T0256499

Information and Instructions

This shop manual contains several sections each covering a specific group of wheel type tractors. The Tab Index on the preceding page can be used to locate the section pertaining to each group of tractors. Each section contains the necessary specifications and the brief but terse procedural data needed by a mechanic when repairing a tractor on which he has had no previous actual experience.

Within each section, the material is arranged in a systematic order beginning with an index which is followed immediately by a Table of Condensed Service Specifications. These specifications include dimensions, fits, clearances and timing instructions. Next in order of arrangement is the procedures paragraphs.

In the procedures paragraphs, the order of presentation starts with the front axle system and steering and proceeding toward the rear axle. The last paragraphs are devoted to the power take-off and power lift systems. Interspersed where needed specifications pertaining to wear limits.

HOW TO USE THE INDEX

Suppose you want to know the procedure for R&R (remove and reinstall) of the engine camshaft. Your first step is to look in the index under the main heading of ENGINE until you find the entry "Camshaft." Now read to the right where under the column covering the tractor you are repairing, you will find a number which indicates the beginning paragraph pertaining to the camshaft. To locate this wanted paragraph in the manual, turn the pages until the running index appearing on the top outside corner of each page contains the number you are seeking. In this paragraph you will find the information concerning the removal of the camshaft.

More information available at haynes.com
Phone: 805-498-6703

J H Haynes & Co. Ltd.

Haynes North America, Inc

ISBN-10: 0-87288-095-8
ISBN-13: 978-0-87288-095-5

FO-31, 8S1, 14-152

SHOP MANUAL

FORD

SERIES 2000-3000-4000
(COVERS MODELS PRIOR TO 1975)

Tractor Serial Number, along with manufacturing and production code numbers and tractor model number, will appear on implement mounting pad at right front corner of transmission (directly behind engine starter). Numbers will be stamped on top of pad, on mounting face of pad, or partially on top of and partially on mounting face of pad. Refer to following explanation of the numbers that will appear at this location:

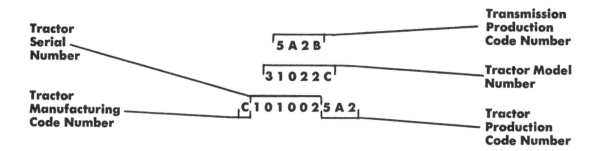

Engine Serial Number, along with tractor size identification, engine type identification and engine production code number, will appear on either the left or right pan rail of cylinder block casting approximately at mid-length of engine. Refer to following explanation of the numbers that will appear at this location.

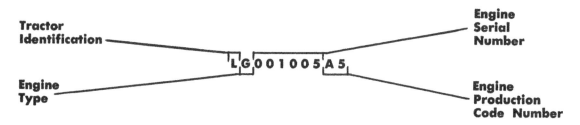

Tractor Identification:		Engine Type:
Before 7-20-68	After 7-19-68	
L—2000	B—2000	D—Diesel Engine
N—3000	C—3000	G—Gasoline Engine
P—4000	D—4000	P—LP-Gas Engine

The following tractor models prior to 1975 are covered in this manual:

2100 All Purpose
2110 L.C.G. (Low Center Gravity)
3100 All Purpose
4100 All Purpose
4110 L.C.G.
4140 S.U.
4200 Row Crop

INDEX (By Starting Paragraph)

CONDENSED SERVICE DATA

General	TRACTOR SERIES (Before 7/20/68)			TRACTOR SERIES (After 7/19/68)		
	2000	3000	4000	2000	3000	4000
Engine Make	Own	Own	Own	Own	Own	Own
No. of Cylinders	3	3	3	3	3	3
Bore—Inches, Non-Diesel	4.2	4.2	4.4	4.2	4.2	4.4
Diesel	4.2	4.2	4.4	4.2	4.2	4.4
Stroke—Inches, Non-Diesel	3.8	3.8	4.2	3.8	3.8	4.4
Diesel	3.8	4.2	4.4	3.8	4.2	4.4
Displacement—Cubic Inches,						
Non-Diesel	158	158	192	158	158	201
Diesel	158	175	201	158	175	201
Compression Ratio, Non-Diesel	8.0:1	8.0:1	8.01:1	7.8:1	7.8:1	7.8:1
Diesel	16.5:1	16.5:1	16.5:1	16.5:1	16.5:1	16.5:1

CONDENSED SERVICE DATA (CONT).

	TRACTOR SERIES (Before 7/20/68)			TRACTOR SERIES (After 7/19/68)		
Tune-Up	2000	3000	4000	2000	3000	4000
Compression, Gage Lbs. @ Cranking Speed of 200 RPM:						
Non-Diesel (All Spark Plugs Out) ...	115-150	115-150	115-150	105-140	115-150	115-150
Diesel	420-510	420-510	420-510	410-490	420-500	420-500
Max. Allowable Variation @ 200 RPM:						
Non-Diesel	25	25	25	25	25	25
Diesel	50	50	50	50	50	50
Firing Order	1-2-3	1-2-3	1-2-3	1-2-3	1-2-3	1-2-3
Valve Tappet Gap—Intake, Hot	0.015	0.015	0.015	0.015	0.015	0.015
Valve Tappet Gap—Exhaust, Hot	0.018	0.018	0.018	0.018	0.018	0.018
Valve Face Angle—Degrees	44	44	44	44	44	44
Valve Seat Angle—Degrees	45	45	45	45	45	45
Ignition Timing			See Paragraph 180			
Injection Timing	19° BTDC	19° BTDC	19° BTDC	19° BTDC	19° BTDC	19° BTDC
Spark Plug Make	Autolite	Autolite	Autolite	Autolite	Autolite	Autolite
Spark Plug Model	AG-5	AG-5	AG-5	AG-5	AG-5	AG-5
Engine Low Idle RPM	600-650	600-650	600-650	600-700	600-700	600-700
Engine High Idle RPM, Non-Diesel	2065-2115	2285-2335	2395-2445	2065-2165	2285-2385	2385-2495
Diesel	2175-2225	2175-2225	2395-2445	2225-2275	2225-2275	2425-2475
Engine Rated RPM, Non-Diesel	1900	2100	2200	1900	2100	2200
Diesel	2000	2000	2100	2000	2000	2200
Battery Terminal Grounded	Negative	Negative	Negative	Negative	Negative	Negative

Sizes-Capacities-Clearances

	2000	3000	4000	2000	3000	4000
Crankshaft Journal Diameter			3.3714-3.3722			
Crankpin Diameter			2.749-2.750			
Camshaft Journal Diameter			2.3895-2.3905			
Piston Pin Diameter			1.4997-1.5000			
Valve Stem Diameter—Intake			0.3711-0.3718			
Valve Stem Diameter—Exhaust			0.3701-0.3708			
Main Bearings Running Clearance			0.0022-0.0045			
Rod Bearings Running Clearance,						
Aluminum Bearings			0.0025-0.0045			
Copper Lead Bearings			0.0017-0.0038			
Camshaft Bearings Running Clearance			0.001-0.003			
Crankshaft End Play			0.004-0.008			
Camshaft End Play			0.001-0.007			
Piston Skirt to Cylinder Clearance			See paragraph 105			
Cooling System, Quarts	13.2	13.8	14.0	13.2	13.8	14.0
Crankcase, Quarts—With Filter	7	8	8	7	7	9
Transmission, Quarts:						
4-Speed	6.5	6.5	....	6.5	6.5	....
6-Speed	....	....	....	14.5	13.8	....
8-Speed	17½	17	17	14.5	13.8	13.8
10-Speed (Select-O-Speed)	12	13¼	13¼	12	12	13.3
Final Drive & Hydraulic, Quarts	23½	20½	26¼	24.6	24.6	33.9
Steering Gear Housing, Quarts	0.85	0.85	0.66	0.85	0.85	0.66
Power Steering System, Quarts	2.31	2.31	2.0	2.3	2.3	2.35

FRONT SYSTEM AND STEERING

(MODELS 2100, 2110, 3100, 4110 & 4140 S.U.)

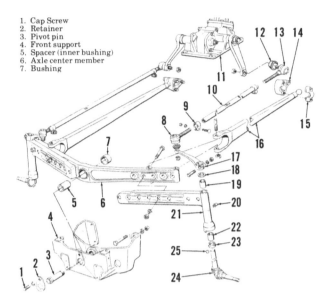

1. Cap Screw
2. Retainer
3. Pivot pin
4. Front support
5. Spacer (inner bushing)
6. Axle center member
7. Bushing

Fig. 1–Exploded view of 2100 and 3100 front axle and related parts used on models prior to production date 8-70. Radius rods on later models attach to axle extensions. Models 2110, 4110 and 4140 are similar. Spacer (5) is available in three lengths; refer to text for selection.

8. Front drag link end
9. Clamp
10. Drag link
11. Steering gear assy.
12. Dust cover
13. Rear drag link end
14. Radius rod ball spacer
15. Radius rod cap
16. Radius rod
17. Steering arm
18. Dust Seal
19. Spindle bushing, upper
20. Grease fitting
21. Axle extension, L.H.
22. Spindle bushing, lower
23. Thrust bearing
24. Spindle
25. Steering arm key

FRONT AXLE ASSEMBLY AND STEERING LINKAGE
Models 2100-2110-3100-4110 4140 S.U.

1. **SPINDLE BUSHINGS.** To renew the spindle bushings (19 and 22—Fig. 1), proceed as follows: Support front of tractor and disconnect steering arms (17) from wheel spindles (24). Slide spindle out of the axle extension (21). Drive old bushings from front axle extension and install new ones using a piloted drift or bushing driver. New bushings will not require final sizing if

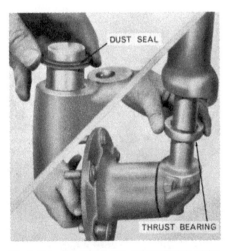

DUST SEAL

THRUST BEARING

Fig. 2–Top view shows proper installation of the dust seal on upper end of spindle before installing steering arm. Lower view shows installation of thrust bearing on spindle before installing spindle in axle extension.

not distorted during installation. Renew thrust bearing (23) if worn or rough. Refer to Fig. 2 for dust seal and thrust bearing installation.

2. **AXLE CENTER MEMBER, PIVOT PIN AND BUSHINGS.** To remove the axle center member (6—Fig. 1), on models prior to production date 8-70, support front of tractor, unbolt radius rods and axle extensions from axle center member and swing the axle extensions and wheel assemblies away from tractor.

NOTE: On models after production date 7-70, the radius rods are attached to axle extensions instead of the axle center member and generally do not need to be disconnected.

Remove cap screw (1) and retainer (2), then unscrew front axle pivot pin (3). Withdraw axle center member from either side of tractor. Remove spacer (5) when so equipped and press out bushing (7).

The front axle support spacer (5) used on models prior to production date 8-70 is available in three lengths: Red, 2.235-2.240; Green, 2.245-2.250 and Grey, 2.255-2.260. Select a spacer that will provide a close fit between pivot pin bosses of front support.

To reassemble, press new bushing into axle center member. If used, insert previously selected spacer into bushing and be sure spacer is a free fit in bushing.

On all models, insert axle center member into front support, then install

pivot pin and tighten pin to a torque of 300-320 Ft.-Lbs. Install retainer (2) and tighten cap screw (1) to a torque of 75 Ft.-Lbs. If removed, tighten radius rod to axle bolts and axle extension bolts to a torque of 130-160 Ft.-Lbs.

3. **FRONT SUPPORT.** To remove front support, proceed as follows: Remove engine hood, grille, lower hood to front support bolts and unbolt radiator from front support. Place floor jack under front end of transmission and take weight of tractor from front axle. Remove front axle pivot pin as in paragraph 2. Remove bolts retaining front support casting to engine and lower the front support to floor. When reinstalling, tighten the front support bolts to a torque of 200-240 Ft.-Lbs. for models prior to production date 8-70, or 180-220 Ft.-Lbs. for later models.

4. **DRAG LINKS AND TOE-IN.** Drag link ends are non-adjustable automotive type and renewal procedure is evident. Refer to Fig. 1 for manual steering tractors and to Fig. 5 for power steering.

Front wheel toe-in should be ¼ to ½-inch; vary the length of each drag link an equal amount to obtain proper toe-in.

MANUAL STEERING GEAR
Models 2100-2110-3100-4110

5. **ADJUSTMENT.** To adjust the steering gear, first be sure that gear housing is properly filled, then disconnect both drag links from steering arms and proceed as follows:

6. **WORMSHAFT END PLAY.** To check wormshaft end play, first loosen the lock nuts (1—Fig. 3) on the sector shaft adjusting screws (8) and back screws out two full turns. The wormshaft should turn freely without perceptible end play. Shims (24) are available in thicknesses of 0.0024 (brass), 0.005 (steel) and 0.010 (steel); also, a 0.005 thick paper gasket is available. Install at least one paper gasket and brass or steel shims as required. Tighten the steering shaft cover retaining cap screws to a torque of 25 Ft.-Lbs. Renew wormshaft bearings as outlined in paragraph 9 if end play is perceptible with paper gasket, but without any metal shims installed.

After checking or adjusting wormshaft end play, readjust sector shaft end play as follows:

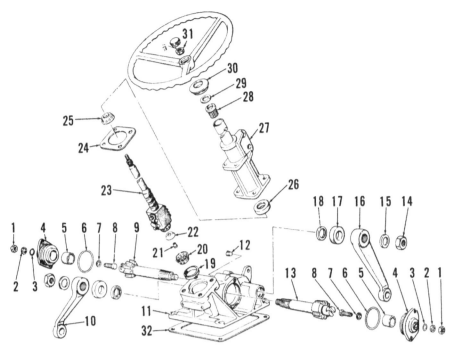

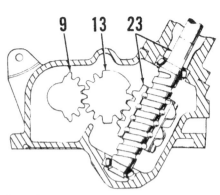

Fig. 4–Cross-sectional view of manual steering gear assembly showing correct installation of sector gears and wormshaft assembly. Refer to Fig. 3 for exploded view and parts identification.

Fig. 3–Exploded view of 2000 and 3000 manual steering gear assembly. Wormshaft (23) end play is adjustable by varying thickness of shims (24); sector shaft end play is adjusted by screws (8). Select a shim (7) that will provide zero to 0.002 clearance between head of adjusting screw (8) and slot in sector shaft (9 or 13).

1. Lock nut	10. Steering arm, L.H.	19. Bearing cup	26. Bearing cup
2. Washer	11. Housing	20. Bearing cone &	27. Steering column
3. Packing	12. Oil level & filler plug	rollers	28. Bushing
4. Side covers	13. Sector gear (double)	21. Bearing retainer	29. Dust seal
5. Bushing	14. Nut	eyelet	30. Grommet
6. "O" ring	15. Washer	22. Bearing retainer	31. Steering wheel nut
7. Shim	16. Steering arm, R.H.	23. Wormshaft assembly	32. Gasket (Select-O-
8. Adjusting screw	17. Felt dust seal	24. Shims	Speed transmissions
9. Sector gear (single)	18. Oil seal	25. Bearing cone &	only)
		rollers	

7. SECTOR SHAFT END PLAY. Before adjusting sector shaft end play, be sure that wormshaft is properly adjusted as outlined in paragraphs 5 and 6, then proceed as follows:

Turn the steering wheel to mid or straight ahead position. With the lock nuts on both sector shaft adjusting screws loosened and the adjusting screw on forward sector shaft (left side) backed out several turns, turn sector shaft adjusting screw (right side) in until there is no perceptible end play in rear sector shaft (13). Tighten lock nut while holding adjusting screw. Turn adjusting screw on left side of unit (front sector shaft) in until there is no perceptible end play of front sector shaft (9). Hold the adjusting screw and tighten lock nut.

Adjustment of steering gear is correct if a pull of 1 to 2¾ lbs. is required at outer edge of steering wheel to turn the steering gear over center position. When adjustment of sector shafts is correct, reconnect the drag links to steering arms.

8. R&R STEERING GEAR ASSEMBLY. To remove the steering gear assembly, proceed as follows: Disconnect battery ground cable and remove the steering wheel. Remove the sheet metal covers at each side of steering gear. Remove screws retaining instrument panel, disconnect ground terminal at left side of panel and rotate panel up out of opening in sheet metal. Disconnect wiring from instrument panel and remove the panel assembly. Remove the light switch and choke or diesel fuel shut-off cable. If equipped with Select-O-Speed transmission, refer to paragraphs 245 and 246 for removal of controls. Remove engine hood and the sheet metal surrounding fuel tank. Shut off the fuel supply, disconnect fuel supply line and diesel fuel return line from tank, then remove fuel tank assembly. Thoroughly clean the steering gear and surrounding area. Disconnect drag links from steering arms, then unbolt and remove steering gear assembly from transmission housing. Drain lubricant from steering gear housing if unit is to be disassembled. Note: On models with Select-O-Speed transmission, take care that no dirt or foreign material enters transmission housing while removing steering gear assembly and place a cover over opening in transmission while steering gear is removed.

To reinstall the steering gear assembly, first place a new gasket on opening in transmission housing (Select-O-Speed transmission models only), then reinstall unit by reversing removal procedure. Refill gear housing with SAE 90 E.P. gear lubricant (Ford specification M2C53-A).

9. OVERHAUL. Major overhaul of the steering gear unit necessitates removal of the unit from tractor as outlined in paragraph 8. After removing unit, refer to Fig. 3 and proceed as follows:

Remove steering arm retaining nuts (14) and pull steering arms (10 and 16) from sector shafts (9 and 13). Sector shafts and side covers (4) can be removed as a unit after removing cover retaining screws. To separate shaft and cover, remove lock nut (1) and thread adjusting screw (8) into cover until threads clear. Unbolt steering shaft cover (27) from gear housing (11) and remove cover, shaft and ball nut assembly. Do not disassemble the ball nut and steering shaft assembly (23) as component replacement parts are not available. If steering shaft and/or ball nut are damaged, renew the complete assembly. The need and procedure for further disassembly and/or overhaul is evident from inspection of unit.

The bushings in gear housing for the sector shafts are not serviced separately from housing; install a new housing if bushings are excessively worn. Bushings (5) in side covers (4) are renewable separately from side covers.

Shims (7) on the adjusting screws (8) are available in thicknesses of 0.063, 0.064, 0.066 and 0.069. When reassembling, select a shim that will provide zero to 0.002 clearance between adjusting screw head and slot in sector shaft.

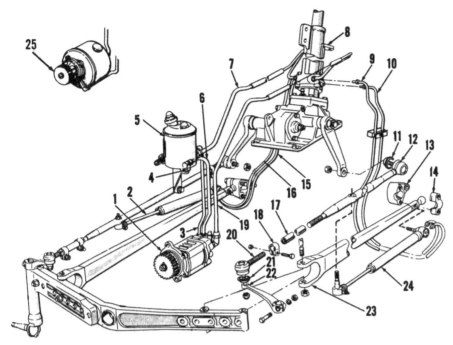

Fig. 5–Drawing of 2000 and 3000 power steering system. For exploded view of steering gear and valve assembly (8), refer to Fig. 12. Exploded view of power steering pump (1) is shown in Fig. 7, late pump (25) in Fig. 10.

1. Power steering pump (early)
2. Steering cylinder, R.H.
3. Tube, pump to reservoir
4. Tube clamp
5. Reservoir
6. Tube, pump to reservoir
7. Return tube
8. Steering gear & valve assy.
9. Front inner tube, L.H.
10. Rear outer tube, L.H.
11. Dust cover
12. Drag link assembly
13. Radius rod ball spacer
14. Radius rod ball cap
15. Rear outer tube, R.H.
16. Front inner tube, R.H.
17. Drag link tube
18. Clamp
19. Pressure tube
20. Drag link front end
21. Dust seal
22. Steering arm
23. Radius rod
24. Steering cylinder, L.H.
25. Steering pump (late)

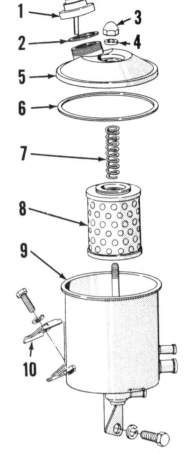

Fig. 6–Exploded view of power steering reservoir and filter assembly used on early models.

1. Dipstick
2. Gasket
3. Nut
4. Sealing washer
5. Cover
6. Gasket
7. Spring
8. Filter element
9. Reservoir
10. Clamp

To reassemble, insert the steering shaft and ball nut unit into gear housing with gear teeth on ball nut forward, then install steering shaft housing (27) with proper number of shims as outlined in paragraph 6. Insert the sector shafts, with their adjusting screws and correct thickness shims, into gear housing so that sector gear teeth and ball nut teeth are timed as shown in Fig. 4. Place new "O" rings (6—Fig. 3) on side covers (4) and pull the side covers into place with sector shaft adjustment screws. Install and tighten the side cover retaining cap screws to a torque of 25 Ft.-Lbs. and adjust the sector shaft end play as outlined in paragraph 7. Install new packing rings (3—Fig. 3) on adjusting screws, install flat washers (2) and tighten the adjusting screw lock nuts (1) while holding adjusting screws in proper position.

Refill gear housing with SAE 90 EP gear lubricant to level with plug (12) opening (approximately 1¾ pints). When reinstalling steering arms, tighten the retaining nuts (14) to a torque of 115-125 Ft.-Lbs.

POWER STEERING SYSTEM
Models 2100-2110-3100-4110-4140

CAUTION: The maintenance of absolute cleanliness of all parts is of the utmost importance in the operation and servicing of the hydraulic power steering system. Of equal importance is the avoidance of nicks or burrs on any of the working parts.

10. **FLUID AND BLEEDING.** Recommended power steering fluid is Ford M2C41-A oil. Maintain fluid level to full mark on dipstick (1—Fig. 6); reservoir assembly (5—Fig. 5) on early models is located under engine hood directly behind radiator. On late models, reservoir is mounted on pump (25). After each 600 hours of operation, it is recommended that the filter element (8 —Fig. 6 or 3—Fig. 10) be renewed, any oil in reservoir be removed with suction gun and that the reservoir be refilled to full mark on dipstick with new, clean oil.

The power steering system is self-bleeding. When unit has been disassembled, refill with new oil to full mark on dipstick, then start engine and cycle the system several times by turning the steering wheel from lock to lock. Recheck fluid level and add as required. System is fully bled when no more air bubbles appear in reservoir as system is being cycled.

11. **CHECKING SYSTEM PRESSURE AND FLOW.** The power steering pump assembly incorporates a pressure relief valve and a flow control valve. System relief pressure and flow should be as follows:

	Early Pump	Late Pump
Pressure, psi	650	650
Flow, gpm @ 1000 rpm	3.5	2.74

To check system relief pressure, install a "T" fitting in pressure line (19—Fig. 5) at pump connection and connect a 0-1000 psi gage to fitting. With the engine running at 1000 RPM and the front wheels turned against lock, gage reading should be 600-700 psi.

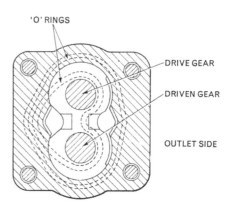

Fig. 8–Drawing showing correct installation of gears and bearing blocks in pump body. View is from rear (flanged) end of pump body.

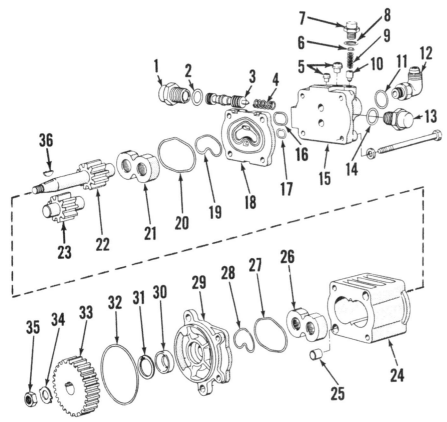

Fig. 7–Exploded view of early power steering pump. Note that flow control valve spring (4) and small tip end of valve (3) is towards side of rear cover (15) containing the relief valve assembly (items 6 through 10). Refer to Fig. 9 for model 4200 shaft seals.

1. Cap plug	10. Pressure relief valve	20. Outer seal ring	30. Seal (except model
2. "O" ring	11. Seal ring	21. Bearing block	4200)
3. Flow control valve	12. Outlet elbow	22. Drive gear & shaft	31. Locating ring
4. Spring	13. Cap plug	23. Driven gear & shaft	32. "O" ring
5. Tubing seats	14. "O" ring	24. Pump body	33. Drive gear
6. Shims	15. Rear cover	25. Bolt (dowel) rings (2)	34. Tab washer
7. Cap plug	16. "O" ring	26. Bearing block	35. Nut
8. "O" ring	17. "O" ring	27. Outer seal ring	36. Woodruff key
9. Spring	18. Rear plate	28. Inner seal ring	
	19. Inner seal ring	29. Front cover	

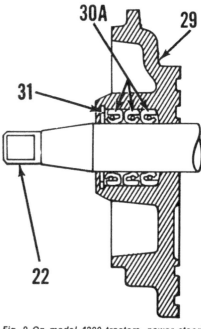

Fig. 9–On model 4200 tractors, power steering pump is fitted with three shaft seals (30A) instead of single seal (30–Fig. 7) used on other models. Lips of two inner seals face inward, outer seal lip is toward retaining snap ring (31). Pump front cover is (29); drive shaft is (22).

CAUTION: When checking system relief pressure, hold the steering wheel against lock for only long enough to observe gage reading; pump may be damaged if steering wheel is held in this position for an excessive length of time.

To adjust the pressure on early models, remove pressure relief valve cap (7 —Fig. 7) and add or remove shims (6) as required. If adding shims under the pressure relief valve cap will not increase system pressure, remove and overhaul power steering pump as outlined in paragraph 12.

On late models, pump must be disassembled to adjust opening pressure. Refer to paragraph 12 and Fig. 10. Shims (25) control system pressure and are available in thicknesses of 0.010, 0.015 and 0.060. A change of 0.005 in total shim pack thickness will alter system pressure about 35 psi. Late pump does not contain a flow control valve.

12. R&R AND OVERHAUL PUMP. Thoroughly clean pump, lines and surrounding area. Disconnect lines from pump and allow fluid to drain. Cap all openings to prevent dirt from entering pump or lines, then unbolt and remove pump assembly from engine front plate. When reinstalling pump, use new sealing "O" ring and tighten retaining bolts to a torque of 23-29 Ft.-Lbs. Reconnect lines, fill and bleed system as in paragraph 10.

On early models, refer to exploded view of pump in Fig. 7 and disassemble pump as follows: Scribe an assembly mark across pump covers and body. Straighten tab on washer (34) and remove nut (35). Pull drive gear (33) from pump shaft and remove key (36). Remove the four through-bolts and separate rear cover assembly (15), plate (18), body (24) and front cover (29). Remove bearing blocks (21 and 26) and gears (22 and 23) from pump as a unit. Remove caps (1, 7 and 13) from

rear cover (15) and withdraw flow control valve (3), pressure relief valve (10) and related parts. Remove snap ring (31) and the oil seal (30) from front cover. Clean all parts in a suitable solvent, air dry, then lightly oil all machined surfaces.

Inspect bearing blocks (21 and 26) for signs of seizure or scoring on face of journals. (When disassembling bearing block and gear unit, keep parts in relative position to facilitate reassembly.) Light score marks on faces of bearing blocks can be removed by lapping bearing block on a surface plate using grade "O" emery paper and kerosene.

Fig. 10–Exploded view of power steering pump with integral reservoir used on most late models.

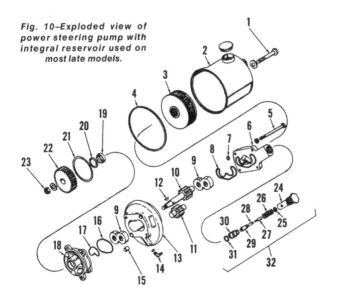

1. Bolt
2. Reservoir
3. Filter
4. Gasket
5. Through-bolt
6. Cover
7. O-ring
8. Seal ring
9. Bearing block
10. Driven gear
11. Follow gear
12. Woodruff key
13. Body
14. Outlet elbow
15. Ring dowel
16. Seal ring
17. Seal ring
18. Flange housing
19. Oil seal
20. Snap ring
21. O-ring
22. Drive gear
23. Shaft nut
24. Valve body
25. Shim pack
26. Spring
27. Spring guide
28. Seal
29. Valve head
30. Valve seat
31. O-ring
32. Relief valve

Fig. 11–When installing reservoir, align indent in reservoir with positioning lugs on pump body as shown.

Examine body for wear in gear running track. If track is worn deeper than 0.0025 on inlet side, body must be renewed. Examine pump gears for excessive wear or damage on journals, faces or teeth. Runout across the gear face to tooth edge should not exceed 0.001. If necessary, the gear journals may be lightly polished with grade "O" emery paper to remove wear marks. The gear faces may be polished by sandwiching grade "O" emery paper between gear and face of scrap bearing block, then rotating the gear. New gears are available in matched sets only.

When reassembling pump, install all new seals, "O" rings and sealing rings. On models with three shaft seals (Fig. 9), the two inner seals are installed lips first and outer seal lip out to serve as a dust seal.

Install flow control valve (3), spring (4) and plugs (1 and 13) with new "O" rings (2 and 14). Install pressure relief valve (10), spring (9) and plug (7), being sure that all shims (6) are in plug and using new "O" ring (8). Assemble pump gears to bearing blocks (use Fig. 8 as a guide, if necessary) and insert the unit into pump body. Be sure the two bolt rings (hollow dowels) are in place in pump body, then position the front cover on body. Place the rear plate (18—Fig. 7) at rear of body and install rear cover. Tighten the four cap screws (throughbolts) to a torque of 13-17 Ft.-Lbs. Install the pump drive gear key, drive gear, tab washer and nut. Tighten the nut to a torque of 55-60 Ft.-Lbs. and bend tab of washer against flat on nut.

On models with integral pump and reservoir refer to Fig. 10. Clean pump and surrounding area and disconnect pump pressure and return lines. Remove the two cap screws securing pump to engine front cover and lift off

pump and reservoir as a unit. Drain the reservoir and remove through-bolt (1), reservoir (2) and filter (3).

Relief valve cartridge (32) can now be removed if service is indicated. For access to shims (25), grasp seat (30) lightly in a protected vise and unscrew body (24). Shims (25) are available in thicknesses of 0.010, 0.015 and 0.060. Starting with the removed shim pack, substitute shims thus varying total pack thickness, to adjust opening pressure. Available shims permit thickness adjustment in increments of 0.005 and each 0.005 in shim pack thickness will change opening pressure about 35 psi. If parts are renewed, the correct thickness can only be determined by trial and error, using the removed shim pack as a guide.

To disassemble the pump, bend back tab washer and remove shaft nut (23), drive gear (22) and key (12). Mark or note relative positions of flange housing (18), pump body (13) and cover (6); then remove pump throughbolts (5). Keep parts in their proper relative position when disassembling pump unit. Pump gears (10 & 11) are available in a matched set only. Bearing blocks (9) are available separately but should be renewed in pairs if renewal is because of wear. Bearing blocks should also be renewed with gear set if any shaft or bore wear is evident.

When reassembling the pump, tighten through bolts (5) to a torque of 25 Ft.-Lbs. and drive gear nut (23) to a torque of 30-40 Ft.-Lbs.

13. **CONTROL VALVE AND STEERING GEAR ASSEMBLY.** The power steering system is of the linkage booster type which utilizes a control valve combined with the steering shaft and gear assembly as an integral unit. Refer to Fig. 12 for an exploded view of the assembly. Adjust-

ment and service information for the unit is contained in the following paragraphs 14 through 17.

14. ADJUST STEERING GEAR. The steering wormshaft (21—Fig. 12) is carried in needle roller bearings (20 and 22) which require no adjustment. End thrust of the wormshaft is utilized to operate the control valve spool.

To adjust sector gears, loosen the lock nuts (1) and back both adjusting screws (8) out two full turns. Turn adjusting screw on rear sector shaft (13) in until there is no perceptible backlash of rear sector shaft gears. Tighten the lock nut while holding adjusting screw in this position. With the rear sector shaft adjusted, turn the adjusting screw on front sector shaft (9) in until there is no perceptible backlash of front sector shaft gears, then tighten lock nut while holding adjusting screw in this position.

15. R&R STEERING GEAR AND CONTROL VALVE ASSEMBLY. To remove the steering gear and power steering control valve assembly, proceed as follows:

16. Disconnect battery ground cable and remove the steering wheel. Drain power steering reservoir and disconnect cylinder line at outer side of either cylinder to drain oil from the control valve. Remove the sheet metal covers at each side of steering gear. Remove screws retaining instrument panel, disconnect ground terminal at left side of panel and rotate panel up out of the opening in sheet metal. Disconnect

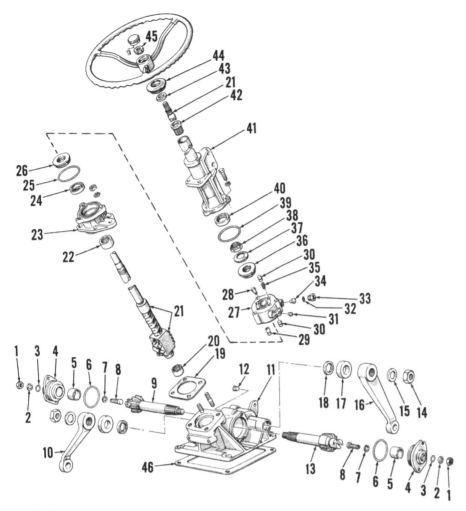

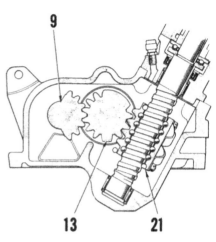

Fig. 13–Cross-sectional view of steering gear assembly showing correct installation (timing) of sector gears and wormshaft; refer to Fig. 12 for exploded view of unit and for parts identification.

Fig. 12–Exploded view of power steering gear and valve assembly. Turning effort on steering wheel, or reaction of front wheels through steering linkage and gears, causes wormshaft (21) to move up or down, thereby actuating the control valve (27).

1. Lock nut	12. Oil level & filler plug	24. Oil seal	36. Thrust bearing assy.
2. Washer	13. Sector gear, double	25. "O" ring	37. Pre-load spring washer
3. Packing	14. Nut	26. Thrust bearing assy.	38. Lock nut
4. Side covers	15. Washer	27. Valve assembly	39. "O" ring
5. Bushing	16. Steering arm, R.H.	28. Tube seat	40. Oil seal
6. "O" ring	17. Felt dust seal	29. Reaction plungers (2)	41. Steering column
7. Shim	18. Oil seal	30. Thrust plungers (6)	42. Upper bushing
8. Adjusting screw	19. Paper gasket (0.005)	31. Tube seat	43. Dust seal
9. Sector gear, single	20. Needle roller bearing	32. "O" ring	44. Grommet
10. Steering arm, L.H.	21. Wormshaft assembly	33. Return union	45. Steering wheel nut
11. Housing	22. Needle roller bearing	34. Check valve	
	23. Adapter assembly	35. Plunger springs (3)	

wiring from instrument panel and remove the panel assembly. Remove the light switch and choke or diesel shut-off cable. If equipped with Select-O-Speed transmission, refer to paragraphs 245 and 246 for removal of controls. Remove hood and the sheet metal surrounding fuel tank. Shut off the fuel supply valve, disconnect fuel supply line and diesel fuel return line from tank, then remove the fuel tank assembly.

17. Thoroughly clean the steering gear, power steering line connections and surrounding area. Disconnect drag links from steering arms and disconnect power steering lines from control valve. Unbolt and remove the steering gear and control valve assembly from

top of transmission housing. Note: On models with Select-O-Speed transmission, take care that no dirt or other foreign material enters the transmission when removing or reinstalling the steering gear and cover the opening in transmission housing while steering gear unit is removed. Drain lubricating oil from gear housing.

Reinstall steering gear unit by reversing removal procedure. On Select-O-Speed transmission models, place a new gasket over opening in transmission housing before reinstalling steering gear. Refill and bleed the power steering system as outlined in paragraph 10. Refill steering gear with SAE 90 E.P. gear lubricant.

18. RENEW CONTROL VALVE UPPER SEAL. If power steering fluid leaks from steering column housing, the control valve upper seal (40–Fig. 12) is leaking and can be renewed as follows:

Follow general procedures outlined in paragraph 16, except that the fuel lines do not need to be disconnected nor does the fuel tank need to be removed. Thoroughly clean the steering gear and contol valve assembly. Remove the cap screws retaining steering column (41) to steering gear and remove the steering column. Remove the seal from lower end of steering column with OTC945-2 puller and slide hammer. Drive the bushing (42) from upper end of steering column with a 1-inch diameter dowel.

19. Fabricate a seal guide from an 18-inch length of ¾-inch thin wall electrical conduit by counterboring the inside diameter to 27/32-inch for a depth of 3 inches. Install new seal in lower end of steering column using a 1 ³⁄₈ inch diameter driver. Install seal so that spring and lip will be downward (towards contol valve). Place counter-bored end of seal guide through seal and install the steering column over steering shaft with new "O" ring in groove in steering column. Install the steering column retaining cap screws to a torque of 25 Ft.-Lbs. and remove the seal guide. Install new bushing in top end of steering column, then reinstall removed parts. Refill and bleed power steering as outline in paragraph 10.

20. OVERHAUL CONTROL VALVE. With steering gear and control valve assembly removed from tractor as outlined in paragraphs 15

through 17, proceed as follows:

Scribe a line across steering column, valve housing and adapter to facilitate reassembly. Remove return line union (33—Fig. 12) from valve housing, discard "O" ring (32) and unscrew the check valve (34). Remove the cap screws retaining steering column and remove column from steering shaft. Carefully unstake lock nut (38) and temporarily install steering wheel on shaft to hold shaft while unscrewing nut. Remove the steering wheel and nut and position steering gear unit on bench so that steering shaft is in horizontal position. Slide spring washer (37), upper thrust bearing (36), control valve assembly (27) and lower thrust bearing (26) from steering shaft. Note: Take care that control valve spool, plungers (30) and check valves (29) do not drop out of valve housing as it is removed from steering shaft. Unbolt and remove adapter (23) from steering gear housing.

Note which end of the control valve spool has the identification groove (Fig. 15), then slide spool from housing. Remove the six plungers (30—Fig. 12), three springs (35) and the two check valves (29), noting which hole contained the check valves. If any of the brass seats (28 and/or 31) for the pump pressure line and the steering cylinder lines are damaged, remove them from valve housing by threading a tap into the seat, clamping tap in vise and pulling the housing from the seat. Remove seal (40) and bushing (42) from steering column. Remove seal (24) from adapter (23).

Inspect all removed parts for damage or undue wear and renew any not ac-

ceptable for further service. Install new "O" rings (25, 32 & 39), seals (24 and 40) and lock nut (38) when reassembling. To reassemble, proceed as follows:

21. Install new seal (24) in adapter (23) with lip of seal up (towards control valve). Install the adapter over steering shaft and to steering gear housing with a new 0.005 thick paper gasket (19) and tighten the retaining nuts to a torque of 25 Ft.-Lbs. Place new "O" ring (25) in groove in top of adapter. Place the small race, bearing retainer and large race over steering shaft and against shoulder on shaft. Lubricate valve spool and insert in housing with identifying groove in same direction as when removed. Insert the check valves in same bore from which they were removed and with open ends inward. Lubricate and insert the six plungers in the remaining bores with a spring between each set of plungers. Carefully place the assembled valve unit on steering shaft next to thrust bearing with the offset cylinder ports upward as shown in Fig. 16. Place large race of upper thrust bearing on shaft against valve, then install bearing retainer and small race. Place spring washer on shaft against small thrust bearing race with cup side of washer towards the race. Install a new lock nut, tighten the nut so that all end play is removed, then loosen the nut 1/6-turn and stake the nut to locating slot in steering shaft as shown in Fig. 16. Reinstall the steering column as outlined in paragraph 19.

22. OVERHAUL STEERING GEAR. With control valve assembly and

adapter removed from top of steering gear as outlined in paragraph 20, refer to Fig. 12 and proceed as follows:

Remove the nuts (14) and with suitable pullers, remove steering arms (10 and 16). Remove the lock nuts (1) from sector shaft adjusting screws (8) and remove cap screws retaining side covers (4) to steering gear housing. With screwdriver, turn adjusting screws in to push side covers from sector shafts and the steering gear housing. Withdraw the sector shafts (9 and 13) and adjusting screws from housing. Remove steering shaft and ball nut assembly (21). Drive the seals (18) from steering gear housing.

Carefully inspect all parts and renew any that are damaged or show undue wear. Sector shaft bushings in steering gear housing are not serviced separately from housing. Bushings in side covers may be renewed or a new side cover and bushing assembly may be installed if bushings are worn. Steering shaft and ball nut assembly is serviced as a complete unit only. Renew needle bearings (20 and 22) in steering gear housing and adapter if loose or worn; drive or press on lettered (trade mark) end of bearing cage only.

Reassemble unit as follows: Install seals (18) in gear housing with lips of seals to inside. Install new seal (24) in adapter (23) with lip of seal upward (toward control valve). Place steering shaft and ball nut assembly in gear housing with gear teeth on ball nut forward, then install the adapter with a new 0.005 thick paper gasket (19). Note: Take care not to damage seal in adapter when installing the adapter over steering shaft; wrap threads, sharp shoulders, etc., with plastic tape. Tighten the adapter retaining nuts to a torque of 25 Ft.-Lbs. Insert adjusting

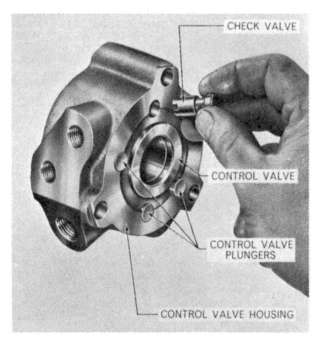

Fig. 14–View showing upper side of control valve and spool assembly; install the three sets of control valve plungers and springs and the two check valves in the bores in housing as shown.

Fig. 15–Insert control valve spool in housing so that identification notch will be towards top end of steering shaft.

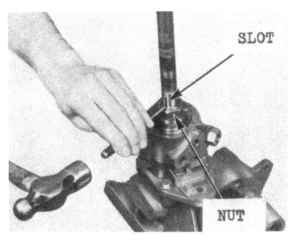

Fig. 16—When valve retaining nut is properly positioned as explained in text; stake nut to slot in shaft as shown.

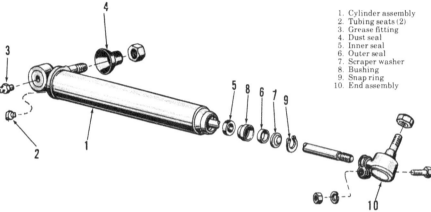

1. Cylinder assembly
2. Tubing seats (2)
3. Grease fitting
4. Dust seal
5. Inner seal
6. Outer seal
7. Scraper washer
8. Bushing
9. Snap ring
10. End assembly

Fig. 17—Exploded view of power steering cylinder used on 2100, 2110, 3100 and 4110 models. Piston rod seals can be renewed but cylinder, piston and piston rod are not serviced except as a complete cylinder assembly.

tighten retaining nuts to a torque of 115-125 Ft.-Lbs. Reinstall control valve as outlined in paragraph 21, steering column as in paragraph 19 and reinstall the steering gear and control valve assembly as outlined in paragraph 17.

23. **POWER CYLINDERS.** Power cylinders are right and left assemblies and should be identified prior to removal. Place a drain pan under cylinder to be removed and disconnect the hydraulic lines. Remove nuts from front and rear ball socket assemblies and bump from mounting holes. Clear cylinder of oil by moving piston rod back and forth.

Loosen clamp and unscrew forward end assembly (10—Fig. 17) from piston rod. Remove snap ring (9) and pull rod to end of travel. Then, remove the scraper washer (7), wiper seal (6), aluminum bushing (8) and inner seal (5). (If seal did not come out with piston rod, remove with awl or other sharp pointed tool.)

If cylinder or piston rod is damaged, the complete power cylinder must be renewed as only the rod end assembly and seal kit are available for service. Reassemble by reversing disassembly procedure. Seals should be soaked in power steering fluid prior to installation. Carefully install inner seal over piston rod with lip of seal inward, then, using a dull pointed tool, work outer lip of seal into proper position in cylinder. Install the aluminum bushing, outer seal with lip outward, washer with scraper lip outward and then install retaining snap ring and forward end assembly. Reinstall power cylinder with ports upward, reconnect cylinder lines and fill and bleed the system as outlined in paragraph 10. Note: Forward end assembly should be screwed far enough onto piston rod so that piston will reach end of stroke before front wheel spindles contact their stops. Piston rod should not flex when cylinder is fully extended.

screw (8) in slot of rear sector gear (13) (gear with double teeth) with a shim (7) of thickness that will give zero to 0.002 clearance between head of adjusting screw and slot in sector shaft. Shims are available in thicknesses of 0.063, 0.064, 0.066 and 0.069. Insert the assembled sector gear and adjusting screw in gear housing so that teeth of sector gear and ball nut are meshed as shown in Fig. 13. Install a new "O" ring (6—Fig. 12) on side cover (4), lubricate the "O" ring and pull side cover onto sector gear with adjusting screw. In-

stall the side cover retaining cap screws and tighten them to a torque of 25 Ft.-Lbs. Turn adjusting screw in until there is no perceptible end play of sector shaft, then install new packing ring (3), washer (2) and lock nut (1). Hold the adjusting screw in proper position while installing and tightening lock nut. Select a shim (7) for the front sector shaft (9), then install front sector shaft as previously outlined for rear sector shaft using Fig. 13 as a guide for timing front sector to rear sector shaft teeth. Install steering arms and

FRONT SYSTEM AND STEERING

(MODEL 4100 ALL PURPOSE)

FRONT AXLE ASSEMBLY AND STEERING LINKAGE

All Models

24. **SPINDLE BUSHINGS.** To renew the spindle bushings (21 and 23 —Fig. 19), support front of tractor and

disconnect steering arms (13 and 19) from wheel spindles (26). Slide the wheel and spindle assemblies (remove wheels from hubs if desired for clearance) out of the front axle extensions (22). Drive old bushings from front axle extensions and install new bushings

with piloted drift or bushing driver. New bushings are final sized and should not require reaming if carefully installed. Renew thrust bearing (24) if worn or rough. Refer to Fig. 18 for correct installation of thrust bearing and spindle dust seal.

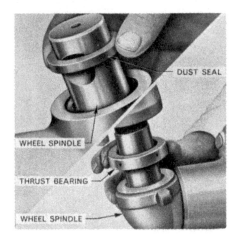

Fig. 18—Top view shows proper installation of dust seal on spindle after spindle is installed in axle extension. Lower view shows proper installation of thrust bearing on spindle prior to installation spindle in axle extension.

1. Plug
2. Front trunnion bracket
3. Bushing
4. Shims (0.002, 0.005 and 0.015)
5. Thrust washer
6. Axle center member
7. Thrust washer
8. Bushing
9. Connecting rod
10. Clamp
11. Connecting rod ends
12. Dust seals
13. Right steering arm
14. Front support
15. Steering gear assy.
16. Drag link rear end
17. Clamp
18. Drag link front end
19. Left steering arm
20. Dust seal
21. Upper spindle bushing
22. Left axle extension
23. Lower spindle bushing
24. Thrust bearing
25. Woodruff key
26. Left wheel spindle

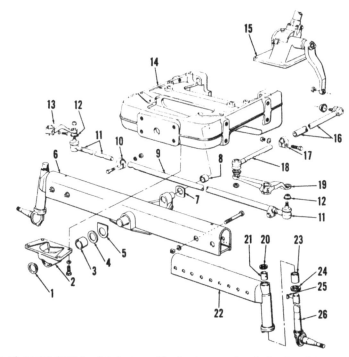

Fig. 19—Exploded view of model 4100 front axle assembly, front support and steering linkage. Manual steering linkage is shown; refer to Fig. 35 for power steering system. Refer to Fig. 21 for exploded view of the steering gear assembly (15) that is used on both manual steering and power steering models.

25. AXLE CENTER MEMBER, PIVOT PINS AND BUSHINGS. To remove axle center member (6—Fig. 19), support front of tractor, remove bolts from clamps (10) on connecting rod (9) and bolts that retain axle extensions (22) in center member. Slide the axle extension, spindle and wheel assemblies out of center member. Support center member and remove trunnion (2), then remove center member from front support. Note: When removing trunnion, take care not to lose or damage shims (4).

Pivot pins are an integral part of the front axle center member and are renewable by installing new center member only. Pivot pin bushing (3) in trunnion and bushing (8) in front support should be renewed if worn. Also, renew thrust washers (5 and/or 7) if worn.

Reinstall front axle center member by reversing removal procedure. Install shims (4) as required to eliminate end play of axle center member between front support and trunnion. Shims are available in thicknesses of 0.002, 0.005 and 0.015. Place all shims between front thrust washer (5) and trunnion (2).

26. FRONT SUPPORT. To remove front support, proceed as follows: Drain cooling system and remove hood, grille and radiator. Support front end of tractor. Disconnect drag link from end assembly (18—Fig. 19) and remove trunnion (2). Roll the front axle assembly away from tractor, attach hoist to front support, then unbolt and remove front support from engine. Install by reversing the removal procedure. Tighten the front support to engine bolts to a torque of 200-240 Ft.-Lbs. Install shims (4) between thrust

washer (5) and trunnion (2) to eliminate end play of front axle on pivot pins. When reconnecting drag link, refer to paragraph 27.

27. DRAG LINK, CONNECTING ROD AND TOE-IN. Front wheel spindle arm connecting rod ends and drag link ends are of the non-adjustable automotive type and procedure for renewing same is evident.

Toe-in should be correct for each tread width position when connecting rod clamp bolt is placed in corresponding notch of rod end assembly. If toe-in is not within limits of 0 to ¼-inch, check for bent or excessively worn parts.

On models prior to production date 8/70, length of drag link between steering gear arm and right front spindle arm should be adjusted for different tread width positions as follows: When right front axle extension is in innermost position or extended 2 inches (one adjustment hole), drag link should be in shortest position. Then, lengthen drag link one notch for each additional 2 inch extension of right front axle.

**STEERING GEAR UNIT
All Manual Steering Models;
Early Models With Power
Steering**

28. Refer to Fig. 21 for exploded view of steering gear unit used on all manual steering models and early tractors with power steering.

29. ADJUSTMENT. If there is no perceptible end play of either the steering (worm) shaft (6—Fig. 21) or the rocker shaft (22) and a pull at outer edge of steering wheel of 1 to 2¾ lbs. is required to turn gear unit past midposition (with drag link or power steering cylinder disconnected from steering arm), adjustment can be considered correct. Although usually performed only during reassembly of gear unit, adjustments for wormshaft and rocker shaft end play can be made to correct excessive end play or turning effort. With unit removed from tractor as outlined in paragraph 32, proceed as follows:

30. WORMSHAFT END PLAY. Remove side cover (25—Fig. 21) and inspect condition of unit. If no obvious damage or excessive wear is noted, add or remove shims (14) and gaskets (15) so that the wormshaft turns freely, but has no perceptible end play. Approximate shim and gasket thickness can be determined by installing steering column without shims or gaskets and measuring resulting gap between steering column and gear housing. Tighten column retaining nuts finger tight and measure gap at several points with feeler gage as shown in Fig. 25. Paper gaskets are 0.010 thick and steel shims are available in thicknesses of 0.003 and 0.010. Use one gasket on each side of shim pack and, on final assembly, apply a light coat of sealer to gaskets. Tighten steering

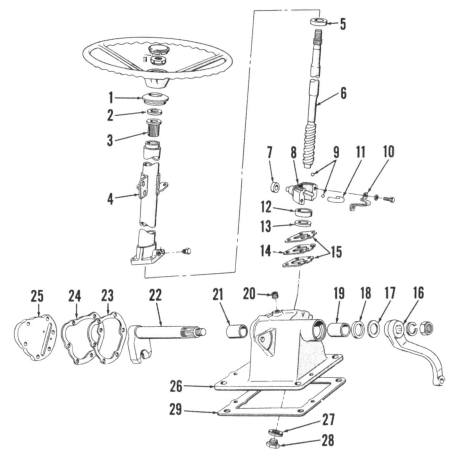

Fig. 22–View of steering gear assembly with side cover removed. Roller moves in slot in side cover.

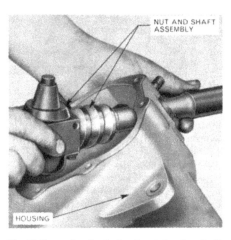

Fig. 23–Removing the ball nut and steering shaft assembly from gear housing.

Fig. 21–Exploded view of steering gear unit of the type used on all manual steering models and on early models with power steering. Thirty-four 1/3-inch diameter steel balls are used in the assembly; twenty are used for steering shaft bearings and 14 are used in the ball nut (8) and tube (11) recirculating groove.

1. Grommet	8. Ball nut	16. Steering arm
2. Dust Seal	9. Steel balls (3/8-in.)	17. Dust seal
3. Bushing	10. Retainer	18. Oil seal
4. Steering column	11. Tube	19. Bushing
5. Upper bearing race	12. Lower bearing race	20. Plug
6. Wormshaft	13. Spacer	21. Bushing
7. Roller	14. Shims	22. Rocker shaft
	15. Gasket	23. Shims

24. Gasket	27. Sealing washer
25. Side cover	28. Plug
26. Gear housing	29. Gasket (Select-O-Speed only)

column retaining nuts to a torque of 25-35 Ft.-Lbs.

31. ROCKER SHAFT END PLAY. First, adjust wormshaft end play as outlined in paragraph 30, then proceed as follows:

Be sure rocker shaft and ball nut are in mid position and roller is in place as shown in Fig. 22, then install side cover (25—Fig. 21) without shims (23) or gaskets (24). Tighten the retaining nuts and cap screws equally finger tight, then measure gap between side cover and steering housing at several points with feeler gage. Average gap measurement is approximate thickness of shims and gaskets required. Shims are 0.005 thick. Use one gasket on each side of shim pack and, on final assembly, apply a light coat of sealer to gaskets.

32. R&R STEERING GEAR ASSEMBLY. To remove the steering gear assembly, proceed as follows: Disconnect battery ground cable and remove the steering wheel. Remove the sheet metal covers at each side of steering gear. Remove screws retaining instrument panel, disconnect ground terminal at left side of panel and rotate panel up out of opening in sheet metal. Disconnect wiring from instrument panel and remove the panel assembly. Remove the light switch and choke or diesel shut-off cable. If equipped with Select-O-Speed transmission, refer to paragraphs 245 and 246 for removal of controls. Remove engine hood and the sheet metal surrounding fuel tank. Shut off the fuel supply, disconnect fuel supply line and diesel fuel return line from tank, then remove fuel tank assembly. Thoroughly clean the steering gear and surrounding area. Disconnect

drag link or power steering cylinder from steering arm, then unbolt and remove steering gear assembly from the transmission housing. Drain lubricant from steering gear housing if unit is to be disassembled. Note: On models with Select-O-Speed transmission, take care that no dirt or foreign material enters transmission housing while removing steering gear assembly and place a cover over opening in transmission housing while steering gear is removed.

To reinstall steering gear assembly, place a new gasket on transmission housing (Select-O-Speed transmission models only), then reinstall steering gear assembly by reversing removal procedure. Refill gear housing with SAE 90 E.P. gear lubricant (Ford specification M-2C53-A).

33. OVERHAUL STEERING GEAR UNIT. With the steering gear assembly removed from tractor as outlined in paragraph 32, refer to exploded

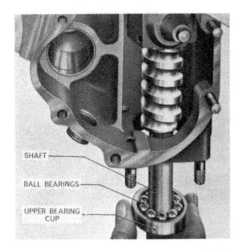

Fig. 24–Installing upper bearing race and ball bearings.

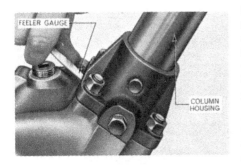

Fig. 25–Measuring clearance between gear housing and steering column housing to determine shim and gasket thickness needed. Shim and gasket thickness required between gear housing and side cover is determined in similar manner.

view of unit in Fig. 21 and proceed as follows:

Remove nut retaining steering arm (16) to rocker shaft (22) and, using suitable pullers, remove arm from shaft. Unbolt and remove the side cover (25), shims (23) and gaskets (24). Remove roller (7) from ball nut (8) and slide the rocker shaft from housing. Unbolt and remove steering column (4), shims (14) and gaskets (15) from gear housing (26). Remove bushing (3) from upper end of steering column. Pull wormshaft (6) upward, then remove upper bearing race (5) and the 10 loose bearing balls (9). Remove the wormshaft and ball nut assembly from gear housing as shown in Fig. 23, then remove the 10 loose bearing balls from gear housing. Unscrew the ball nut assembly from wormshaft and remove the 14 recirculating balls from the nut. Tube (11—Fig. 21) can be removed from nut (8) if necessary. Remove lower bearing race (12), spacer (13), bushings (19 and 21) and oil seal (18) from gear housing (26).

To reassemble, proceed as follows: Install new bushings (19 and 21) using piloted drift or bushing driver, then install new seal (18) with lip to inside of gear housing. Install spacer (13) and lower bearing race (12) in gear housing, then stick the 10 bearing balls in race with grease. Assemble tube (11) to ball nut (8) if removed, then stick the 14 recirculating balls in tube and groove of nut with grease. Thread the ball nut assembly onto wormshaft, then install the shaft and nut assembly in gear housing as in Fig. 23. Carefully insert wormshaft into lower bearing to avoid dislodging any of the bearing balls, then while holding shaft in the bearing, place upper bearing race over shaft and invert the assembly allowing gear housing to rest against end of shaft. Stick the 10 bearing balls in upper race with grease, then push bearing assembly up into housing as shown in Fig. 24. While holding against upper bearing, turn the assembly upright. Install new bushing (3 —Fig. 21) in steering column, then refer to paragraph 30 for wormshaft adjustment and column installation. Insert rocker shaft. Install the steering arm (16) and tighten the retaining nut to a torque of 125 Ft.-Lbs. Place roller on end of ball nut (see Fig. 22) and install side cover with proper shims and gaskets from rocker shaft adjustment as outlined in paragraph 31.

Late Power Steering Models

34. Late power steering models use the integral power assist unit shown schematically in Fig. 26 and exploded in Fig. 27. A piston is built into the shaft ball nut and a cylinder machined into the gearcase housing; and the entire case unit is pressurized by the steering oil. Control is by means of a rotary valve which is built into piston and ball nut unit (12) and is not available separately. The pressure passage to top of piston (P) is internal while lower end is pressurized by external flow through pressure tube (24). The piston and ball nut moves **upward** for a right-hand turn. Manual operation of steering gear is made possible by a check ball (7) located in valve housing which recirculates oil within gear housing when pump is inoperative.

35. **REMOVE AND REINSTALL.** To remove the steering gear assembly, first disconnect battery ground cable and remove steering wheel. Remove sheet metal covers at each side of steering gear. Remove screws retaining instrument panel, disconnect ground terminal (left side) and rotate panel up and out of opening in cowl. Disconnect wiring and remove instrument panel. Remove light switch and

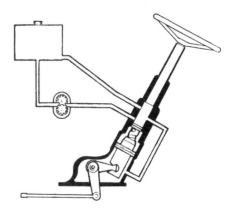

Fig. 26–Schematic view of late power steering gear unit used on some models. Steering cylinder is built on ball nut and gear unit is pressurized. Refer to Fig. 27 for exploded view.

choke (or diesel shut-off cable). On Select-O-Speed Models, refer to paragraphs 245 and 246 for removal of controls. Remove engine hood and cowl. Shut off fuel, disconnect and remove fuel tank. Clean steering gear unit. Disconnect drag link and pump pressure and return lines, then unbolt and lift off steering gear assembly.

NOTE: On Select-O-Speed Models, make sure no dirt or foreign material falls into transmission during removal or while steering gear is off.

To install steering gear, reverse removal procedure. Use a new transmission housing gasket on Select-O-Speed Models. Refill steering gear after complete assembly by cycling power steering, engine running, while keeping pump reservoir filled.

36. **OVERHAUL.** Before disassembling the removed steering gear, temporarily reinstall steering wheel and disconnect external oil feed pipe (24—Fig. 27). Turn steering wheel from lock to lock several times until as much fluid as possible is pumped from housing.

Remove steering arm (15) using Special Tool 1001 or other suitable puller. Remove side cover (23), gasket (20), and rocker shaft end float shim (22). Turn steering shaft until rocker shaft arm is centered in housing opening as shown in Fig. 28, then withdraw rocker shaft.

Remove the four stud nuts securing steering column (1—Fig. 27) and lift off the column and shaft (2). Remove and save shim pack (5). Install the oil seal protecter sleeve (Tool SW 23/1) over steering shaft spline as shown in Fig. 29, gently tap valve housing (6) away from bearing housing (9) and lift off valve housing, saving bearing adjustment shims (8) as valve housing is removed.

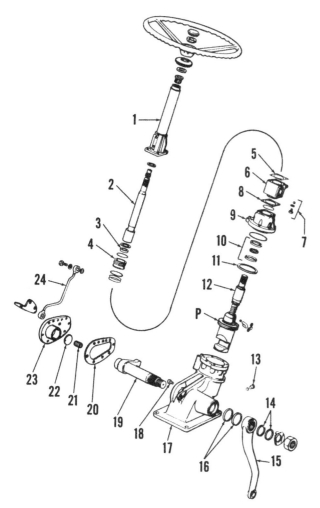

Fig. 27–Exploded view of late power assist steering gear.

1. Steering column
2. Steering shaft
3. Housing seal
4. Bushing sleeve
5. Shim pack
6. Valve housing
7. Check valve
8. Shim pack
9. Bearing housing
10. Bearing
11. Piston ring
12. Ball nut
13. Guide peg
14. Dust seal
15. Steering arm
16. Oil seal
17. Gear housing
18. Wear pin
19. Rocker shaft
20. Gasket
21. Spring
22. Float shim
23. Side cover
24. Pressure tube
P. Piston

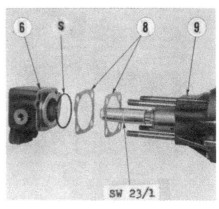

Fig. 29–Oil seal protector sleeve (SW 23/1) should be used as shown. Refer to Fig. 27 for parts identification except for O-ring seal (S).

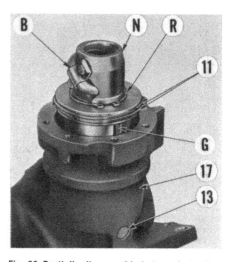

Fig. 30–Partially disassembled view of steering gear. Refer to Fig. 27 for parts identification except for the following:

B. Clamp bracket
G. Locating groove
N. Ball nut
R. Snap ring

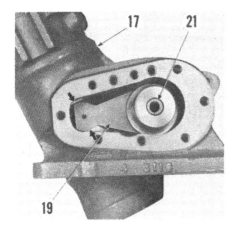

Fig. 28–Rocker shaft can be withdrawn when arm is centered in housing opening as shown. Refer to Fig. 27 for parts identification.

Remove cap screws retaining bearing housing (9—Fig. 27); then, turning splined end of steering shaft counter-clockwise, force bearing housing (9) and bearing (10) up and out of ball nut and main gear housing. Shaft bearing (10) contains fifteen 5/16-inch diameter loose steel balls which are free to fall as the parts are removed. Bearing balls are interchangeable with the 28 steel balls used in steering nut (N—Fig. 30).

Working through side opening, carefully push ball nut (N), piston and associated parts out of main housing. Note that ball nut is prevented from turning in housing by groove (G) which fits over guide peg (13). Be careful not to damage piston rings (11) as piston is withdrawn. Remove clamp bracket (B), transfer tube and the 28 bearing balls from main nut.

Examine all parts for wear or scoring and make sure parts are thoroughly cleaned. Renew all seals, gaskets and O-rings when unit is disassembled, as all parts are under system pressure. O-rings are located on piston guide peg (13); and between bushing sleeves in control valve housing as shown in Fig. 31. Carefully push bushings out top end of housing. Seal (3) must be installed from underside (chamfered) end of bushing sleeve (4), using Special Tool SW23/2 which applies pressure in the groove between the two seal lips. Coat sealing O-rings sparingly with a suitable lubricant and install carefully using Fig. 31 as a guide.

Assemble the steering gear as follows: Install piston rings (11—Fig. 30), if removed, and rotate rings until end gaps are 180° apart. Align groove (G) with locating pin (13) and carefully install piston using a suitable ring compressor. Position ball nut assembly so that rocker shaft arm slot is aligned with main housing side opening (Fig. 28) and install rocker shaft with spring (21) removed. Temporarily install main housing side cover (23—Fig. 27), using a new gasket (20) but omitting end float shim (22) at this time. Install and tighten side cover retaining cap screws (35-45 Ft.-Lbs.); then using a dial indicator, measure and record rocker shaft end float. Remove side cover then reassemble, installing spring (21) and an end float shim (22) which most nearly equals measured end float minus 0.008. Shims (22) are available in thicknesses of 0.050,

Fig. 31–Control valve housing partially disassembled, showing location of bushing sleeves and O-ring seals.

3. Oil seal	L. Lower sleeve
4. Bushing sleeve	M. Middle sleeve
6. Valve housing	S. O-ring seal

0.060, 0.070 and 0.080. Retighten side cover cap screws to 35-45 Ft.-Lbs.

Turn rocker shaft until ball nut and piston unit is at top of its stroke. Install worm shaft and lower half of transfer tube as shown in Fig. 32, then feed in the 28 bearing balls using clean grease. Install upper half of transfer tube and clamp bracket (B—Fig. 30).

Install bearing housing (9—Fig. 27) and tighten retaining cap screws to a torque of 15-20 Ft.-Lbs. Slide lower race of bearing (10) into bearing housing bore, grooved side up, then install the 15 steel balls in bearing groove using clean grease.

Position Oil Seal Protecter Sleeve (SW 23/1) over shaft splines as shown in Fig. 33; then, omitting bearing adjusting shim pack (8—Fig. 27), install control valve housing. Tap housing lightly into place until it bottoms and, using a feeler gage, measure the clearance between bearing housing and valve housing as shown in Fig. 33. Install a shim pack (8—Fig. 27) equal to the measured clearance minus 0.003. Shims (8) are available in thicknesses of 0.004, 0.005, 0.010 and 0.025. Shim pack must be accurate to within 0.0015. Shim pack (8) controls preload of worm gear bearing (10).

With bearing pre-load correctly adjusted, assemble and install steering column, upper steering shaft and bearing washer. Omit shim pack (5) on trial assembly. Make sure steering column is bottomed on steering shaft splines; then measure clearance between steering column flange and valve housing as shown in Fig. 34. Install shim pack (5—Fig 27) equal to the measured clearance PLUS 0.005. Shim (5) is available in 0.005 thickness only. Tighten steering column stud nuts to a torque of 25-30 Ft.-Lbs. Complete the assembly by reversing the disassembly procedure. Tighten ex-

Fig. 32–With piston at top of stroke, install worm shaft (S) and lower half of transfer tube (T), then feed in the 28 loose balls (B) using clean grease. Refer to text.

ternal feed line banjo bolts to 25-30 Ft.-Lbs., steering arm nut to 200-250 Ft.-Lbs. and steering wheel nut to 80-100 Ft.-Lbs. Install steering gear unit as outlined in paragraph 35.

FLUID AND BLEEDING

All Models

CAUTION: The maintenance of absolute cleanliness of all parts is of the utmost importance in the operation and servicing of the hydraulic power steering system. Of equal importance is the avoidance of nicks or burrs on any of the working parts.

37. Recommended power steering fluid is Ford M-2C41A oil. Maintain fluid level to full mark on dipstick on early models with separate reservoir; or bottom of filler neck on late models with integral pump and reservoir. After each 600 hours of operation, it is recommended that filter element be changed and reservoir cleaned.

The power steering system is self-bleeding. When unit has been disassembled, refill reservoir to full level, start and idle engine, and refill if level lowers. Cycle steering gear by turning steering wheel at least five times from lock to lock, maintaining fluid level at or near full mark. System is fully bled when no more air bubbles appear in reservoir and fluid level ceases to lower.

38. **SYSTEM PRESSURE AND FLOW.** Power steering system pressure should be 850 psi for Model 4100 with linkage booster cylinder & valve unit; and 1100 psi for late Model 4100 with integral power assist gear unit.

NOTE: On some early Model 4100 tractors, power steering system pressure was initially adjusted to 1100 psi, but manufacturer recommended reducing the pressure to 850 psi when service was performed on unit.

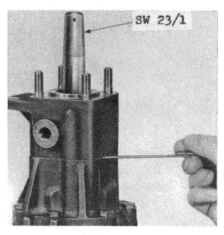

Fig. 33–Using protector sleeve (SW 23/1), install and bottom valve housing without shims; then measure shim pack thickness as shown. Install shims equal to measured clearance minus 0.003.

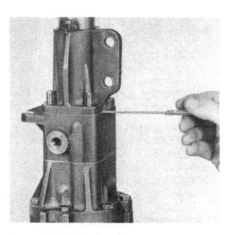

Fig. 34–Make sure steering column is bottomed in steering shaft splines, then measure clearance for upper shim pack as shown. Install shims equal to measured clearance PLUS 0.005.

Early models with separate steering system reservoir were equipped with a flow control valve which maintained a regulated fluid flow of 3.5 gpm at 1000 engine rpm. On late model pumps with integral reservoir, a flow control valve is not used and normal pump flow at 1000 engine rpm is 2.74 gpm.

On all models, pressure and flow can be checked by teeing into pump pressure line. On early models, pressure relief valve cap plug (7—Fig. 7) is externally located and pressure can be adjusted without disassembly. On late models with integral pump reservoir, pump must be removed for relief valve adjustment, refer to paragraph 12.

HYDRAULIC PUMP

All Models

39. Except for relief valve pressure settings, power steering pumps are similar to those used on Models 2000 and 3000, and overhaul procedure is outlined in paragraph 12.

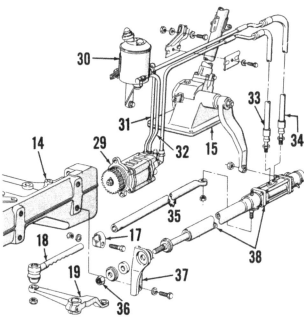

Fig. 35–Power steering system for model 4100 tractor. Front axle and steering linkage is same as for manual steering shown in Fig. 21, except that drag link (35) is shorter and attaches to steering cylinder instead of steering gear arm. Refer to Fig. 36 for later type cylinder rod mounting.

14. Front support
15. Steering gear assy.
17. Clamp
18. Drag link front end
19. Steering arm, R.H.
30. Power steering reservoir
31. Pump to reservoir bypass line
32. Reservoir to pump line
33. Pressure line
34. Return line
35. Drag link
36. Nut
37. Bracket
38. Power cylinder and control valve assy.

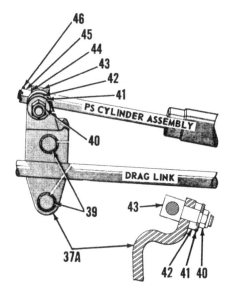

Fig. 36–Views showing installation of latter type power steering cylinder rod mounting bracket. The above parts may be installed on early production tractors if desired.

37A. Bracket
39. Cap screws
40. Nylock nut
41. Washer
42. Buffer pad
43. Connector
44. Washer
45. Castellated nut
46. Cotter pin

POWER CYLINDER AND CONTROL VALVE
Early Model 4100

40. **REMOVE AND REINSTALL.** To remove the power cylinder and control valve assembly, refer to Figs. 35 and 36, then proceed as follows:

Disconnect fluid lines (33 & 34—Fig. 35) from cylinder and allow fluid to drain. Disconnect drag link (35) from cylinder. Remove nut from front end of piston rod and disconnect cylinder from steering gear arm. Then remove cylinder and valve assembly from tractor. To overhaul the unit, refer to paragraphs 41 through 44.

Reinstall power steering cylinder and valve assembly by reversing removal procedure. If removed, tighten cylinder rod bracket to front support cap screws to a torque of 200-250 Ft.-Lbs. On later production units, tighten the connector Nylock nut (40—Fig. 36) to a torque of 80-100 Ft.-Lbs. Refill and bleed the system as outlined in paragraph 10.

41. **RENEW PISTON ROD SEALS.** To renew the piston rod seals, refer to exploded view of the power cylinder and control valve unit in Fig. 37 and proceed as follows:

If not already removed, slide the insulator (2) and rear anchor washer (3) from piston rod. Pry the scraper assembly (5) from end of cylinder and remove snap ring (6), washer (7), retainer (8), seal (9) and seal spreader (10).

Thoroughly clean and inspect the piston rod for scoring or burrs and inspect cylinder for any damage in seal bore. If either are damaged, the com-

plete assembly (11) must be renewed; refer to paragraph 42 for installation procedure. Note: A new cylinder and piston rod assembly includes new seals and scraper.

Install new seal kit by reversing removal procedure. When installing seal (9), work lips of seal into cylinder with dull pointed tool. Drive scraper into cylinder after installing snap ring (6).

42. RENEW POWER CYLINDER UNIT. With power cylinder and control valve assembly removed from tractor as outlined in paragraph 40, proceed as follows:

Cap the openings in control valve housing (27—Fig. 37) and thoroughly clean outside of unit. Remove the four cap screws (48) and separate cylinder from control valve unit; be careful that valve housing (27) does not slide off of valve spool and lay the control valve and ball socket unit aside. Remove spacer (18), shims (17), spring (16) and outer ball seat (14). Slide the ball stud inward from rubber boot (4) and out open end of cylinder. Remove remaining ball seat (14), grease fitting (12) and pin (13) if pin remained with cylinder instead of staying in control valve housing. Clean and inspect the removed parts and renew any not suitable for further use.

To install the new power cylinder and reinstall removed parts, proceed as follows: Place boot (4) over cylinder so that hole in boot is aligned with ball stud hole. Insert one ball seat (14), then insert ball stud (15), threaded end first, into open end of cylinder and extend the threaded end out through hole in

cylinder and boot. Insert second ball seat (14) and shims (17) that were removed plus at least an additional two 0.010 thick shims. Insert spring housing (18) in end of cylinder, but do not install spring (16) at this time. Assemble the cylinder to control valve unit and tighten the four cap screws (48) evenly. This should leave a gap between end of cylinder and control valve housing; if not, remove control valve unit and spring housing frcm cylinder and add additional shims. Reassemble cylinder unit and spring housing to control valve, then measure gap between cylinder and valve housing with feeler gage as shown in Fig. 39. Record the gap measurement, then remove control valve unit and spring housing from cylinder and remove shim thickness equal to the gap measurement plus an additional thickness of 0.008-0.010. Install remaining shims, spring housing and spring (16) in end of cylinder, then assemble control valve unit to cylinder with two new passage sealing "O" rings inserted in grooves of valve housing. Tighten the four cap screws evenly and securely, then reinstall cylinder and valve assembly on tractor as outlined in paragraph 40.

43. **RENEW OR OVERHAUL CONTROL VALVE.** With power cylinder and control valve unit removed from tractor as outlined in paragraph 40, proceed as follows:

Thoroughly clean outside of unit and remove the four cap screws (48—Fig.

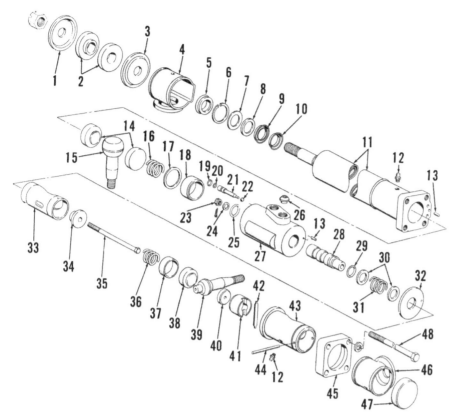

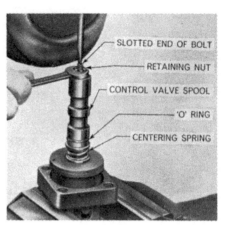

Fig. 38–Valve spool bolt has screwdriver slot in threaded end to facilitate removing and installing spool retaining nut.

Fig. 37–Exploded view of the power cylinder and control valve unit (38–Fig. 35). Cylinder tube, piston and rod are serviced as a complete unit (11) only and cannot be disassembled.

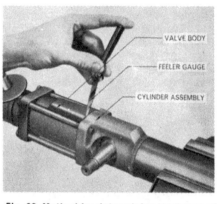

Fig. 39–Method for determining thickness of shims (17–Fig. 37) required for reassembly of cylinder. Refer to text for procedure.

1. Front anchor washer		25. "O" ring	36. Spring
2. Insulators	14. Ball stud cups	26. Tubing seats	37. Spacer
3. Rear anchor washer	15. Ball stud	27. Control valve	38. Ball stud cup
4. Rubber boot	16. Spring	housing	39. Ball stud
5. Scraper	17. Shims (0.002 &	28. Control valve spool	40. Ball stud cup
6. Snap ring	0.010)	29. "O" ring	41. Cup holder
7. Washer	18. Spring housing	30. Washers	42. Retaining clip
8. Retainer	19. "O" ring	31. Spring	43. Sleeve
9. Seal	20. "O" ring	32. Spacing disc	44. Retaining pin
10. Seal spreader	21. Check valve pin	33. Sleeve	45. Clamp
11. Cylinder, piston &	22. Check valve ball	34. Collar	46. Rubber boot
rod assembly	23. Castellated nut	35. Valve spool bolt	47. Cover (cup)
12. Grease fittings	24. Washer		48. Cap screws (4)
13. Dowels			

37). Separate the power cylinder from control valve and slide the valve housing (27) from valve spool. Place the ball socket and valve spool unit upright in vise as shown in Fig. 38. Remove cotter pin from valve spool bolt (35—Fig. 37), hold bolt from turning with screwdriver and remove the castellated nut (23) and flat washer (24). Remove valve spool (28), washers (30), centering spring (31) and spacer disc (32) from bolt. Remove "O" ring (25) from valve housing and "O" ring (29) from valve spool. Remove check valve pin (21) with a stiff wire, then remove check valve ball (22) from valve housing. Remove "O" ring (20) from valve pin.

Thoroughly clean and inspect the control valve parts. If either the spool or housing is damaged beyond further use, a new matched valve and valve housing must be installed.

Reassemble using all new "O" rings as follows: Install "O" ring (20) on check valve pin (21), drop ball (22) in

hole in valve housing, lubricate "O" ring and insert pin, small end first, into hole. Place spacer disc (32), washers (30) and the centering spring (31) on spool bolt. Place new "O" ring (29) in groove on rear end of valve spool and place spool on bolt with small end inside washer and centering spring. Install flat washer (24) and nut (23), turn nut down tight (be sure valve spool enters second washer), then back nut off to first castellation in which retaining cotter pin can be inserted. Be sure that the roll pins (13) are in place in each end of valve housing. Renew tubing seats (26) in valve housing if necessary. Insert "O" ring (25) in groove in forward end of bore in valve housing, lubricate "O" rings and slide the valve housing over valve spool. Be sure pin in housing enters hole in spacer disc. Place new "O" rings (19) in grooves around oil passages in front end of control valve, then assemble valve and ball socket unit to power cylinder. Note: If a new control valve

housing and spool are being installed, it may change requirements for shim thickness (17); refer to paragraph 42. Tighten the four retaining cap screws evenly and securely. Reinstall power cylinder and control valve assembly as outlined in paragraph 40.

44. **OVERHAUL ACTUATING BALL STUD ASSEMBLY.** With control valve spool (28—Fig. 37) removed as outlined in paragraph 43 proceed as follows:

Remove boot (46) and pry lip of cup (47) from groove on rear end of outer sleeve (43). Using a punch, drive out pin (44). Pry out retainer (42) and unscrew ball seat holder (41). Use a screwdriver to hold ball seat against spring (36) while removing ball stud (39). Then, remove the seat (38), spacer (37), spring (36), bolt (35) and collar (34) from inner sleeve, then separate inner and outer sleeves.

Clean and carefully inspect all parts. If there is considerable wear or damage, the ball socket unit (items 33 through 47) is available as a complete assembly.

To reassemble, proceed as follows: Place clamp (45) on outer sleeve (43). Insert collar (34), bolt (35), spring (36), spacer (37), and ball seat (38) in inner sleeve and insert the assembly into outer sleeve. Using a screwdriver to hold seat (38) against spring, insert ball stud (39) through large holes in sleeves. Insert ball seat (40) in holder (41), screw holder into inner sleeve until tight, then back holder out approximately ¼ turn and install new retainer (42). Drive pin (44) into hole through rear end of outer sleeve and install boot (46). Crimp edge of new end cap (47) into groove around outer sleeve. Reassemble control valve and housing onto bolt (35) as outlined in paragraph 43.

FRONT SYSTEM AND STEERING

(MODEL 4200)

45. Model 4200 Rowcrop tractors are equipped with a front pedestal adaptable to a wide adjustable front axle, a single front wheel or a dual wheel tricycle type front end. All versions are equipped with a hydrostatic power steering system which consists of an engine driven pump and separate fluid reservoir, a Saginaw Hydramotor steering unit and two double acting steering cylinders mounted in the front pedestal.

WIDE ADJUSTABLE FRONT AXLE
Model 4200

46. **R&R FRONT AXLE ASSEMBLY.** To remove the wide front axle assembly as a unit, proceed as follows: Straighten tabs of locking plate (7—Fig. 40) and remove the two cap screws retaining center steering arm (8) to steering shaft (34—Fig. 43). Support front of tractor, unbolt the front (4) and rear (16) pivot pin supports from front pedestal and roll the front axle and wheel assembly forward.

Reinstall front axle assembly by reversing removal procedure.

47. **SPINDLE BUSHINGS.** To renew the spindle bushing (24 and 26—Fig. 40), proceed as follows: Support front of tractor and disconnect steering arms (21) from wheel spindles (28). Slide the wheel and spindle assemblies (remove front wheels from hubs if desired for clearance) out of the front axle extension (25). Drive old bushings from front axle extensions and install new bushings with piloted drift or bushing driver. New bushings are final sized and should not require reaming if carefully installed. Renew thrust bearing (27) if worn or rough.

48. **AXLE CENTER MEMBER, RADIUS ROD, PIVOT PINS AND BUSHINGS.** To remove the axle center member (3—Fig. 40), support front of tractor, unbolt axle extensions from center member and swing the axle extension and wheel assemblies aside. Remove retainer (1) and unbolt radius rod (13) from center member, then remove center member from pivot pin. Remove bushing (2) from center member.

With axle center member removed, radius rod can then be removed by removing bolt (15) and pin (17), then lowering radius rod from tractor. To remove radius rod with front axle in place, unbolt radius rod from center member and unbolt rear pivot pin support (16) from front pedestal. Move the unit rearward until radius rod is free of axle center member, then lower unit from tractor. Remove bolt (15), pin (17) and support (16) from radius rod. Note: Be careful not to lose the thrust washers (12). Remove bushing (14) from radius rod.

To renew pivot pins and bushings, proceed as follows: Remove steering arm (8) from steering shaft (34—Fig. 43). Support front end of tractor so as to take weight from pivot pins and support front axle radius rod to keep unit from tipping. Remove retainers (1 and 6—Fig. 40) from front pivot pin and remove bolt (15) from rear pivot pin. Remove both pivot pins, raise tractor to clear front axle and roll the unit forward. Remove bushings (2 and 14) and carefully install new bushings. Bushings are final sized and should not require reaming if not distorted during installation. Renew thrust washers (12) and pivot pins if worn and reassemble by reversing removal procedure.

49. **DRAG LINKS AND TOE-IN.** Drag link (10 and 20—Fig. 40) are of the non-adjustable automotive type and procedure for renewing same is evident.

Toe-in should be correct in each tread width adjustment when clamp bolt of drag link (18) is in proper notch of outer end assembly (20). However, if toe-in is not ¼ to ½-inch, adjust as follows: Remove clamp bolt at outer end and loosen lock nut at inner end of each tie-rod. Then, vary length of each tie-rod an equal amount to obtain correct toe-in when clamp bolts are installed. Retighten lock nuts and clamp bolts when adjustment is correct.

TRICYCLE FRONT SPINDLES
All Models So Equipped

50. Refer to Figs. 41 and 42 for exploded views of the single and dual front spindle assemblies. Either unit can be bolted to the steering shaft (34—Fig. 43). Single wheel spindle must be installed with offset to left side of tractor as shown in Fig. 41 and dual

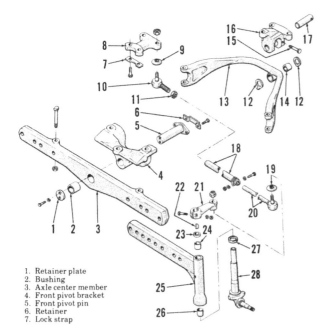

Fig. 40–Exploded view of model 4200 wide adjustable front axle assembly. Unit is interchangeable with single front wheel spindle (Fig. 41) or dual wheel tricycle spindle (Fig. 42) Note: Only the left axle extension, spindle, steering arm and tie rod are shown.

8. Center steering arm
9. Dust seal
10. Tie rod inner end
11. Lock nut
12. Thrust washers
13. Radius rod
14. Bushing
15. Retainer bolt
16. Rear pivot bracket
17. Rear pivot pin
18. Tie rod
19. Dust seal
20. Tie rod outer end
21. Steering arm
22. Woodruff key
23. Dust seal
24. Upper spindle bushing
25. Axle extension
26. Lower spindle bushing
27. Thrust bearing
28. Spindle

1. Retainer plate
2. Bushing
3. Axle center member
4. Front pivot bracket
5. Front pivot pin
6. Retainer
7. Lock strap

wheel spindle must be installed with knife edge to front as shown in Fig. 42 so that proper caster will be obtained. Note: Be sure to deflate tire before attempting to disassemble the single wheel disc (38—Fig. 41) and flange (39).

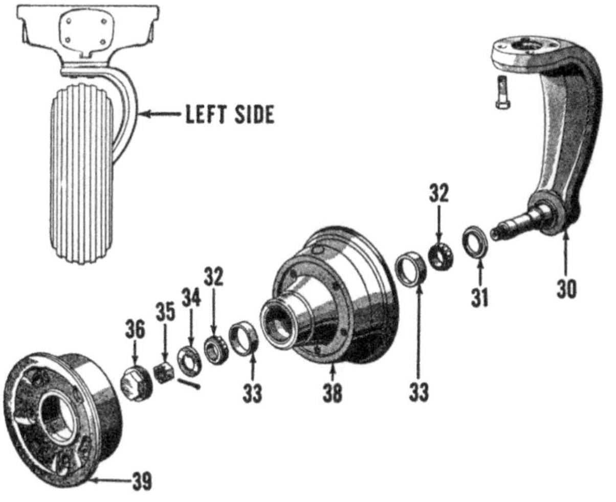

FRONT SUPPORT
Model 4200

51. **R&R FRONT PEDESTAL ASSEMBLY.** First, drain cooling system, then remove engine hood and radiator grille assembly. Disconnect radiator hoses and remove radiator from pedestal. Remove covers from top of pedestal and disconnect power steering lines (17 and 18—Fig. 43) from manifold (39). Support front end of tractor and remove wide front axle assembly as outlined in paragraph 52, or if a tricycle model, unbolt and remove the spindle and wheel assembly from steering shaft. Attach a hoist to front support, then unbolt and remove front support from engine cylinder block and oil pan. Note: Be careful not to lose or damage shims, if any, located on front support to oil pan bolts.

52. Reinstall front pedestal by reversing removal procedure, then priming and refilling the power steering system as outlined in paragraph 54. However, care should be taken that the proper number (thickness) of shims are installed between pedestal and oil pan when pedestal is being installed. The correct shimming procedure is as follows:

Check to be sure the oil pan to transmission bolts are tightened to a torque of 220-300 Ft.-Lbs. Attach pedestal to engine and tighten the four bolts to a torque of 200-240 Ft.-Lbs.; do not install the pedestal to oil pan bolts at this time. Using a feeler gage, determine clearance between pedestal and oil pan at the two bolting points, then install pedestal to oil pan bolts with this total thickness of shims. Shims are available in thicknesses of 0.014-0.015, 0.017-0.018, 0.020-0.022, 0.023-0.025 and 0.026-0.028. Tighten the two pedestal to oil pan bolts to a torque of 200-240 Ft.-Lbs.

53. **OVERHAUL PEDESTAL ASSEMBLY.** With the pedestal removed from tractor as outlined in paragraph 51, proceed as follows:

Remove the nuts from power cylinder rear end assemblies and bump rear ends of cylinders from front support. Remove the cap screws (23—Fig. 43) clamping cylinder front end assemblies in steering shaft arm (22) and remove cylinder from arm. Unbolt manifold (39) from pedestal and lift out manifold, lines and cylinders as a unit.

Fig. 41–Exploded view of tricycle single front wheel spindle and wheel assembly. Unit is interchangeable on model with wide adjustable front axle or dual front wheel and pedestal assembly.

30. Spindle
31. Retainer
32. Bearing cone
33. Bearing cup
34. Washer
35. Nut
36. Hub cap
38. Wheel disc
39. Wheel flange

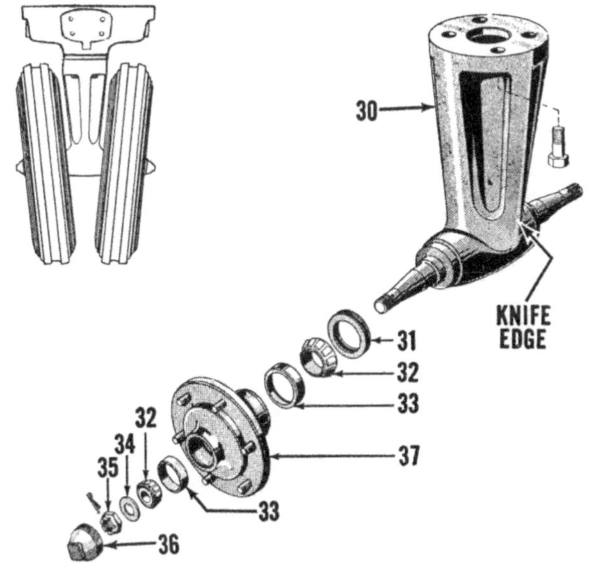

Fig. 42–Exploded view of dual wheel tricycle spindle and wheel hub unit. Note that knife edge of spindle must be towards front of tractor. Unit is interchangeable with wide adjustable front axle or single front wheel.

30. Spindle
31. Retainer
32. Bearing cone
33. Bearing cup
34. Washer
35. Nut
36. Hub cap
37. Hub

Refer to paragraph 65 for service information on the power cylinders.

Pry cap (25) from pedestal and remove cotter pin and castellated nut (26). Bump steering shaft (34) downward out of pedestal and lift out the upper bearing cone and roller assembly (27) and steering arm (22). Remove lower bearing cone and roller assembly (32) from steering shaft and remove bearing cups (28 and 31) and seals (29, 30 and 33) from the pedestal.

To reassemble, drive new bearing cups (28 and 31) into pedestal until firmly seated. Drive or press lower bearing cone and roller assembly (32) onto steering shaft until seated. Install lower seal (5.51 O.D.) (33) into pedestal with lip up. Install the 3.35 O.D. seal (30) with lip down and the 2.25 O.D. seal (29) with lip up. Pack the bearings with grease and while holding steering arm (22) in position, insert steering shaft up through pedestal. Insert upper cone and roller assembly (27) into cup and onto shaft and install nut (26).

Tighten nut so that all steering shaft end play is removed and bearings are slightly preloaded, then install cotter pin and cap (25). Reconnect steering cylinders to steering arm and pedestal and install manifold (39) to pedestal retaining cap screws. Reinstall pedestal as outlined in paragraph 52.

POWER STEERING SYSTEM
Model 4200

CAUTION: The maintenance of absolute cleanliness of all parts is of the utmost importance in the operation and servicing of the hydraulic power steering system. Of equal importance is the avoidance of nicks or burrs on any of the working parts.

54. **FLUID AND BLEEDING.** Recommended power steering fluid is Ford M-2C41A oil. Maintain fluid level to full mark on dipstick. After each 600 hours of operation, it is recommended that filter element be changed and reservoir cleaned.

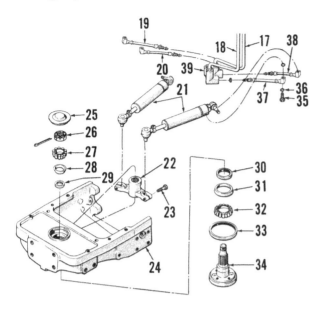

Fig. 43–Exploded view of model 4200 pedestal, steering shaft and steering cylinders. Fluid lines (17 and 18) lead to "Hydramotor" steering unit.

17. Left turn fluid line
18. Right turn fluid line
19. Left turn fluid line
20. Right turn fluid line
21. Power cylinders
22. Steering arm
23. Clamp bolts
24. Front pedestal
25. Cap
26. Castellated nut
27. Bearing cone
28. Bearing cup
29. Grease seal
30. Grease seal
31. Bearing cone
33. Grease seal
34. Steering shaft
35. Drilled bolt
36. "O" rings
37. Right turn fluid line
38. Left turn fluid line
39. Manifold

WILL NOT STEER IN EITHER DIRECTION. The manual steer check ball between pump return and pressure passages in Hydramotor unit may not be seating. Disassemble unit and clean passage with solvent and dry with compressed air. Renew Hydramotor unit if check ball cannot be made to seat.

FRONT WHEELS JERK OR TURN WITHOUT MOVING STEERING WHEEL. Check for sticking rotor vanes, rotor springs out of place or broken, scored pressure plate, scored rotor ring, scored housing, ball check valves in pressure plate leaking, improper assembly causing gap between rotor components. Disassemble the Hydramotor unit, carefully clean and inspect all parts and renew components or complete Hydramotor unit if necessary.

The power steering system is self-bleeding. When unit has been disassembled, refill reservoir to full level, start and idle engine, and refill if level lowers. Cycle steering gear by turning steering wheel fully in each direction at least five times, maintain fluid level at or near full mark. System is fully bled when no more air bubbles appear in reservoir, steering action is steady and solid, and fluid level ceases to lower.

55. SYSTEM PRESSURE AND FLOW. The power steering pump assembly incorporates a pressure relief valve and a flow control valve. System relief pressure should be 1100 psi. All models are equipped with a flow control valve which maintains a regulated flow of 3.5 gpm at 1000 engine rpm.

56. STEERING SYSTEM TROUBLESHOOTING. Refer to the following paragraphs for checking causes of steering system malfunction:

HARD STEERING. Check column bearings and bearings in Hydramotor unit; renew if rough or damaged. Check ring, rotor and vanes for wear and renew the assembly if necessary. Check for sticking control valve spool or blocking spool in Hydramotor; clean valves or renew Hydramotor unit as required.

EXCESSIVE WHEEL DRIFT. Check blocking spool spring and guide assembly and renew if spring is broken. Check for leakage past blocking valve; if excessive, renew Hydramotor unit. Check seals on steering cylinder pistons and renew pistons and/or cylinder inner tubes as required.

STEERING WHEEL TURNING UNAIDED. Check the Hydramotor unit for sticking control valve spool, broken valve spool spring, acuator

shaft binding or torque shaft (inside actuator shaft) broken. Clean spool and bore or renew Hydramotor unit as required.

STEERING WHEEL SLIPPAGE. Hydramotor control valve spool scored (renew Hydramotor unit) or rotor seals leaking (renew seals).

EXCESSIVE NOISE. Hydraulic lines vibrating against tractor frame or broken control valve spool spring; insulate lines from tractor or renew Hydramotor unit if spring is broken.

ERRATIC MOVEMENT OF FRONT WHEELS. Check Hydramotor ring, rotor or vanes for scoring, wear or binding condition; renew the ring and rotor assembly if necessary.

57. R&R AND OVERHAUL PUMP. Except for relief valve pressure setting, pumps are similar to early power steering pump used on Models 2000 and 3000, and overhaul procedures are outlined in paragraph 12.

58. HYDRAMOTOR STEERING UNIT. Refer to following paragraphs for information on removal, overhaul and reinstalling the Saginaw Hydramotor power steering unit. Refer to paragraph 56 for troubleshooting information.

59. R&R HYDRAMOTOR UNIT. To remove the steering unit, first remove fuel tank. Then, proceed as follows:

Unbolt and remove the steering wheel and shaft assembly from fuel

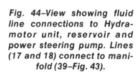

Fig. 44–View showing fluid line connections to Hydramotor unit, reservoir and power steering pump. Lines (17 and 18) connect to manifold (39–Fig. 43).

1. Power steering pump
2. Fluid reservoir
3. Pump to reservoir by-pass line
4. Reservoir to pump suction line
5. Connector
6. Seal
7. "O" ring
8. Pump to Hydramotor pressure line
9. Hydramotor to reservoir return line
10. Pump to Hydramotor rear line
11. Hydramotor to reservoir rear line
12. Adapter connection
13. Steering motor and support unit
14. Right turn rear fluid line
15. Left turn rear fluid line
16. Support bracket
17. Right turn fluid line
18. Left turn fluid line

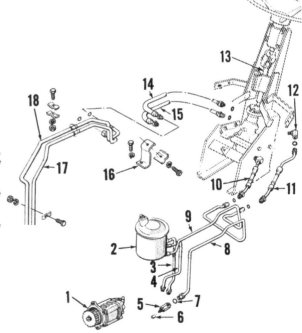

tank and steering support. Disconnect the two cylinder lines and the pump pressure and fluid return lines from steering motor (21—Fig. 45). Unbolt the bracket and motor (21), then remove bracket and motor from support (18).

Reverse removal procedure to reinstall the steering motor unit using new "O" rings when reconnecting fluid lines. Before reinstalling fuel tank, run engine on nurse tank to fill and bleed the power steering system as outlined in paragraph 54. After being sure no fluid leakage is occuring, reinstall fuel tank. CAUTION: The blocking (reaction) valve adjuster (1—Fig. 48) must be in position shown in Fig. 50 before reinstalling fuel tank.

60. R&R BLOCKING SPOOL (RE-ACTION) VALVE. The blocking spool valve and related parts can be removed and reinstalled after the Hydramotor steering unit has been removed as outlined in paragraph 59. Refer to Fig. 48 and proceed as follows:

Remove the lockout adjuster nut (1). Plug (3) and spool valve (4) may now be removed by pushing the plug into bore against spring pressure with screwdriver, then quickly releasing the plug to allow spring to pop it out of bore. Remove plug and, if spool sticks in bore, invert the unit and tap housing (12) with soft faced mallet to jar spool out. Invert unit and allow spring (5) and spring and guide assembly (6) to drop from bore.

Spool is not serviced separately from complete Hydramotor unit. Renew other parts as necessary, or install new Hydramotor unit if spool and/or spool bore in housing are not serviceable.

To reassemble, install parts in bore of housing (12) as shown in exploded view, renewing the "O" ring (2) on plug (3) and tightening adjuster nut to a torque of 10-15 Ft.-Lbs. Note: The adjuster (1) is not accessible after tractor is fully assembled; thus, the adjuster must be in position shown in Fig. 50 when unit is being reinstalled.

61. R&R COVER RETAINING SNAP RING. To remove snap ring (7—Fig. 46) used to retain cover (30) to housing (12), proceed as follows:

With unit removed from tractor as outlined in paragraph 59, check to see that end gap of snap ring is near hole in cover as shown in Fig. 49; if not, bump snap ring into this position with hammer and punch. Insert a pin punch into hole and drive punch inward to dislodge snap ring from groove. Hold punch under snap ring and pry ring from cover with screwdriver. Usually, the coil spring (27—Fig. 46) will push housing from cover; if not, bump cover

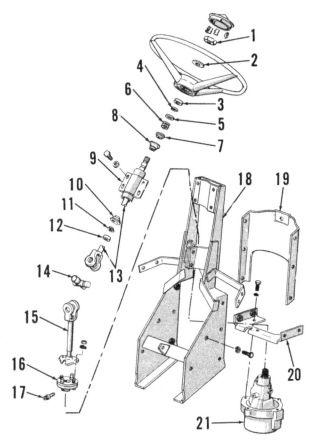

Fig. 45–Exploded view of steering shaft assembly, support and motor mounting bracket. Refer to Fig. 46 for exploded view of the Saginaw Hydramotor steering unit (21).

1. Steering wheel nut
2. Flat washer
3. Dust seal
4. Bearing retaining ring
5. Washer
6. Spring
7. Bearing seat
8. Bearing assembly
9. Steering shaft jacket
10. Bearing assembly
11. Bearing seat
12. Spacer
13. Upper steering shaft
14. Universal joint assembly
15. Lower steering shaft
16. Connecting flange
17. Special bolt
18. Fuel tank and steering motor support
19. Steering motor cover
20. Steering motor bracket
21. Steering motor (Hydramotor)

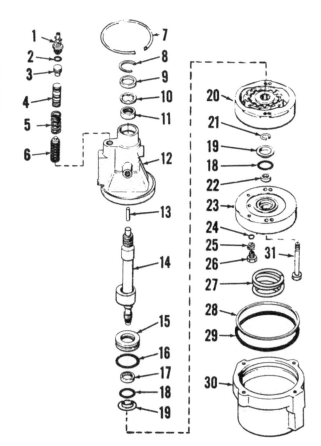

Fig. 46–Exploded view of the Saginaw Hydramotor steering unit. Also refer to Figs. 48 to 61 for photos showing disassembly and reassembly techniques.

1. Blocking valve adjuster
2. "O" ring
3. Plug
4. Blocking valve
5. Spring
6. Spring and guide assembly
7. Snap ring
8. Snap ring
9. Dust seal
10. Oil seal
11. Needle bearing
12. Housing
13. Dowel pins (2)
14. Actuator shaft assembly
15. Bearing support
16. "O" ring
17. Needle bearing
18. "O" ring
19. Rotor seal ring
20. Ring, rotor & vane assy.
21. Snap ring
22. Needle bearing
23. Pressure plate assembly
24. Check valve balls (2)
25. Check valve springs (2)
26. Retaining plugs (2)
27. Pressure plate spring
28. Back-up ring
29. "O" ring
30. Cover
31. Cap screws

Fig. 47–Pressure plate retaining cap screw heads must be positioned as shown to allow installation of cover.

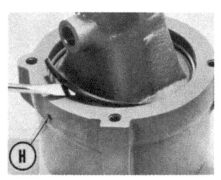

Fig. 49–To remove cover retaining snap ring, drive pin punch through hole in cover to disengage snap ring from groove.

Fig. 51–Using sleeve and arbor press to push housing into cover to allow installation of the cover retaining snap ring (7).

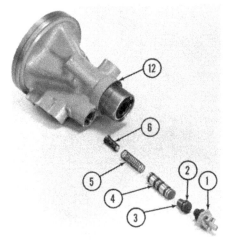

Fig. 48–Exploded view of Hydramotor housing and blocking valve components. Blocking valve can be removed without disassembly of Hydramotor.

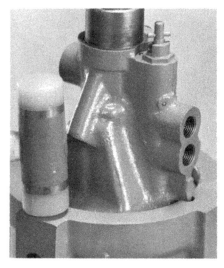

Fig. 50–In event coil pressure plate spring does not push cover from housing, tap alternate mounting lugs on cover with soft face mallet as shown above.

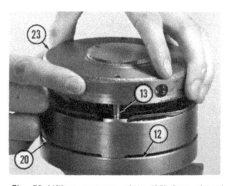

Fig. 52–Lifting pressure plate (23) from dowel pins (13).

1. Lockout
2. "O" ring
3. Plug
4. Blocking valve
5. Spring
6. Spring & guide assy.
12. Housing

loose by tapping around edge with mallet as shown in Fig. 50.

To reinstall the cover retaining snap ring, housing must be held in cover against pressure from spring. It is recommended that the unit be placed in an arbor press and the housing be pushed into cover with a sleeve as shown in Fig. 51. CAUTION: Do not push against end of shaft (14—Fig. 46). Place snap ring over housing before placing unit in press. Carefully apply force on housing with sleeve until flange on housing is below snap ring groove in cover. Note that lug on housing must enter slot in cover. If housing binds in cover, do not apply heavy pressure; remove unit from press and bump cover loose with mallet as shown in Fig. 50. When housing has been pushed far enough into cover, install snap ring in groove with end gap near hole in the cover as shown in Fig. 49.

62. OVERHAUL HYDRAMOTOR STEERING UNIT. With the unit removed from tractor as outlined in paragraph 59 and the cover retaining snap ring removed as outlined in paragraph 61, proceed as follows:

Clamp flat portion of hydramotor housing in a vise and remove cover (30 —Fig. 46) by pulling upward with a twisting motion. Remove the pressure plate spring (27), then remove cap screws (51) and lift off the pressure plate (23) as shown in Fig. 52. Remove the dowel pins (Fig. 53), then remove snap ring (21) from shaft (14) with suitable snap ring pliers and screwdriver; discard the snap ring. Pull pump ring and rotor assembly (20) up off of shaft as shown in Fig. 54. Tap outer end of shaft with soft faced mallet as shown in Fig. 55 until bearing support (15) can be removed, then carefully remove the actuator shaft assembly from housing as shown in Fig. 57. Note: As the actuator shaft and control valve spool assembly is a factory balanced unit and is not serviceable except by renewing

the complete Hydramotor steering unit, it is recommended that this unit not be disassembled.

Carefully clean and inspect the removed units. Refer to paragraph 60 for information on the blocking valve assembly. If the housing control valve bore or blocking valve bore are deeply scored or worn, or if the blocking valve spool or the actuator shaft and control valve spool assembly are damaged in any way making the unit unfit for further service, a complete new Hydramotor steering unit must be installed. If these components (housing, blocking valve and actuator assembly) are serviceable, proceed with overhaul as follows:

Remove the check valve retainers (26—Fig. 46), springs (25) and check valve balls (24) from pressure plate (23) and blow passages clear with compressed air. Renew the pressure plate assembly if check valve seats or face of plate are deeply scored or damaged.

Fig. 53–Removing dowel pins (13) from motor ring and housing. Then, remove the snap ring (21) retaining rotor to actuator shaft (14).

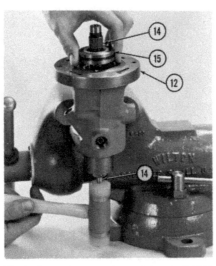

Fig. 55–Tap on outer end of actuator shaft (14) to bump bearing support (15) from the housing (12).

Fig. 57–Removing the actuator shaft assembly from housing. Be careful not to cock control valve spool in bore of housing.

Fig. 54–Lifting the motor ring, rotor and vane assembly (20) from actuator shaft (14) and housing (12).

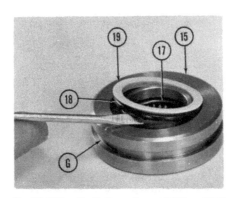

Fig. 56–Lifting the Teflon rotor seal (19) and "O" ring (18) from bearing support (15). Needle bearing (17) is serviced separately from bearing support. Groove (G) is for support sealing "O" ring (16–Fig. 46). Identical seals (18 and 19) are used in pressure plate.

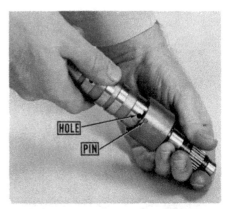

Fig. 58–Pin in actuator sleeve must be engaged in hole in end of control valve spool before actuator assembly is installed. If spool cannot be pulled out of sleeve, pin is engaged.

Renew needle bearing (22), springs (25) and/or check valve balls (24) if damaged and pressure plate is otherwise serviceable. Note: Drive or press on lettered (trademark) end of bearing cage when installing new needle bearing.

If bearing support (15) is otherwise serviceable, a new needle bearing (17) may be installed; drive or press only on lettered end of bearing cage.

Remove snap ring (8), dust seal (9) and oil seal (10) and inspect the needle bearing (11); renew needle bearing if worn or damaged. Press only on lettered (trademark) end of bearing cage when installing new bearing. Install oil seal with lip towards inside (needle bearing), then install dust seal and retaining snap ring.

If the rotor ring, rotor or vanes are worn, scored or damaged beyond further use, or if any of the vane springs are broken, renew the unit as a complete assembly (20). If unit was disassembled and is usable, reassemble as

follows: Place ring on flat surface and place rotor inside ring. Insert vanes with rounded edges out in the rotor slots aligned with large inside diameter of ring, turn the rotor ¼-turn and insert remaining vanes. Hook the springs behind the vanes with a screwdriver as shown in Fig. 60, then turn the assembly over and hook springs behind the vanes on opposite side of rotor.

To reassemble the Hydramotor unit, place housing (with needle bearing, seals and snap ring installed) in a vise with flat (bottom) side up. Check to be sure that pin in actuator is engaged with hole in valve spool; if spool can be pulled away from actuator as shown in Fig. 58, push the spool back into actuator and be sure pin is engaged in one of the holes in spool. Then, lubricate spool and shaft and carefully insert the assembly into bore of housing. Place bearing support, with outside "O" ring and needle bearing installed,

on shaft and carefully push the support into housing as shown in Fig. 59. Insert rotor sealing "O" ring and rotor seal into bearing support. Place ring and rotor assembly on shaft and housing with chamfered outer edge of ring up (away from housing). Install a new rotor retaining snap ring and insert the dowel pins through ring into housing. Using heavy grease, stick the "O" ring and rotor sealing ring into pressure plate, then install the pressure plate assembly over shaft, pump ring and rotor assembly and the two dowel pins. Install the two pressure plate retaining cap screws (31—Fig. 46), tighten them to 96-120 In.-Lbs. of torque, then back off the two cap screws ½ to ¾ turn. Be sure bolt head corners are within perimeter of cover as shown in Fig. 47. NOTE: A minimum torque of 10 In.-Lbs. must remain on each cap screw after they have been backed off. Place the coil spring on top of pressure plate. Install new "O" ring and backup ring

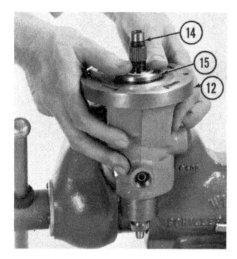

Fig. 59–When pushing bearing support (15) into housing (12), be careful not to damage the "O" ring on outside groove of support.

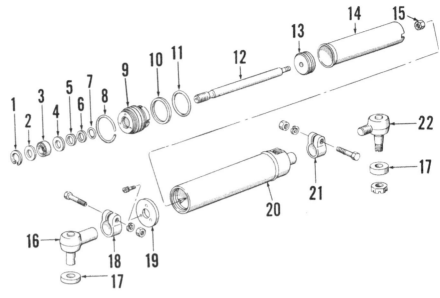

Fig. 62–Exploded view of a model 4200 power steering cylinder. Refer to Fig. 43 for view showing mounting of the power cylinders in front pedestal. Plate (19) and snap ring (8) retain packing gland (9) in cylinder.

1. Snap ring	7. "O" ring	13. Piston & seal assy.	18. Clamp
2. Washer	8. Snap ring	14. Cylinder inner tube	19. Retaining plate
3. Outer seal	9. Packing gland	15. Lock nut	20. Cylinder
4. Washer	10. Back-up ring	16. Cylinder front end	21. Clamp
5. Washer	11. "O" ring	assembly	22. Cylinder rear end
6. Back-up ring	12. Piston rod	17. Dust cover	assembly

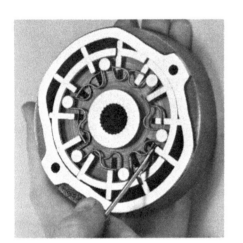

Fig. 60–Be sure all vane springs are engaged behind the rotor vanes. Springs can be pried into place with screwdriver as shown.

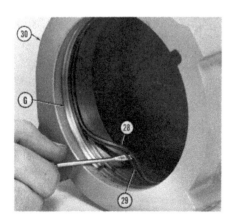

Fig. 61–"O" ring (28) and back-up ring (29) are installed in cover (30); be sure back-up ring is to outside (open side) of cover. Groove (G) is for cover retaining snap ring.

in second groove of cover (See Fig. 61), lubricate the rings and push cover down over the pressure plate and ring. While holding the cover on the assembly, place the unit in an arbor press and insert the cover retaining snap ring as outlined in paragraph 61.

63. **POWER CYLINDERS.** Two double-acting power cylinders are located in front pedestal on Model 4200. Refer to Fig. 43 for view showing cylinder location and to Fig. 62 for exploded view of the cylinder. Refer to following paragraphs for servicing information.

64. **R&R POWER CYLINDERS.** Drain cooling system and remove hood, grille and radiator from front pedestal. Disconnect the power steering lines (17 and 18—Fig. 43) from the manifold (39). Place a support under front end of tractor so that weight is taken from front wheels, attach a hoist to front pedestal, then unbolt pedestal from oil pan and engine and roll the front assembly forward. Remove any cover plates necessary to gain access to cylinders and disconnect lines (19, 20, 37 and 38) from cylinders. Remove the nuts retaining cylinder rear end assemblies in each side of pedestal and remove the cap screws (23) clamping cylinder front ends in steering arm (22). Then, bump the cylinder ends out of pedestal and the steering arm and remove cylinders from pedestal.

To reinstall power cylinders, reverse the removal procedure. Fill and bleed the power steering system as outlined in paragraph 54.

65. **OVERHAUL POWER CYLINDERS.** With the cylinders removed as outlined in paragraph 64, refer to exploded view in Fig. 62 and proceed as follows:

Remove clamp (18) bolt and unscrew cylinder front end assembly (16) from piston rod. Remove the screws retaining plate (19) to packing gland (9) and remove retaining plate. Push packing gland inward to expose snap ring (8), remove snap ring and bump the packing gland out of the cylinder with piston rod and piston. Remove cylinder inner tube (14) from cylinder or piston as case may be. Unscrew nut (15) and piston (13) from inner end of piston rod and slide packing gland from rear end of rod. Remove snap ring (1), washers (2, 4 and 5), outer seal (3), back-up ring (6) and "O" ring (7) from packing gland inside diameter and remove the back-up ring (10) and "O" ring (11) from outside groove on gland.

Renew piston rod (12), gland (9) and/or the cylinder inner tube (14) if excessively worn or scored. Reassemble cylinder using new piston and seal assembly (13) and cylinder gland seal kit (includes items 1 through 11, except gland, which are not serviced separately). To reassemble, proceed as follows:

Lubricate the piston rod and slide the seal kit components on rod from rear (piston) end in following order: snap ring (1), large O.D. washer (2), outer seal (3) with scraper lip towards snap ring, large O.D. washer (4) (same as 2), small O.D. washer (5), back-up ring (6) and "O" ring (7). Slide the gland (9) onto rear end of piston rod over the "O" ring, etc., and install the retaining snap ring (1) in front end of gland. Install new piston and lock nut on piston rod. Install new back-up ring (10) and "O" ring (11) on gland, slide the cylinder inner tube (14) over piston and rear (inner) end of gland, then insert the assembly into cylinder outer tube (20) and retain with snap ring (8). Install retainer plate (19) and tighten the three plates to gland screws securely. Install new cylinder end assemblies (16 and/or 22) if required and reinstall cylinder(s) in pedestal as outlined in paragraph 64.

ENGINES AND COMPONENTS

R&R ENGINE WITH CLUTCH
All Models

66. To remove engine and clutch assembly, first remove the front axle, front support or pedestal, radiator and radiator shell assembly as outlined in paragraph 77 or 80. Then, proceed as follows:

Remove battery, battery tray and support brackets. If engine is to be disassembled, drain oil pan. Disconnect wiring from starting motor, generator, ignition distributor or diesel thermostart unit, oil pressure sender switch and temperature gage unit. Shut off fuel supply valve, remove fuel supply line and the diesel excess fuel return line. Disconnect throttle linkage, choke wire or diesel stop wire and the proofmeter drive cable. On models with piston type hydraulic pump, remove panel from below left side of fuel tank, disconnect hydraulic lines from pump and remove pump from engine. Remove any engine accessories as required for indicated engine overhaul, attach hoist to engine, then unbolt and remove engine and clutch assembly from transmission.

To reinstall engine, reverse removal procedure. Refer to Fig. 63 for engine to transmission bolt tightening torques. Bleed the diesel fuel system as outlined in paragraph 129 or 131, fill and bleed the power steering system as in paragraph 10, 37 or 54 and piston type hydraulic pump as outlined in paragraph 361.

ENGINE COMPRESSION PRESSURES
All Models

67. The Ford Motor Company recommends that compression pressures be checked only at cranking speeds. Non-

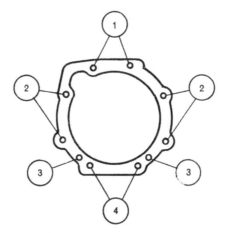

Fig. 63–When reinstalling engine to transmission housing, tighten bolts shown at locations in drawing to the following torque specifications:

1. 220-300 Ft.-Lbs.
2. 125-140 Ft.-Lbs.
3. 35-50 Ft.-Lbs. (engines with stamped steel oil pan)
4. 220-300 Ft.-Lbs. (engines with cast iron oil pans)

diesel engine compression pressures should be checked with all spark plugs removed, the throttle in wide open position and with the choke knob in. Note: Insufficient cranking speed can cause low compression pressure readings. For this reason, the battery should be well charged and the starting system functioning properly to assure a cranking speed of 150 to 200 RPM.

The following table gives compression pressure ranges and maximum allowable cylinder-to-cylinder variation for a cranking speed of approximately 200 RPM:

Non-Diesel Engines:
Compression Pressure
Range .... 115-150 psi
Max. Allowable Variation ..... 25 psi

Diesel Engines:
Compression Pressure
Range .... 420-510 psi
Max. Allowable Variation ..... 50 psi

R&R CYLINDER HEAD
All Models

68. To remove cylinder head, proceed as follows: Remove vertical exhaust muffler, if so equipped, and remove the engine hood. On models with horizontal exhaust system, disconnect exhaust pipe from exhaust manifold. Remove the exhaust manifold, battery, battery tray and the battery support brackets. Drain cooling system and remove the upper radiator hose.

On diesel models, disconnect the air intake hose from intake manifold, shut off the fuel supply valve and remove the fuel filter assembly from intake manifold. Remove the fuel injectors as outlined in paragraph 149. Remove the intake manifold.

On gasoline models, shut off fuel supply valve, disconnect the air inlet hose, fuel line and throttle rod from carburetor, disconnect the vacuum line from intake manifold and distributor and remove the intake manifold and carburetor as a unit. Remove the spark plugs.

Disconnect ventilation tube from rocker arm cover and remove rocker arm cover, rocker arms assembly and push rods. Remove remaining cylinder head bolts and remove cylinder head from engine block.

NOTE: If cylinder head gasket failure has occured, check cylinder head and block mating surfaces for flatness. Maximum allowable deviation from flatness is 0.006 overall or 0.003 in any six inches. If cylinder head is not within flatness specified or is rough, the surface may be lightly remachined. If the cylinder block is not within flatness specified, it may be machined providing the distance between top of pistons at top dead center and top surface of cylinder block is not less than 0.002 after machining. Also, install cylinder head to block without gasket, install rocker arm supports, washers and all head bolts finger tight. Then using feeler gage, measure clearance between underside of bolt heads and cylinder head or rocker arm supports. If clearance is 0.010 or more, cut threads of bolt hole deeper in block with a ½-inch, 13 UNC-2A tap.

69. Two different thickness cylinder head gaskets have been used. On tractors built before mid 1969, the head gasket has a compressed thickness of 0.029. Later gaskets have a compressed thickness of 0.037. Because of construction differences, the earlier gasket should not be used on later engines.

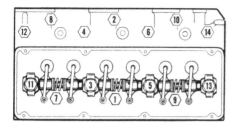

Fig. 64–Drawing showing cylinder head bolt tightening sequence for three cylinder engines.

Fig. 66–Valve guides are integral with the cylinder head. Reamers in 0.003, 0.015 and 0.030 oversize are available (Nuday tool No. SW-502) for repairing worn guides to fit oversize valve stems. Refer to text.

When reassembling, do not use gasket sealer or compound and be sure that gasket is properly positioned on the two dowel pins. Using the sequence in Fig. 64, tighten cylinder head bolts in three steps as follows:

First Step . . . . .50-60 Ft.-Lbs.
Second Step . . . . .80-90 Ft.-Lbs.
Final Step . . . . 105-115 Ft.-Lbs.

Note: Torque values given are for lubricated threads; tighten cylinder head bolts **only** when engine is **cold**. Adjust valve gap as indicated in paragraph 72. Complete the reassembly of engine by reversing disassembly procedure. Tighten the intake manifold bolts to a torque of 23-28 Ft.-Lbs. the exhaust manifold bolts to a torque of 25-30 Ft.-Lbs. With reassembly complete, bleed the diesel fuel system as outlined in paragraph 129 or 131, start engine and bring to normal operating temperature. With engine running at slow idle speed, readjust valve gap to 0.015 on intake valves and 0.018 on exhaust valves. The cylinder head bolts should be retorqued and the valve gap readjusted after 50 hours of operation.

VALVES, STEM SEALS AND SEATS
All Models

70. Exhaust valves are equipped with positive type rotators and an "O" ring type seal is used between valve stem and rotator body. Intake valve stems are fitted with umbrella type oil seals. Both the intake and exhaust valves seat on renewable type valve seat inserts that are a shrink fit in cylinder head. Inserts are available in oversizes of 0.010, 0.020 and 0.030 as well as standard size.

Intake and exhaust valve face angle is 44 degrees and valve seat angle is 45 degrees resulting in a 1 degree interference angle. Renew valve if margin is less than 1/32-inch after valve is refaced. Desired valve seat width is 0.080-0.102 for intake valves and 0.084-0.106 for exhaust valves. Seats can be narrowed and centered by using 30 and 60 degree stones. Total seat runout should not exceed 0.0015.

Desired stem to guide clearance is 0.001-0.0024 for intake valves and 0.002-0.0034 for exhaust valves. New (standard) stem diameter is 0.3711-0.3718 for intake valves and 0.3701-0.3708 for exhaust valves. Valves with 0.003, 0.015 and 0.030 oversize stems are available as well as reamers (Nuday tool No. SW502) for enlarging valve guide bore to 0.003, 0.015 or 0.030 oversize.

NOTE: Although valves for gasoline and diesel engines are dimensionally the same, exhaust valve material is different. For this reason, gasoline and diesel exhaust valves should not be interchanged. Exhaust valves for gasoline engines may be identified by an 0.18 diameter depression in center of top face of valve head whereas exhaust valves for diesel engines have a flat top face on valve head.

VALVE GUIDES AND SPRINGS
All Models

71. Intake and exhaust valve guides are an integral part of the cylinder head and are non-renewable. Standard valve guide diameter is 0.3728-0.3735. Valve guides may be reamed to 0.003, 0.015 or 0.030 oversize and valves with oversize stems installed if stem to guide clearance is excessive. Desired stem to guide clearance is 0.001-0.0024 for intake valves and 0.002-0.0034 for exhaust valves. A valve stem guide reamer kit (Nuday tool No. SW502 as shown in Fig. 66) is available.

Intake and exhaust valve springs are interchangeable. Valve spring free length should be 2.15 inches. Springs should exert a force of 61 to 69 pounds when compressed to a length of 1.74 inches, and a force of 125-139 pounds when compressed to a length of 1.32 inches. Valve springs should also be

checked for squareness by setting spring on flat surface and checking with a square; renew spring if clearance between top end of spring and square is more than 1/16-inch with bottom end of spring against square. Also, renew any spring showing signs of rust or erosion.

ADJUSTMENT
All Models

72. **TAPPET GAP ADJUSTMENT.** Recommended initial (cold) tappet gap adjustment is 0.017 for intake valves and 0.021 for exhaust. Recommended setting at operating temperature is 0.015 for intake valves and 0.018 for exhaust valves.

Valves can be statically adjusted using the two-position method and Figs. 67 and 68 as a guide, proceed as follows: Turn crankshaft until "TDC" flywheel timing mark is aligned with timing pointer, then check the two front (No. 1 Cylinder) rocker arms. If rocker arms are loose, No. 1 cylinder is on compression stroke; adjust the valves shown in Fig. 67. If front rocker arms are tight, No. 1 cylinder is on exhaust stroke; adjust valves shown in Fig. 68. In either event, complete the adjustment by turning the crankshaft one complete revolution and using the appropriate alternate diagram.

EXHAUST VALVE ROTATORS
All Models

73. The positive type valve rotators require no maintenance, but each exhaust valve should be observed while engine is running to be sure the valve rotates slightly. Renew the rotator on any exhaust valve that fails to turn.

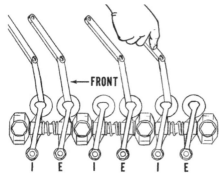

Fig. 67–With TDC timing mark aligned and No. 1 piston on compression stroke, adjust the four indicated valves. Turn crankshaft one complete turn until TDC timing mark again aligns and refer to Fig. 68.

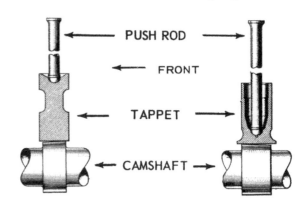

Fig. 71–Views showing previous and current types of push rods, tappets and camshafts for diesel engines.

PREVIOUS TYPE   **CURRENT TYPE**

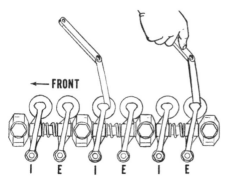

Fig. 68–With TDC timing mark aligned and No. 1 piston on exhaust stroke, the two indicated valves may be adjusted. Refer also to Fig. 67.

VALVE TAPPETS (CAM FOLLOWERS)
Gasoline Engines

74. The tappets used in gasoline engines are fabricated cylindrical or cast webbed cylindrical type of furnace hardened iron. Desired tappet to bore clearance for fabricated tappets is 0.0006-0.0021. Desired clearance for cast tappets is 0.0005-0.0023. Tappet bore diameter is 0.990-0.991. Tappets may be removed from above with magnet after removing cylinder head as outlined in paragraph 68.

Diesel Engines

75. On early production diesel engines, barrel type tappets were used; late production engines are equipped with semi-mushroom type tappets. Refer to Fig. 71. Tappet bore diameter in cylinder block is 0.990-0.991. Tappet diameter is 0.9887-0.9895; desired tappet to bore clearance is 0.0005-0.0023.

The early barrel type tappets can be removed from above with a magnet after removing cylinder head as outlined in paragraph 68. To remove the late semi-mushroom type tappets, first

remove camshaft as outlined in paragraph 87.

Lobe width on camshaft used with early barrel type tappets is 0.545-0.575; if renewing early camshaft with later type having lobe width of 0.825-0.855, the late semi-mushroom type tappets and longer push rods should also be installed. Early type barrel tappets require push rods of 10.63-10.67 inches in length; push rods used with the late semi-mushroom type tappets are 12.15-12.19 inches long.

ROCKER ARMS
All Models

76. To remove rocker arms, lift hood, swing battery tray out, remove rocker arm cover and unscrew, but do not remove, the four cylinder head bolts that retain rocker arm assembly to cylinder head. Lift out the rocker arm assembly and head bolts as a unit.

To disassemble, withdraw the cylinder head bolts. Rocker arm to shaft clearance should be 0.002-0.004. Shaft diameter is 1.000-1.001; rocker arm inside diameter is 1.003-1.004. Renew rocker arm if clearance is excessive or

if valve contact pad is worn more than 0.002. Torque required to turn valve adjustment screw in rocker arm should be 9 to 26 Ft.-Lbs.; renew rocker arm and/or screw if torque required to turn screw is less than 9 Ft.-Lbs.

When reassembling, be sure notch (N—Fig. 72) in one end of rocker arm shaft is up and towards front end of engine; this will correctly place the rocker arm oiling holes. Back each rocker arm adjusting screw out two turns, then tighten all retaining bolts evenly until valve springs are compressed and rocker arm supports are snug against the cylinder head; then, tighten all bolts to a torque of 95-105 Ft.-Lbs.

NOTE: A few engines were built with two thin hardened washers under the valve rocker arm shaft support bolt heads instead of the simple 0.16-0.17 thick washer (2—Fig. 72) which is listed in the Ford Tractor Parts Catalog for this location. When one of these engines is serviced, it is very important when reassembling, that either the correct washer is installed or both the thinner washers are reinstalled.

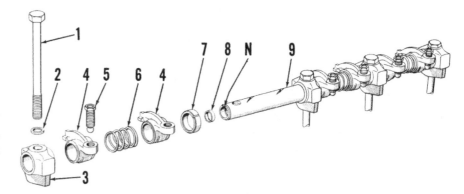

Fig. 72–Exploded view of rocker arm assembly. Cylinder head bolts (1) are used to retain the rocker arm supports. Notch (N) in end of shaft must be installed upward and to front of engine. Refer to note in text concerning washers (2).

N. Notch
1. Cylinder head bolts
2. Flat washers
3. Rocker arm supports
4. Rocker arms
5. Adjusting screws
6. Springs
7. Spacers
8. Shaft end plugs
9. Rocker arm shaft

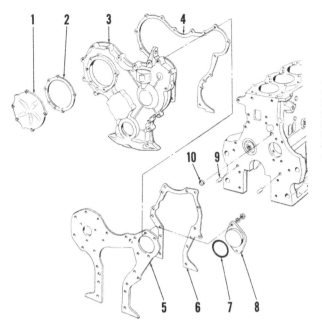

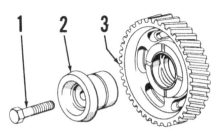

Fig. 73–View of timing gear cover and engine front plate installation. On non-diesel engines, governor assembly is used instead of plate (1). On models without power steering, plate (8) covers hole in engine front plate (5).

1. Cover plate
2. Gasket
3. Timing gear cover
4. Gasket
5. Engine front plate
6. Gasket
7. "O" ring
8. Cover plate
9. Dowel pins
10. Oil gallery plug

Fig. 75–Camshaft drive gear and adapter shaft as removed from front end of cylinder block. The bushing for gear (3) is not available separately.

1. Retainer bolt
2. Adapter shaft
3. Camshaft drive gear

After rocker arm shaft is installed, adjust valve clearance, cold, to 0.017 on intake valves and 0.021 on exhaust valves. After engine is assembled, start and bring to normal operating temperature and adjust valve clearance as outlined in paragraph 72.

Reinstall rocker arm cover with new gasket and tighten retaining bolts to a torque of 10-12 Ft.-Lbs.

R&R TIMING GEAR COVER
Models 2100, 2110, 3100, 4110 & 4140SU

77. To remove timing gear cover, remove hood, drain cooling system and disconnect radiator and front mounted air cleaner hose. On models with transmission oil cooler, remove grille and disconnect cooler tubes from radiator lower tank. Disconnect battery ground

cable and headlight wires. Unbolt radiator shell upper support bracket from shell. Support front end of tractor under front of transmission housing. Disconnect drag links from spindle arms and unbolt radius rods from axle. Support front ends of radius rods and drag links on power steering models. Drive wedges between front support and axle, then unbolt front support from engine and roll the front assembly away from tractor.

78. With radiator and front support removed, proceed as follows: Remove gasoline engine fuel pump, fuel pump push rod and on all models, remove fan belt and generator front mounting bolt. Disconnect non-diesel governor linkage. Drain and remove engine oil pan. Remove cap screw and washers from front end of crankshaft. Using suitable pullers and shaft protector,

remove the crankshaft pulley. Then, unbolt and remove the timing gear cover along with non-diesel governor cover and lever assembly or diesel fuel injection pump drive gear cover.

With timing gear cover removed, the crankshaft front oil seal and dust seal can be renewed as outlined in paragraph 97.

To reinstall timing gear cover, reverse removal procedure. Tighten timing gear cover retaining cap screws to a torque of 13-18 Ft.-Lbs. Tighten crankshaft pulley retaining cap screw to a torque of 130-145 Ft.-Lbs. Install oil pan as outlined in paragraph 101 or 102.

79. Reattach front axle and radiator assembly to engine and tighten the retaining bolts to a torque of 200-240 Ft.-Lbs. Tighten radius rod foot bolts to a torque of 130-160 Ft.-Lbs.

Models 4100 and 4200

80. To remove the timing gear cover, remove engine hood, drain cooling system and disconnect radiator and air cleaner hoses. On model 4100, disconnect battery ground strap and the headlight wires. On models so equipped, remove grille and disconnect transmission oil cooler tubes from radiator lower tank.

On model 4200, disconnect power steering fluid lines at manifold in front pedestal. On model 4100, disconnect steering drag link from left spindle arm and unbolt power steering cylinder bracket from front support on models so equipped.

Adequately support front unit with chain and hoist and support tractor under front end of transmission housing. Unbolt pedestal or front axle support from engine and move unit away from tractor.

With the front unit removed, refer to paragraph 78 for removal and installation procedure for timing gear cover. With timing gear cover reinstalled, reverse removal procedure for reinstalling model 4100 front axle support and tighten the retaining support to

Fig. 74–Timing gear backlash can be checked with dial indicator or with feeler gage as shown. Be sure to check backlash at several points around the gear.

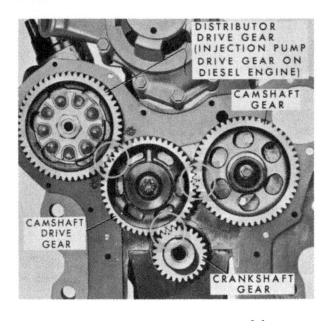

Fig. 76—View showing timing marks on crankshaft gear, camshaft gear, camshaft drive gear and distributor drive gear; timing marks are identical on diesel engines.

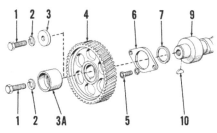

Fig. 77—Drawing showing assembly of camshaft gear to camshaft. Flat washer (3) is used on diesel engines; fuel pump eccentric (3A) is used on gasoline engines.

1. Cap screw	5. Cap screws (2)
2. Lockwasher	6. Thrust plate
3. Flat washer (diesel)	7. Spacer
3A. Eccentric Shaft	9. Camshaft
4. Camshaft gear	10. Woodruff key

engine bolts to a torque of 200-240 Ft.-Lbs., drag link nut to a torque of 55-65 Ft.-Lbs. and the power steering cylinder bracket bolts to a torque of 200-250 Ft.-Lbs. On model 4200, refer to paragraph 52 for shimming procedure when reinstalling front pedestal.

TIMING GEARS
All Models

81. Before removing any gears in the timing gear train, first remove rocker arms assembly to avoid the possibility of damage to piston or valve train if either the camshaft or crankshaft should be turned independently of the other.

The timing gear train consists of the crankshaft gear, camshaft gear, injection pump drive gear or distributor drive gear and a camshaft drive gear (idler gear) connecting the other three gears of the train. Refer to Fig. 76.

Timing gear backlash between crankshaft gear and camshaft drive gear, or between camshaft drive gear and camshaft gear should be 0.001-0.006 on gasoline engines and 0.001-0.009 on diesel engines. Backlash between camshaft drive gear and fuel injection pump drive gear or distributor drive gear should be 0.001-0.012. If backlash is not within recommended limits, renew the camshaft drive gear, camshaft drive gear shaft and/or any other gears concerned.

83. **CAMSHAFT DRIVE GEAR AND SHAFT.** To remove, unscrew the self locking cap screw and remove the camshaft drive gear and shaft (adapter) from front face of cylinder block. Renew shaft and/or gear if bushing to shaft clearance is excessive, or if bearing surfaces are scored. Inspect gear teeth for wear or score

marks; small burrs can be removed with fine carborundum stone.

To reinstall the camshaft drive gear, turn crankshaft so that No. 1 piston is at top dead center on compression stroke and turn camshaft and fuel injection pump drive gear or distributor drive gear so that timing marks point to center of camshaft drive gear location. Place the camshaft drive gear within the other three gears so that all timing marks are aligned as shown in Fig. 76, then install the adapter (shaft) and tighten the self-locking cap screw to a torque of 100-105 Ft.-Lbs.

84. **CAMSHAFT GEAR.** To remove the camshaft gear, remove cap screw (1—Fig. 77), lockwasher (2) and washer (3) on diesel models or fuel pump eccentric (3A) on gasoline models; then, pull gear from shaft. Gear should be a hand push fit on shaft. With gear removed, inspect drive key (10), thrust plate (6) and spacer (7) and renew if damaged in any way.

To reinstall gear, first install spacer, thrust plate and drive key, then install gear, washer or gasoline fuel pump eccentric, lock washer and gear retaining cap screw. Tighten the cap screw to a torque of 40-45 Ft.-Lbs.

85. **CRANKSHAFT GEAR.** If not removed with timing gear cover and seal assembly, slide the spacer (5—Fig. 78) from crankshaft; then, using remover-replacer (Nuday tool No. SW-501) and insert (Nuday tool No. SW-501-1) or equivalent tool, pull gear from crankshaft. Inspect the gear and crankshaft pulley drive key (3) and renew if damaged in any way.

To reinstall gear, first drive the key (3) into crankshaft keyway until fully seated, then install the gear with timing mark outward using remover-replacer tools as used in removal procedure, or push gear onto shaft with bolt threaded into front end of crankshaft and using nut, large washer and a sleeve.

86. **FUEL INJECTION PUMP OR DISTRIBUTOR DRIVE GEAR.** For diesel models, refer to paragraph 157. On gasoline models, refer to paragraph 167 for information concerning distributor drive unit and governor.

CAMSHAFT AND BEARINGS
All Models

87. To remove camshaft, first remove engine as outlined in paragraph 66 and timing gear cover as outlined in paragraph 78. On gasoline engines and on diesel engines with early barrel type tappets, remove cylinder head as in paragraph 68, then lift tappets (cam followers) from openings in top of cylinder block with a magnet. On diesel

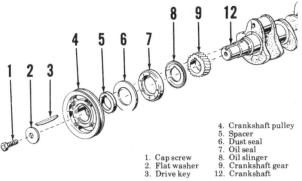

Fig. 78—Crankshaft gear installation is shown for all models. Dust seal (6) and oil seal (7) are pressed into timing gear cover from inside and ride on spacer (5) on late production tractors; on early units, dust seal and oil seal ride on crankshaft pulley hub.

1. Cap screw	4. Crankshaft pulley
2. Flat washer	5. Spacer
3. Drive key	6. Dust seal
	7. Oil seal
	8. Oil slinger
	9. Crankshaft gear
	12. Crankshaft

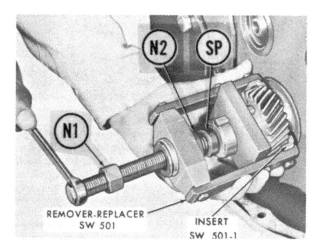

Fig. 79–Using special remover-replacer (Nuday tool No. SW501) to remove crankshaft gear. To install gear, nut (N2) and shaft protector (SP) are removed, threaded shaft is screwed into end of crankshaft and nut (N1) is turned to push gear onto shaft.

reassembly. Remove the cap screw (1—Fig. 80), lock washer (2) and flat washer (3) (diesel models) or the fuel pump eccentric (3A) (gasoline models) and pull the camshaft gear from shaft. Remove the Woodruff key (10), thrust plate (6) and spacer (7). Withdraw camshaft and hydraulic pump drive gear from rear of cylinder block, then remove pump drive gear if necessary.

The camshaft is supported in four bushing type bearings. Check camshaft and bearings against the following values:

Camshaft journal
diameter 2.3895-2.3905
Desired journal to bearing
clearance 0.001-0.003
Camshaft end play 0.001-0.007

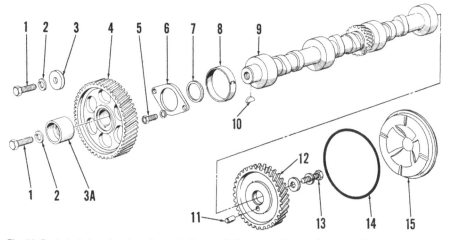

Fig. 80–Exploded view drawing of camshaft, camshaft gear and hydraulic pump drive gear. End play is controlled by thrust plate (6). Renewable camshaft bearings (8) are used in cylinder block for all camshaft journals. Washer (3) is used on diesel models; fuel pump eccentric (3A) is used on gasoline models.

1. Cap screw	5. Cap screws (2)
2. Lockwasher	6. Thrust plate
3. Flat washer	7. Spacer
3A. Fuel pump eccentric	8. Camshaft bearing
4. Camshaft gear	

9. Camshaft	13. Cap screw
10. Woodruff key	14. "O" ring (early)
11. Dowel pin	15. Hydraulic pump
12. Hydraulic lift pump	drive gear cover
drive gear	

If excessive bearing wear is indicated, the bearings can be removed and new bearings installed with bearing driver (Nuday tool No. SW 506) and handle (Nuday tool No. N6261-A) or equivalent tools. Note: It will be necessary to invert engine and remove oil pan to remove and install bearings. Pay particular attention that the oil holes in bearings are aligned with the oil passages in cylinder block. New bearings are pre-sized and should not require reaming if carefully installed.

Reinstall camshaft by reversing removal procedure. Tighten the hydraulic pump drive gear retaining cap screw to a torque of 40-45 Ft.-Lbs., tighten thrust plate cap screws to a torque of 12-15 Ft.-Lbs. and tighten the camshaft gear retaining cap screw to a torque of 40-45 Ft.-Lbs. Place hydraulic drive gear cover "O" ring (models prior 8-67) in groove in cylinder block, lubricate the "O" ring and push cover into place as shown in Fig. 83. Note: Be sure that camshaft gear is timed as shown in Fig. 76.

engines with late semi-mushroom type tappets, remove rocker arm assembly and push rods, then invert engine assembly to allow cam followers to fall away from camshaft. Note: On diesel engines, measure push rods before removing cylinder head; if push rods are approximately 10½ inches long, engine is equipped with early barrel type tappets. If diesel engine push rods are approximately 12 5/32 inches long, engine is equipped with semi-mushroom type tappets. Remove clutch, flywheel and engine rear plate. Remove the hydraulic pump drive gear cover from rear end of cylinder block (models not equipped with piston type pump) and push the camshaft rear cover plate from cylinder block with punch (Fig. 81). Remove the oil filter and oil pump drive gear (Fig. 95); on diesel engine, the oil pump drive shaft should be removed with gear on engine inverted in stand.

Using a feeler gage or dial indicator, measure camshaft end play. Desired end play is 0.001-0.007. If end play is excessive, renew thrust plate during

Fig. 81–Camshaft rear cover plate can be removed with a punch after removing flywheel, rear engine plate and hydraulic pump.

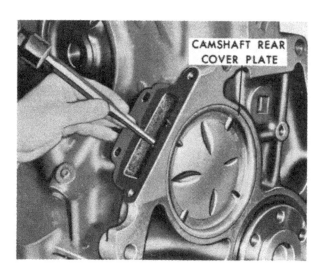

CAMSHAFT REAR COVER PLATE

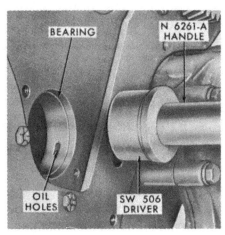

Fig. 82–View showing special tool for removing and installing camshaft bearings. Be sure that bearings are installed with oil holes aligned with oil holes in cylinder block.

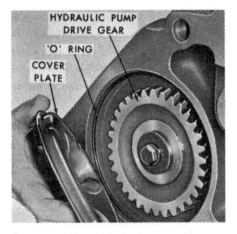

Fig. 83–Installing the hydraulic pump drive gear cover plate (15–Fig. 80). Note: "O" ring is no longer used.

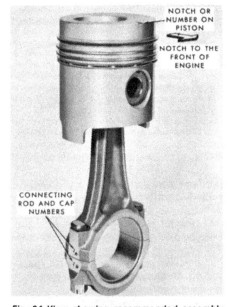

Fig. 84–View showing recommended assembly of piston to connecting rod; however, connecting rod can be installed on crankshaft with numbers facing either way. Piston must be installed with notch or number to the front of engine.

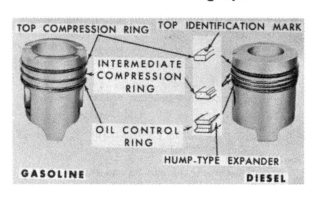

Fig. 85–Views showing proper installation of piston rings on non-diesel (gasoline) and diesel pistons.

CONNECTING ROD AND PISTON UNITS

All Models

88. Connecting rod and piston units are removed from above after removing the cylinder head and oil pan. Be sure to remove top ridge from cylinder bores before attempting to withdraw the assemblies.

Connecting rod and bearing cap are numbered to correspond to their respective cylinder bores. When renewing the connecting rod, be sure to stamp the cylinder number on new rod and cap.

When reassembling, it is important that the identification number or notch in top face of piston is towards the front end of engine. It is standard practice to assemble connecting rod to piston with cylinder numbers to right side of engine (away from camshaft); however, the rod is symmetrical and rod can be installed with numbers in either direction (away from or towards camshaft) without affecting performance or durability of engine.

When installing connecting rod cap, be sure that bearing liner tangs, and the cylinder identification number, of rod and cap are towards same side of engine. Tighten the connecting rod nuts to a torque of 60-65 Ft.-Lbs.

PISTON RINGS

Non-Diesel Engines

89. Each cam ground piston is fitted with two compression rings and one oil control ring. Top compression ring and oil control ring are chrome plated. Top compression ring is barrel face type and must be installed with identification mark up. Second compression ring has straight face with inner bevel and must be installed with groove on inside diameter up. The oil control ring and oil ring expander may be installed either side up.

Piston ring sets are available in oversizes of 0.020, 0.030 and 0.040 as well as standard size. The standard size rings are to be used with both grades of standard size pistons and with 0.004 oversize pistons. Refer to the following specifications for checking piston ring fit:

Ring End Gap:
Top compression ring 0.012-0.038
Second compression ring .. 0.012-0.035
Oil control ring 0.013-0.038
Ring Side Clearance In Groove:
Top compression ring .. 0.0029-0.0046
Second compression ring 0.0025-0.0045
Oil control ring 0.0024-0.0041

Diesel Engines

90. Pistons are fitted with three compression rings and one oil control ring. The two top compression rings and the oil control ring are chrome plated. Top compression ring is barrel face type and must be installed with identification mark up. Second and third compression rings have straight face with bevel on inside diameter and must be installed with beveled side up. The oil control ring and expander may be installed with either side up.

Piston ring sets are available in oversizes of 0.020, 0.030 and 0.040 as well as standard size. The standard size rings are to be used with both grades of standard size pistons and also with 0.004 oversize pistons. Refer to the following specifications for checking piston ring fit:

Ring End Gap:
Top compression ring 0.012-0.038
Second & third compression
   rings 0.012-0.035
Oil control ring 0.013-0.038
Ring Side Clearance In Groove:
Top compression ring .. 0.0044-0.0061
Second & third compression
   rings 0.0039-0.0056
Oil control ring 0.0024-0.0041

PISTONS AND CYLINDERS
All Models

91. Non-diesel engines have cam ground aluminum alloy pistons with a cast iron insert containing the top ring groove. Diesel engines have trunk type aluminum alloy pistons with a continuous skirt and a Ni-Resist insert containing the top ring groove. Early design pistons were straight in the upper

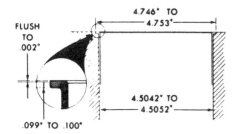

Fig. 86–Top of installed sleeve flange should be flush to 0.002 above gasket surface of block as shown.

Fig. 87–Three different crown heights (H) have been used in 4.4 inch bore gasoline engines; and two different crown heights used on 4.4 inch bore diesel engines. If in doubt, measure crown height to be sure correct piston is being installed.

land area; whereas, later production and service pistons are tapered in the upper land area.

Cylinder bores in engine block are normally unsleeved; however, repair sleeves have been factory installed in engines with 4.2 inch diameter cylinders. In factory installation, all cylinders are sleeved, and block is stamped "SB" on left side oil pan flange.

Series 4000 tractors (4.4 bore) have available for service a thin wall cast sleeve which is not used in production and which cannot be overbored.

Pistons, pins and rings are available in standard size and 0.004 oversize for use in standard bores, and oversizes of 0.020, 0.030 and 0.040 for use in rebored blocks.

92. **INSTALLING SLEEVES.** To install sleeves in unsleeved 4.2 inch bore engines, finish-bore the block to an inside diameter of 4.358. Selectively fit the sleeves to provide, as nearly as possible the optimum 0.004 interference fit for all sleeves.

Thoroughly clean and dry the sleeve and bore and carefully apply a 3-inch band of LOCTITE Sleeve Retainer (green) to top of bore. Install sleeve, outside chamfered edge down and press into position until top of sleeve is flush to 0.001 above block surface. Allow block to cure overnight before completing the assembly. Re-examine the installation and mill the block if necessary, to provide correct sleeve height. NOTE: DO NOT remove more than 0.005 from top surface. Finish bore and hone the sleeve after milling, to 4.200-4.2024 for standard piston or to the appropriate oversize if oversize piston is installed. Refer also to paragraph 93 for additional data on piston and ring installation.

The flanged, thin-wall cast sleeves used in 4.4 inch diameter bore engines can be used only with standard size or 0.004 oversize pistons, and cannot be re-bored. Thin wall sleeves are a light press fit in engine block and should be installed as follows:

First make sure that sleeve and

block bore are clean and dry. Chill the sleeves for 15 minutes in dry ice and push the sleeve as far as possible into block bore. Seat the sleeve, if necessary, using a puller and suitable installing plate. Top of installed sleeve flange should be flush to 0.002 above gasket surface of block as shown in Fig. 86. Machining dimensions for installing sleeves in unbored block are shown in Fig. 86.

93. **FITTING PISTONS.** Recommended method for fitting pistons is as follows: Before checking piston fit, deglaze cylinder wall using a hone or deglazing tool. Using a micrometer, measure piston diameter at centerline of and at right angle to the piston pin bore. Then, using an inside micrometer, measure cylinder bore diameter at a distance of 2¾ inches from top of cylinder block crosswise with the block. Subtract the piston diameter from the cylinder bore diameter; the resulting piston to cylinder bore clearance should be within the following specification for proper piston fit:

Non-Diesel Engines:
4.2 Inch Bore 0.0027-0.0037
4.4 Inch Bore 0.0032-0.0042
Diesel Engines:
4.2 Inch Bore 0.0075-0.0085
4.4 Inch Bore 0.0080-0.0090

Note: Two grades of standard pistons are available. Grade "D" standard size pistons (color coded blue) are larger in diameter than grade "B" standard size piston (color coded red). If grade "D" piston fits too loosely in standard cylinder bore, try a 0.004 oversize piston. If not possible to fit the 0.004 oversize

piston by honing, rebore cylinder to 0.020 oversize.

NOTE: After honing or deglazing cylinder bore, wash bore thoroughly with hot water and detergent until a white cloth can be rubbed against cylinder wall without smudging, then rinse with cold water, dry thoroughly and oil to prevent rusting.

All pistons used in 4.4 inch bore engines are not interchangeable. In April, 1968, Series 4000 gasoline engine stroke was increased from 4.2 to 4.4 inches. At that time piston crown height (measured from top of piston pin as shown at H—Fig. 87) was decreased from 1.992 to 1.774. Installation of the wrong piston could result in piston to cylinder head contact or decreased compression.

PISTON PINS
All Models
94. The 1.4997-1.5000 diameter floating type piston pins are retained in the piston pin bosses by snap rings and are available in standard size only. Piston pin should have a clearance of 0.0005-0.0007 in connecting rod bushing and a clearance of 0.0003-0.0005 in piston bosses. After installing new piston pin bushings in connecting rods, the oil hole in bushing must be drilled as shown in Fig. 88 and the bushings final sized with a spiral expansion reamer to obtain the specified pin to bushing clearance. Bushing inside diameter should then be 1.5003-1.5006. When assembling, identification notch or number in top face of piston must be to front end of engine and the identification number on rod and cap towards right side of engine

(away from camshaft); however, the rod is symmetrical and can be installed either way without affecting engine performance or durability.

CONNECTING RODS AND BEARINGS

All Models

95. Connecting rod bearings are precision type, renewable from below after removing the oil pan and connecting rod bearing caps. When removing bearing caps, note which way that cylinder identification numbers are placed on each rod assembly. It is standard practice that the numbers face away from the camshaft side of engine; however, the connecting rods are symmetrical and may be placed with identification numbers to either side of engine. It is very important that the pistons be placed in cylinder bore with the identification number or notch in top face of piston towards front end of engine; therefore, do not attempt to re-align rod identification numbers if they are to camshaft side of engine.

Crankpin bearing liners may be of two different materials, copper-lead or aluminum-tin alloy. The bearings will have an identification marking as follows:

Copper-lead . . . . PV or G
Aluminum-tin . . . . G and AL

Standard size bearing liners of each material are available in two different thicknesses and are color coded to indicate as follows:

Copper-lead bearing thickness:
Red 0.0943-0.0948
Blue 0.0947-0.0952
Aluminum-tin alloy bearing thickness:
Red 0.0939-0.0944
Blue 0.0943-0.0948

In production, connecting rods and crankshaft crankpin journals are color coded to indicate bore and journal diameters as follows: Connecting rod bore diameter:

Red 2.9412-2.9416
Blue 2.9416-2.9420
Crankpin journal diameter:
Red 2.7500-2.7504
Blue 2.7496-2.7500

When installing a new crankshaft and the color code marks are visible on connecting rods, the crankpin bearing liners may be fit as follows: If the color code markings on both rod and crankshaft crankpin journal are red, install two red bearing liners; if both color

Fig. 88—Oil hole in connecting rod bushing must be drilled after bushing is installed, but before reaming or honing bushing to fit pin.

code markings are blue, install two blue coded bearing liners. If color code marks on rod and crankpin do not match (one is red and the other is blue) install one red and one blue bearing liner. Note: Be sure that both bearing liners are of the same material; that is, either both are copper-lead or both are aluminum-tin alloy. If color code mark is not visible on connecting rod or crankshaft, bearing fit should be checked with plasti-gage for the proper clearance according to bearing material as follows:
Crankpin journal to bearing liner clearance:
    Copper-lead bearings   0.0017-0.0038
    Aluminum-tin
    bearings 0.0025-0.0046
As well as being available in either red-coded or blue-coded standard size, bearing liners are also available in undersizes of 0.002, 0.010, 0.020, 0.030 and 0.040. When installing undersize crankpin bearing liners, the crankpin must be reground to one of the following exact undersizes:

Bearing Undersize	Crankpin Journal Dia.
0.002	2.7476-2.7484
0.010	2.7396-2.7404
0.020	2.7296-2.7304
0.030	2.7196-2.7204
0.040	2.7096-2.7104

NOTE: When regrinding crankpin journals, maintain a 0.12-0.14 fillet radius and chamfer oil hole after journal is ground to size.

When reassembling, tighten the connecting rod nuts to a torque of 60-65 Ft.-Lbs.

CRANKSHAFT AND MAIN BEARINGS

All Models

96. Crankshaft is supported in four main bearings. Crankshaft end thrust is controlled by the flanged main bearing liner which is used on the second main journal from front. Before removing main bearing caps, check to see that they have an identification number so that they can be installed in same position from which they are removed.

Main bearing liners may be of two different materials, copper-lead or aluminum-tin alloy. The bearings will have an identification marking to indicate bearing material as follows:
Copper-lead . . . . PV or G
Aluminum-tin alloy . . . . G and AL
Standard size bearing liners are available in two different thicknesses and are color-coded to indicate thickness as follows:
Red 0.1245-0.1250
Blue 0.1249-0.1254

In production, main bearing bores in block and main bearing journals on crankshaft are color coded to indicate bore and journal diameter as follows:
Main bearing bore diameter:
Red 3.6242-3.6246
Blue 3.6246-3.6250
Main journal diameter:
Red 3.3718-3.3723
Blue 3.3713-3.3718

When installing a new crankshaft and the color code marks are visible in crankcase at main bearing bores, new main bearing liners may be fit as follows: If the color code marks on bore and journal are both red, install two red coded bearing liners; if both marks are blue, install two blue coded bearing liners. If color code mark on bore is not the same as color code on journal (one is blue and the other is red), install one red coded bearing liner and one blue coded bearing liner. Note: Be sure both liners used at one journal are of the same material however, copper-lead bearing liners may be used on one or more journals with aluminum-tin alloy liners on the remaining journals. If color code marks are not visible at the main bearing bores in block, check bearing fit with plasti-gage and install red or blue, or one red and one blue liner to obtain a bearing journal to liner clearance of 0.0022-0.0045. Note: Recommended clearance is for either copper-lead or aluminum-tin alloy bearing material.

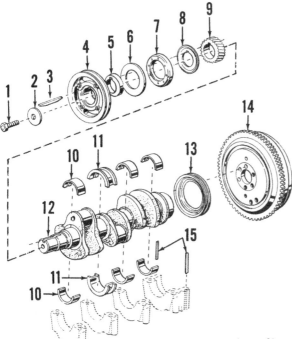

Fig. 89–Exploded view showing crankshaft, bearings and related parts for three cylinder engine.

1. Cap screw
2. Flat washer
3. Drive key
4. Crankshaft pulley
5. Spacer
6. Dust seal
7. Oil seal
8. Oil slinger
9. Crankshaft gear
10. Main bearing liners
11. Thrust bearing liners
12. Crankshaft
13. Rear oil seal
14. Flywheel assembly
15. Rear bearing cap seals

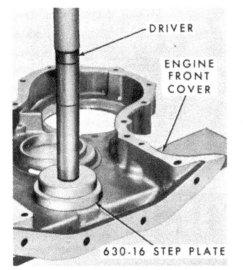

Fig. 90–Installing the crankshaft front oil seal in engine timing gear cover.

Fig. 91–Installing rear oil seal using special installation tool.

As well as being available in either red-coded or blue-coded standard size, new main bearing liners are also available in undersizes of 0.002, 0.010, 0.020, 0.030 and 0.040. When installing undersize main bearing liners, the crankshaft journals must be reground to one of the following exact undersizes:

Bearing Undersize	Main Journal Diameter
0.002	3.3694-3.3702
0.010	3.3614-3.3622
0.020	3.3514-3.3522
0.030	3.3414-3.3422
0.040	3.3314-3.3322

NOTE: When regrinding crankshaft main bearing journals, maintain a fillet radius of 0.12-0.14 and chamfer oil holes after journal is ground to size.

When reinstalling main bearing caps, proceed as follows: Be sure the

Fig. 92–Installing rear main bearing cap and side seals.

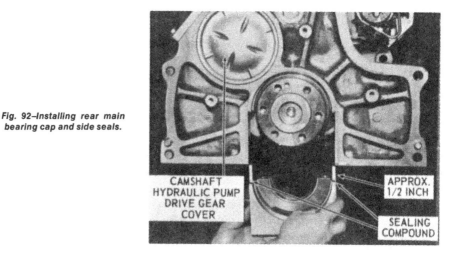

bearing bores and rear main bearing oil seal area are thoroughly clean before installing bearing liners. Be sure tangs on bearing inserts are in the slots provided in cylinder block and bearing caps. Refer to paragraph 99 for rear cap side seals. Tighten bearing cap bolts to a torque of 115-125 Ft.-Lbs.

CRANKSHAFT OIL SEALS
All Models

97. **FRONT OIL SEAL.** Crankshaft front oil seal is mounted in timing gear cover and cover must be removed to renew seal. Timing gear cover removal procedure is outlined in paragraph 77 or 80. To renew seal, drive dust seal and oil seal out towards inside of timing gear cover. Install new dust seal in timing gear cover first, then using a seal driver (OTC No. 630-16 step plate or equivalent), install new oil seal in cover with spring loaded lip towards inside of cover. Refer to Fig. 90.

On early production engines, the crankshaft front seal rides on the crankshaft pulley hub; later production engines are fitted with a revised crankshaft pulley and pulley spacer so that the oil seal rides on the spacer instead of pulley hub. Carefully inspect pulley hub (early production) or pulley spacer (later production) for wear at seal contact surface and renew pulley or spacer if wear or scoring is evident. Note: later production crankshaft pulley and spacer can be installed on early production engine if desired.

98. **REAR OIL SEAL.** Crankshaft rear oil seal can be renewed after removing clutch assembly, flywheel and engine rear plate. Pry old seal from bore in cylinder block and rear main bearing cap and thoroughly clean crankshaft seal journal. Apply a light coat of high temperature grease to seal bore, crankshaft seal journal, lip of seal and the outside diameter of seal. Install new seal with a 4⅞ inch I.D. sleeve so that rear face of seal is 0.060

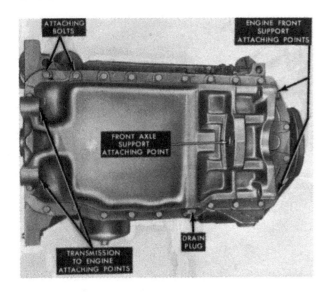

Fig. 93–Bottom view of cast iron oil pan used on industrial, rowcrop, and utility models. Engine front support attaching points require a definite shimming procedure which is outlined in appropriate front support removal paragraph.

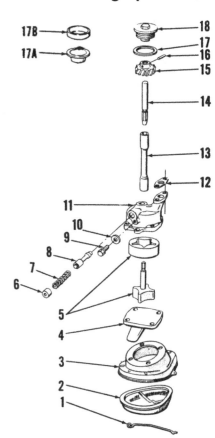

Fig. 96–Exploded view of oil pump, floating shaft and drive gear assembly used on all models. Refer to Fig. 95 for removal of early type stop (18), drive gear (15) and shaft assembly. Floating shaft (13) may be removed from either below or above.

1. Retainer clip	11. Pump body
2. Screen	12. Gasket
3. Screen cover	13. Floating shaft
4. Pump cover	14. Drive shaft
5. Rotor set	15. Drive gear
6. Plug	16. Pin
7. Relief valve spring	17. Gasket
8. Relief valve	17A. Shaft stop (late)
9. Cap screw	17B. Cup plug (late)
10. Retainer washer	18. Shaft stop

Fig. 95–Removing the oil pump drive gear and shaft assembly. Floating shaft connecting the drive gear and shaft assembly to oil pump rotor can be removed at this time. Refer to Fig. 96.

below flush with rear face of block. Using a dial indicator, check runout of rear face of seal; runout should not exceed 0.015. A special seal installation tool (Nuday tool No. SW 520) is available; using three flywheel bolts, press seal in with tool until flange on tool bottoms and tighten the bolts to 25 Ft.-Lbs. Then, remove bolts and seal installation tool. Note: Some early models may have a stop installed in front of seal; if so remove and discard stop before installing new seal.

99. To install new rear main bearing cap side seals, remove the cap and proceed as follows: Place side seals in grooves of cap so they extend about ½-inch from top surface (see Fig. 92). Apply sealing compound to top face of cap making sure chamfers on edge of cap are covered. Insert and lubricate the bearing liner, then install cap and tighten retaining bolts to torque of 115-125 Ft.-Lbs. Tap side seals into block so that they extend about 0.005-0.15 from block and cap. With rear cap and side

seals installed, soak side seals with penetrating oil and install rear oil seal as in paragraph 98.

FLYWHEEL
All Models

100. The flywheel can be removed after splitting tractor between engine and transmission and removing the clutch or Select-O-Speed torque limiting clutch. Flywheel can be installed in one position only. On models with clutch, the flywheel retaining cap screws also retain the clutch shaft pilot bearing retainer. When reinstalling flywheel, tighten the retaining cap screws to a torque of 100-110 Ft.-Lbs.

Starter ring gear is installed from front face of flywheel; therefore, the flywheel must be removed to renew the ring gear. Heat the gear to be removed with a torch from front side of gear and knock gear off of flywheel. Heat new gear evenly until gear expands enough to slip onto flywheel. Tap gear all the way around to be sure it is properly seated and allow it to cool. Note: Be sure to heat gear evenly; if any portion of gear is heated to a temperature higher than 500° F., rapid wear will result. Ring gears for diesel engines have 128 teeth; those for gasoline engines have 162 teeth.

OIL PAN (SUMP)
Models 2100, 2110, 3100, 4100, 4110 & 4140SU

101. The oil pan can be unbolted and removed without removing other components. Be sure that gasket surfaces of engine block, timing gear cover and pan are clean and that pan rim is flat. If removed, reinstall timing gear cover prior to reinstalling oil pan. Cement gasket to pan rim with thin film of

gasket sealer. Tighten retaining cap screws finger tight and, if engine is removed, be sure that pan does not extend beyond rear face of block. Starting from middle of pan, tighten cap screws to a torque of 20-25 Ft.-Lbs.

Model 4200

102. On models with cast iron pan (see Fig. 93), the front support assembly must be removed prior to removing oil pan. Refer to paragraph 80 for model 4200. With front support off, remove the two oil pan to transmission bolts, place a floor jack under oil pan, unbolt pan from engine and lower pan to floor.

To reinstall pan, proceed as follows: If removed, reinstall timing gear cover prior to reinstalling oil pan. Be sure all gasket surfaces are clean and cement gasket to pan with thin film of gasket sealer. Lift pan into place with floor

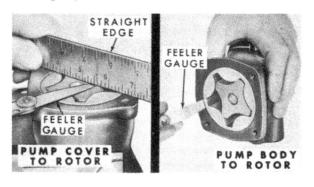

Fig. 97–Checking pump cover to rotor clearance and pump body to rotor clearance. Refer to text for specifications. Refer also to Fig. 98.

jack and reinstall pan to engine retaining cap screws finger tight only. Tighten, then loosen the two oil pan to transmission bolts. Then, starting from center of oil pan, tighten pan to engine cap screws to a torque of 30-35 Ft.-Lbs. Tighten the oil pan to transmission bolts to a torque of 220-300 Ft.-Lbs., then reinstall front support, front axle and radiator assembly as outlined in paragraph 80.

OIL PUMP AND RELIEF VALVE
All Models

103. To remove oil pump, first remove oil pan as outlined in paragraph 101 or 102, then remove the two retaining cap screws and remove pump from cylinder block. Refer to Fig. 96 for exploded view of oil pump and the oil pump drive gear assembly. The floating drive shaft (13) will usually be removed with the pump.

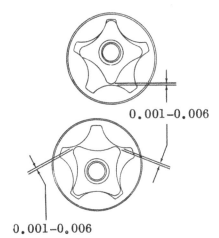

0.001-0.006

0.001-0.006

Fig. 98–Checking pump rotor to rotor clearance; renew rotor set if clearance exceeds 0.006. Refer also to Fig. 97.

To disassemble pump, remove clip (1) and oil screen (2), then remove screws retaining screen cover (3) and pump cover (4) to pump body (11). Remove the covers and pump rotor set (5), noting which direction outer rotor was placed in pump body. Remove retainer screw (9) and if necessary, thread a self-tapping screw into plug (6) and pull plug from pump body. Remove the spring (7) and oil pressure relief valve (8).

Check the pump for wear as shown in Figs. 97 and 98. Pump cover to rotor clearance (rotor end play) should be 0.001-0.0035; pump body to rotor clearance should be 0.006-0.011; and rotor clearance should be 0.001-0.006 when measured as shown in Fig. 98. Renew the rotor set and/or pump body if clearances are excessive. Renew pump cover plate if excessively worn or scored. Relief valve spring should exert a force of 10.7 to 11.9 pounds when compressed to a length of 1.07 inches. Engine oil pressure should be 60-70 psi at 1000 engine RPM.

Assemble the oil pump and reinstall by reversing removal and disassembly procedure. Tighten the oil pump retaining cap screws to a torque of 23-28 Ft.-Lbs. Note: Prime the oil pump by immersing in clean oil and turning the rotor shaft prior to installing pump in engine.

GASOLINE FUEL SYSTEM

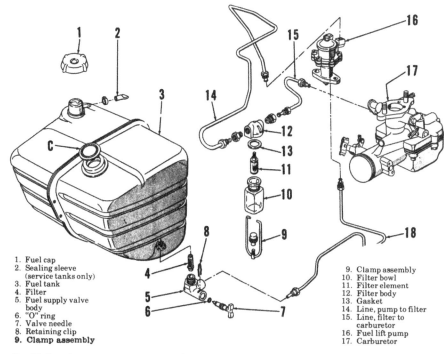

1. Fuel cap
2. Sealing sleeve (service tanks only)
3. Fuel tank
4. Filter
5. Fuel supply valve body
6. "O" ring
7. Valve needle
8. Retaining clip
9. Clamp assembly
9. Clamp assembly
10. Filter bowl
11. Filter element
12. Filter body
13. Gasket
14. Line, pump to filter
15. Line, filter to carburetor
16. Fuel lift pump
17. Carburetor

Fig. 99–Drawing showing typical gasoline fuel system. Refer to Fig. 100 for exploded view of fuel pump and to Fig. 101 for exploded view of carburetor assembly. Note filter (4) which is accessible only by removing fuel supply valve (tap) (5). An accessory kit is also available to install a renewable element type filter in line between fuel tank and fuel pump. Cap (C) is used on model 2000 without fuel gage only.

FILTERS AND SCREENS
All Models

104. The gasoline fuel system incorporates four separate fuel screens and filters. A filter screen (4—Fig. 99) in the fuel tank is accessible by disconnecting fuel supply line (18) and unscrewing the fuel supply valve (5) from tank. A second screen (3— Fig. 100) is located in the fuel pump assembly and is accessible after removing cover (1). A disc type filter element (11—Fig. 99) is located in the sediment bowl; when reinstalling element, tighten finger tight only. A screen is incorporated in the carburetor inlet fitting (41—Fig. 101). An accessory kit is also available for installing an additional renewable element type filter in line (18—Fig. 99) between fuel tank and fuel pump.

FUEL PUMP
All Models

105. Refer to Fig. 100 for exploded view of the diaphragm type fuel pump used on all models. Fuel pump assembly may be removed after disconnecting lines and unbolting lower body

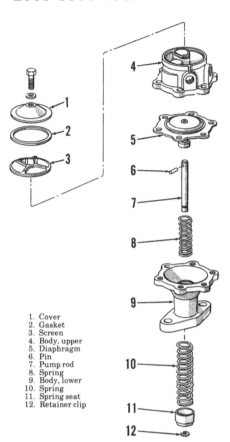

1. Cover
2. Gasket
3. Screen
4. Body, upper
5. Diaphragm
6. Pin
7. Pump rod
8. Spring
9. Body, lower
10. Spring
11. Spring seat
12. Retainer clip

Fig. 100–Exploded view of fuel lift pump assembly. Note filter screen (3). Pump is actuated by eccentric attached to front end of engine camshaft via a push rod.

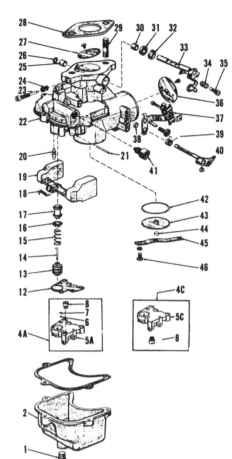

Fig. 101–Exploded view of the Holley carburetor assembly. Metering cluster cover (A) is used on 2000, 3000 and 4000; check valve ball (6) and retainer (7) is used on early production units only. Note filter screen in inlet fitting (41). Refer also to Fig. 102.

1. Drain plug	23. Idle fuel needle
2. Float bowl	24. Spring
4A. Metering cluster	25. Bushing
assy. (early)	26. Plug
4C. Metering cluster	27. Throttle plate
assy. (late)	28. Gasket
5A. Cluster cover (early)	29. Mounting studs
5C. Cluster cover (late)	30. Bushing
6. Check ball (early	31. Seal
units only)	32. Seal retainer
7. Retainer	33. Throttle shaft
8. Main jet	34. Spring
12. Gasket	35. Idle speed screw
13. Accelerating pump	36. Choke disc assy.
piston	37. Choke cable bracket
14. Link	38. Packing
15. Spring	39. Spring
16. Spring seat	40. Choke shaft
17. Vacuum piston	41. Inlet fitting & screen
18. Float spring	42. Gasket
19. Float assembly	43. Air horn plug
20. Inlet needle	44. Sintered bronze plug
21. Float hinge pin	45. Retainer spring
22. Carburetor body	46. Retainer screw

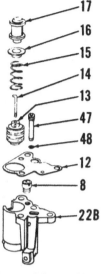

Fig. 102–On late models, accelerating pump cylinder is in metering valve body (22B) instead of carburetor body (22–Fig. 101).

8. Main jet	16. Spring seat
12. Gasket or diaphragm	17. Vacuum piston
13. Accelerating pump	22B. Metering body
piston	47. Pump valve
14. Link	48. Gasket
15. Spring	

Fig. 103–View showing correct installation of float spring (18); float lever is (19).

CARBURETOR
All Models

106. Holley carburetors are used on all gasoline engines. Refer to Fig. 101 for carburetor exploded view and component parts for different models.

Very early production carburetors were equipped with a metering cluster cover (5A) having a ball check valve (6) and valve retainer (7). When servicing one of these units, be sure that valve and retainer are reinstalled in the cluster cover.

Early production carburetors were equipped with a metering cluster cover (5A) having main jet (8) installed on top side of cover. If flooding during hillside operation is encountered with one of these units, a kit that includes new type cover (5C) with main jet on lower side and a float spring (18) should be installed. Note: Kit does not include main jet (8); main jet from old cover can

be installed in new cover. Spring (18) is installed to assist float lift; refer to Fig. 103 for spring installation. Later production carburetors include these changes.

Float level is measured as shown in Fig. 104. For early production units not having float spring (18—Fig. 101), measurement should be 61/64 to 1 1/64 inch. For models having hillside flooding correction kit installed, or on

late production units, measurement should be from 27/32 to 29/32-inch. Adjust float level by bending lever that contacts float valve (20).

Float valve (inlet needle) seat is an integral part of the carburetor body (22). As the body is not serviced separately, a new carburetor must be installed if seat or body is worn or damaged beyond further use.

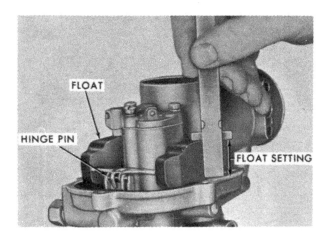

Fig. 104–Measuring float setting on Holley carburetor; bend float tang if necessary to correct float level.

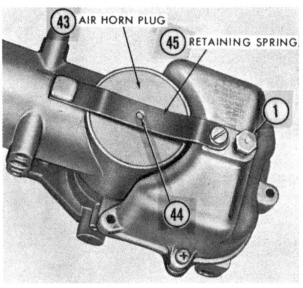

Fig. 105–It is very important that the air horn plug and sintered bronze plug (44) are properly installed to prevent dust entry into engine air intake. Drain plug is (1).

LP-GAS FUEL SYSTEMS

(Model 4000 So Equipped)

107. A factory installed Century LP-Gas fuel system is available on some model 4000 tractors; refer to Fig. 106 for exploded view of the system. A Century model STF filter-fuelock assembly (13), model H converter (vaporizer-regulator) (35) and model 3C-705-WD carburetor (31) are used.

The model STF filter-fuelock is a solenoid operated fuel shut-off device which closes off the fuel supply to the converter when ignition switch is in "OFF" position. Turning the ignition switch to "ON" position energizes the solenoid and opens the fuelock.

The model H converter is equipped with a solenoid coil operated plunger which opens the secondary valve allowing fuel to flow to the carburetor

when the solenoid is energized. A push button switch located near the instrument panel (primer button) energizes the solenoid and is used instead of a conventional choke valve within the carburetor.

NOTE: Do not push primer button except prior to starting engine, then depress button only the one or two seconds required to fill vapor hose to carburetor with fuel. Be sure throttle is open before operating primer button. The primer is used for starting purposes as the carburetor is not equipped with a conventional choke mechanism.

The model 3C Carburetor is equipped with idle speed, idle fuel mixture and power fuel mixture adjustments.

TROUBLE SHOOTING

108. When malfunction of the LP-Gas fuel system is encountered, refer to the following paragraphs for possible causes of trouble.

109. **NO FUEL TO CARBURETOR.** First, check fuel tank to be sure it is not empty, then close and slowly open the fuel tank shut-off valve to be sure the safety excess flow valve is not closed. If gas does not flow after shut-off valve is opened, loosen the fuel line at converter inlet. If no fuel is present, malfunctioning filter-fuelock is indicated; refer to paragraph 118. If fuel escapes, malfunctioning converter is indicated; refer to paragraph 119.

110. FAULTY FILTER-FUELOCK. Trouble may be caused by:
1. Fuel filter plugged.
2. Fuelock not opening; check for:
  A. No electricity to coil.
  B. Defective coil.

111. FAULTY CONVERTER. Trouble may be caused by:
1. Primary regulator not working because of:
  A. Primary orifice plugged from excessive amount of thread sealer used on assembly.
  B. Primary regulator assembled without spring.
2. Secondary regulator not working because of:
  A. Valve seat stuck to orifice.
  B. Lever far out of adjustment.
  C. Primer plunger too short to open valve.
  D. Inoperative primer due to defective push button switch, defective coil or incorrect wiring.

112. **FUEL LEAKING THROUGH CONVERTER.** Check the primary regulator pressure. If normal, leakage is through secondary regulator. If primary regulator pressure is excessive, leakage is due to primary regulator. Overhaul of converter is given in paragraph 119.

113. FAULTY PRIMARY REGULATOR. Trouble may be caused by:
1. Valve seat dirty or defective.
2. Diaphragm broken.
3. Valve lever distorted.
4. Spring or washers incorrectly installed.

114. FAULTY SECONDARY REGULATOR. If primary pressure is normal, leakage through secondary regulator could be caused by:
1. Dirt on valve seat.
2. Spring missing.
3. Lever incorrectly adjusted.
4. Primer plunger too long.
5. Primer actuated at all times.

115. **CONVERTER FREEZING.** Note: Repeated converter freezing can loosen back screws or distort back

cover. Converter freezing could be caused by:

1. Water lines restricted.
2. Internal fuel leak.
3. Low engine coolant level.
4. Defective engine water pump.
5. Loose engine fan belt.
6. Defective hose or fittings.
7. Air lock; remove top water hose from the converter and bleed system.
8. Engine head gasket leaking.

CARBURETOR ADJUSTMENT

116. First, adjust slow idle fuel mixture and idle speed as follows: Refer to Fig. 108 and loosen the drag link locknuts (18L and 18R). With the engine running and throttle control lever in slow idle position, adjust idle fuel mixture by rotating the knurled barrel (16) to lengthen or shorten the drag link. When smoothest engine idle operation is obtained, adjust engine idle speed by turning the idle speed set screw (13—Fig. 107) in or out to obtain engine slow idle speed of 600-650 RPM. Then, readjust drag link as necessary to obtain smoothest engine operation at slow idle speed and tighten the drag link locknuts. Note that one locknut (18L—Fig. 108) has left hand threads and the other (18R) has right hand threads.

Power (load) adjustment is made by rotating the spray bar (1—Fig. 107). The end of the spray bar is marked "R" for rich and "L" for lean mixture. Adjustment must be made with tractor connected to a PTO dynamometer or using an exhaust analyzer. When using dynamometer, rotate spray bar to obtain maximum horsepower. When using an exhaust analyzer, rotate spray bar to obtain a wide open throttle reading of 14.3 on LP-Gas scale; be sure to closely follow analyzer operating instructions. It will not be necessary to readjust idle speed and fuel mixture as they are not affected by spray bar adjustment.

R&R AND OVERHAUL

117. **CARBURETOR.** To remove the carburetor, disconnect the vapor hose, air cleaner hose and governor rod, then unbolt and remove carburetor from intake manifold.

Prior to disassembling, note position of throttle lever and drag link assemblies; it is possible to reassemble the carburetor with components reversed. Exploded view of carburetor in Fig. 107 shows correct assembly; however, unit is shown upside down. Refer to Fig. 108 for completely exploded view of the drag link assembly.

Completely disassemble the carburetor and clean all parts in a suitable

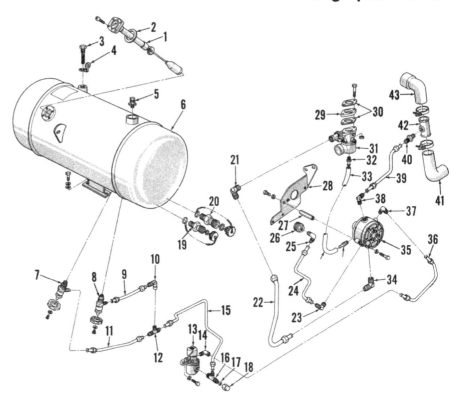

Fig. 106–Exploded view of the LP-Gas system used on model 4000 tractor. Refer to Fig. 109 for exploded view of filter-fuelock assembly (13); to Fig. 107 for exploded view of carburetor (31) and to Fig. 110 for exploded view of converter (vaporizer-regulator) (35).

1. Fuel gage	12. "Tee"	22. Tube	33. Fuel hose
2. Gasket	13. Filter-fuelock	23. Elbow	34. Elbow
3. Relief valve	14. Elbow	24. Tube, water	35. Converter
4. Raincap	15. Tube, fuel	25. Elbow	36. Tube, fuel
5. 80% fill valve	16. "Tee"	26. Adapter	37. Elbow
6. Fuel tank	17. Gasket	27. Spacers	38. Elbow
7. Supply valve	18. Cap	28. Mounting bracket	39. Tube, water
8. Supply valve	19. Liquid-vapor valve	29. Spacer	40. Connector
9. Tube	20. Vapor return-fill	30. Gaskets	41. Radiator hose
10. Elbow	valve	31. Carburetor	42. Adapter
11. Tube	21. Elbow	32. Fitting	43. Radiator hose

Fig. 107–Exploded view of Century LP-Gas carburetor assembly. View is with unit turned upside down, and with air horn removed. Refer to Fig. 108 for fully exploded view of drag link assembly.

1. Spray bar	8. Stop pin	15. Clamp screw	22. Washer
2. Throttle shaft	9. Retaining screw	16. Drag link body	23. Metering valve body
3. Throttle disc	10. Metering valve lever	17. Spring	24. Metering valve seal
4. Seal retainer	11. Spring	18. Jam nut	25. Metering valve
5. Shaft seals	12. Ball studs	19. Drag link end	26. Spring
6. Bushings	13. Idle speed screw	20. Ball socket bearings	27. Plug
7. Throttle body	14. Throttle lever	21. Metering valve lever	28. Gasket

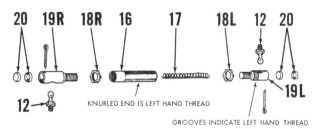

Fig. 108–Exploded view of carburetor metering valve drag link. "L" suffix on call-out indicates left hand thread; "R" indicates right hand thread. To adjust idle fuel mixture, loosen jam nuts (18L and 18R) and turn body (16).

KNURLED END IS LEFT HAND THREAD

GROOVES INDICATE LEFT HAND THREAD

12. Ball studs
16. Drag link body
17. Spring
18. Jam nuts
19. Drag link ends
20. Ball stud bearings

solvent. Blow out passages with compressed air. Thoroughly inspect all parts and renew any that are damaged or excessively worn. Always renew gaskets and throttle shaft seals when reassembling and reinstalling carburetor. Readjust carburetor as outlined in paragraph 116.

118. **FUELOCK AND FILTER.** With engine running, shut off the fuel supply valve at fuel tank and run engine until fuel is exhausted from lines. Then, turn ignition switch to "OFF" position, disconnect solenoid lead wire and the fuel lines, then unbolt and remove the fuelock and filter unit.

Refer to exploded view of unit in Fig. 109. To clean the filter, remove the screws retaining bowl (12) to housing (8), then remove bowl and filter.

Remove the screw (1), then lift cover (2) and coil (3) from plunger guide (4) and separate coil from cover. Remove the screws retaining plunger guide to housing and remove the guide, plunger spring (5) and plunger (6). Discard the "O" ring (7). Renew the plunger and, if

distorted or damaged in any way, also renew the spring. Place new "O" ring in housing, then place the spring in plunger and slide plunger upward into guide and install the guide and plunger assembly on housing. Reinstall filter element and bowl, attach air hose to bowl inlet and check unit for leaks using a soap solution.

If necessary to renew coil, pull old coil from cover and insert new coil, then install coil and cover on plunger guide. Reinstall fuelock and filter as-

Fig. 110–Exploded view of Century model H converter (vaporizer-regulator) used on model 4000 LP-Gas system. Diaphragm assemblies are riveted together and are serviced as a complete assembly only. Check valve assembly (39) screws into cover (8) at (CV). Primer solenoid (1 through 6) is serviced as complete assembly only.

CV. Check valve port
1. Cover
2. Solenoid coil
2G. Ground wire
2S. Solenoid lead
3. "O" ring
4. Plunger
5. Spring
6. Plate
7. Gasket
8. Cover
9 to 13. Secondary diaphragm assembly
14. Gasket
15. Secondary valve lever
16. Pin
17. Valve
18. Pin
19. Spring
20. Converter body
21. Gasket
22. Cover
23. Plug
24. Primary valve lever
25. Pin
26. Valve
27. Gasket
28. Valve body
29. Retainer
30 to 36. Primary diaphragm assembly
37. Spring
38. Cover
39. Check valve assembly

sembly on tractor and check connections for leaks with soap solution.

119. **CONVERTER.** With engine running, shut off fuel supply valve at tank and run engine until fuel is exhausted from system. Then, shut off ignition switch and proceed as follows:

Drain the cooling system and disconnect coolant lines from converter. Disconnect primer solenoid lead and the fuel lines from converter, then unbolt and remove converter unit from spacers on mounting bracket.

To disassemble converter, refer to exploded view in Fig. 110 and proceed as follows: Remove the primer solenoid unit (1 through 7) from front cover (8) as an assembly, then unscrew check valve (39). Remove front cover (8), then lift off secondary diaphragm (11) assembly, remove the two screws retaining valve lever pin (16) and remove lever (15) and spring (19). Remove the two screws retaining inlet valve housing (28) to converter and remove

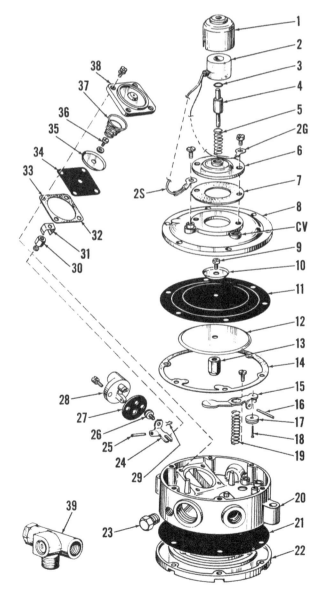

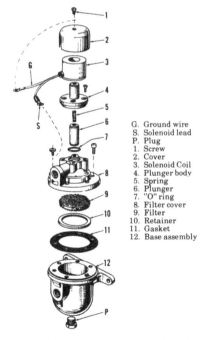

G. Ground wire
S. Solenoid lead
P. Plug
1. Screw
2. Cover
3. Solenoid Coil
4. Plunger body
5. Spring
6. Plunger
7. "O" ring
8. Filter cover
9. Filter
10. Retainer
11. Gasket
12. Base assembly

Fig. 109–Exploded view of fuel filter-fuel-lock assembly. Ground wire (G) is connected to plunger body (4) by screw (1).

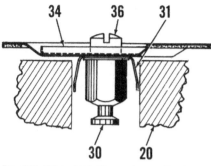

Fig. 111—When installing primary diaphragm assembly into converter body, legs of damper spring (31) must touch against sides of cavity in body (20).

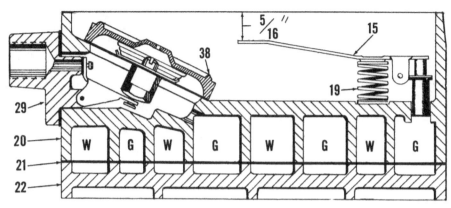

Fig. 115—Cross-sectional drawing of converter body, primary regulator and secondary regulator valve assembly. Distance between machined face of body (20) and valve lever (15) should be 5/16-inch as shown. Hot engine coolant in passages marked "W" vaporizes the LP-Gas in passages marked "G". Refer to Fig. 110 for parts identification.

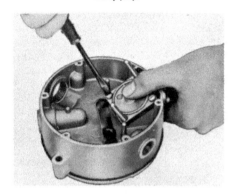

Fig. 112—When installing primary diaphragm cover, remove one alignment pin and install one retaining screw at a time.

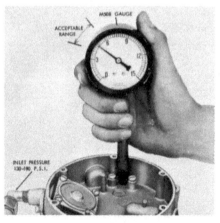

Fig. 113—Checking primary regulator pressure; refer to text.

Fig. 114—Checking secondary valve lever adjustment with Century gage No. 2V-01. Alternate method is to use straight edge and steel rule; refer to Fig. 115.

the housing and inlet valve as an assembly. Remove the square diaphragm cover (38) and primary diaphragm (34) assembly. Remove the back cover (22) and discard gasket (21).

Do not disassemble the diaphragm units as they are serviced as complete assemblies only. Remove primary valve lever pin (25), remove clip (29) and discard the valve seat (26) and gasket (27). Thoroughly clean all metal parts in a suitable solvent and air dry. Inspect all parts for wear, damage or distortion and discard any questionable items. Be sure all mating joints are flat; lap, machine or renew any warped or uneven castings.

To reassemble, proceed as follows: Lay converter body face down, screw allignment pins into back cover retaining holes and install new gasket and cover over the pins (Borg Warner part No. M-501). Install cover retaining screws in holes not having alignment pins, then remove the pins and install remaining screws. Tighten the center screws first, then alternate from side to side when tightening outer screws. Tighten the screws securely, then turn converter body face up.

Place the primary diaphragm assembly over opening in converter body making sure the damper spring legs contact the flat sides as shown in Fig. 111, then install four alignment pins (Borg Warner part No. M-501) through holes in diaphragm. Place spring (37—Fig. 110) on diaphragm plate, then install cover (38) over the springs and alignment pins. Hold the cover down securely, remove an alignment pin and install a retaining screw until all four screws are installed and securely tightened, then release pressure from cover.

Install a new valve seat (26) in primary valve lever (24) and secure with clip (29). Place new gasket (27) on valve body (28), install lever with pin (25), then carefully install the valve

assembly on converter body making sure lever straddles the diaphragm link (30) and tighten retaining screws securely.

Connect an air hose to converter inlet and regulate air pressure to 130-180 psi. Hold a pressure gage (Borg Warner part No. M-508 or equivalent) against secondary valve venturi in converter body as shown in Fig. 113. Pressure gage reading should read 4 to 6 psi. If pressure creeps upward, the primary valve is leaking and must be reworked. Lift gage slightly allowing some air to escape; pressure should drop slightly, then return to original reading when gage is pressed back against venturi.

Install new secondary valve seat (17 —Fig. 110) on lever (15) with pin (18), bend pin over (do not hammer) and clip off excess length of pin. The seat should fit loose enough to be self-aligning against venturi. Place spring (19) in pocket in converter body, then install pin (16) through lever and install the assembly with the two retaining screws. Open valve by hand and allow it to snap closed several times to align seat with orifice, then using special gage (Borg Warner part No. 2V-01) or straight edge and rule, measure lever height as shown in Fig. 114. Bend lever as necessary to obtain the 5/16-inch distance between lever and machined face of converter body.

Reconnect air supply to converter inlet. Plug one water fitting opening and apply soap bubble to other water opening; any continuous growth of bubble will indicate leakage past back cover gasket. Immerse entire unit in water or check all joints including the secondary valve with soap solution and correct any leakage noted before proceeding further.

Install three alignment pins (M-501) in alternate holes in converter body and slide gasket (14—Fig. 110) down

Fig. 116–Use alignment pins when installing gasket, secondary diaphragm assembly and cover.

Fig. 117–Pull up on diaphragm with pliers while tightening cover retaining screws.

over the pins with gasket ears over the primary cover (38) mounting screws. Mount the secondary diaphragm assembly on pins with button against valve lever. Place cover (8) on pins so that it will be right side up when mounted on tractor and install three retaining screws. Remove the alignment pins and install the other three screws. Refer to Fig. 117 and tighten cover retaining screws while pulling up on diaphragm screw (9—Fig. 110) with pliers. Reinstall primer coil assembly and connect to 12-volt battery to check

primer operation. Apply air pressure to converter inlet and check for leakage at vapor outlet with soap bubble. Actuate primer either manually or electrically to check for fuel flow out of vapor outlet; volume will not be great but escaping air should be audible. Reinstall converter by reversing removal procedure. Be sure to bleed any trapped air from the converter by loosening top water hose.

DIESEL FUEL SYSTEM

The diesel fuel system consists of three basic components: the fuel filters, injection pump and injection nozzles. When servicing any unit associated with the fuel system, the maintenance of absolute cleanliness is of utmost importance. Of equal importance is the avoidance of nicks or burrs on any of the working parts.

Probably the most important precaution that service personnel can impart to owners of diesel powered tractors is to urge them to use an approved fuel that is absolutely clean and free from foreign material. Extra precaution should be taken to make certain that no water enters the fuel storage tanks.

TROUBLESHOOTING
All Diesel Models

120. If the engine will not start, or does not run properly after starting, refer to the following paragraphs for possible causes of trouble.

121. **FUEL NOT REACHING INJECTION PUMP.** If no fuel will run from line when disconnected from

pump, check the following:
Be sure fuel supply valve is open.
Check the filters for being clogged (Including filter screen in fuel supply valve).
Bleed the fuel filters.
Check lines and connectors for damage.

122. **FUEL REACHING NOZZLES BUT ENGINE WILL NOT START.** If, when lines are disconnected at fuel nozzles and engine is cranked, fuel will flow from connections, but engine will not start, check the following:
Check cranking speed.
Check throttle control rod adjustment.
Check pump timing.
Check fuel lines and connections for pressure leakage.
Check engine compression.

123. **ENGINE HARD TO START.** If the engine is hard to start, check the following:
Check cranking speed.
Bleed the fuel filters.

Check for clogged fuel filters.
Check for water in fuel or improper fuel.
Check for air leaks on suction side of transfer pump.
Check engine compression.

124. **ENGINE STARTS, THEN STOPS.** If the engine will start, but then stops, check the following:
Check for clogged or restricted fuel lines or fuel filters.
Check for water in fuel.
Check for restrictions in air intake.
Check engine for overheating.
Check for air leaks in lines on suction side of transfer pump.

125. **ENGINE SURGES, MISFIRES OR POOR GOVERNOR REGULATION.** Make the following checks:
Bleed the fuel system.
Check for clogged filters or lines or restricted fuel lines.
Check for water in fuel.
Check pump timing.
Check injector lines and connections for leakage.
Check for faulty or sticking injector nozzles.
Check for faulty or sticking engine valves.

NOTE: On models with C.A.V. fuel injection pump, refer also to paragraph 152.

126. **LOSS OF POWER.** If engine does not develop full power or speed, check the following:
Check throttle control rod adjustment.
Check maximum no-load speed adjustment.
Check for clogged or restricted fuel lines or clogged fuel filters.
Check for air leaks in suction line of transfer pump.
Check pump timing.
Check engine compression.
Check for improper engine valve gap adjustment or faulty valves.

127. **EXCESSIVE BLACK SMOKE AT EXHAUST.** If the engine emits excessive black smoke from exhaust, check the following.
Check for restricted air intake such as clogged air cleaner.
Check pump timing.
Check for faulty injectors.
Check engine compression.

FILTERS AND BLEEDING

Model 3000 With Simms Fuel Injection Pump

128. **MAINTENANCE.** On models produced prior to 12-68, the fuel filter head is fitted with two renewable type elements as shown in Fig. 118. On models produced after 11-68, only one filter assembly is used. Water drain plug (cap) should be removed each 50

hours of operation and any water in sediment bowl be drained; remove cap and drain more often than each 50 hours if excessive condensation is noted.

After each 1200 hours (dual element), or 600 hours (single element) of operation, the fuel filter elements should be renewed. Refer to Fig. 119. Unscrew the cap screw at top side of filter head and remove the filter element and sediment bowl as shown. Clean the sediment bowl and reinstall with new element and rubber sealing rings; be sure rings are placed as shown in Fig. 119. Bleed the diesel fuel system as outlined in paragraph 129.

129. **BLEEDING.** Bleeding procedure will remain basically the same regardless of whether dual or single filters are used. The following is based on the tractor being equipped with dual filters and when bleeding models with single filter ignore reference to rear and front filter bleed screws.

Refer to Fig. 120. Open rear bleed screw on filter head and actuate hand (priming) lever on fuel lift pump until fuel flowing from bleed screw is free of bubbles. Close the rear bleed screw and open front bleed screw on filter head and actuate primer lever until bubble free fuel flows from opening. Then, close the fuel filter bleed screw, open

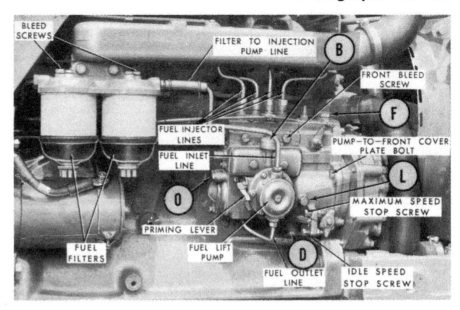

Fig. 120–Side view of a typical engine showing filter and the Simms fuel injection pump. Note priming lever on fuel lift (transfer) pump and bleed screw location on fuel filter head and on fuel injection pump. Simms injection pump used on some model 3000 tractors is similar except for three plunger pump instead of four plunger unit shown. Although not shown, an outlet (overflow) tube is connected at fitting (O). Breather is located at (B), filler plug at (F), oil level plug at (L) and drain plug at (D).

front bleed screw on fuel injection pump and actuate primer lever until fuel flowing from bleed screw is free of bubbles. Close the pump bleed screw while actuating primer lever. Loosen the fuel injector lines at the injectors

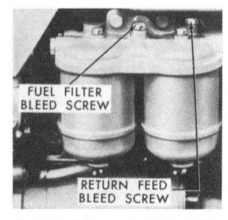

Fig. 121–Bleed screws are located as shown on filter head used with C.A.V. fuel injection pump; refer to Fig. 122 for bleed screw location on the C.A.V. pump assembly.

Fig. 118–After each 50 hours operation, remove water drain plugs from bottom of each fuel filter and allow to drain until only clean fuel flows, then reinstall plugs. The filters used with Simms injection pumps are shown at left; filters at right are used with C. A. V. injection pumps.

Fig. 119–To renew filter elements, unscrew cap screw at top side of filter head, then remove sediment bowl and filter element. Be sure rubber sealing rings are placed as shown when installing new element.

and crank engine until fuel appears at all injectors, then tighten the fuel injector line connections and start engine.

Models With C.A.V. Fuel Injection Pump

130. **MAINTENANCE.** Filter maintenance is the same as outlined for models with Simms fuel injection pump; refer to paragraph 128.

131. **BLEEDING.** Turn on the fuel supply valve at fuel tank and open the bleed screw(s) (See Fig. 121) on fuel filter head; when fuel is flowing freely with no bubbles, close the bleed screw(s) on filter head and open the bleed screw (See Fig. 122) on fuel injection pump. Crank the engine until fuel flows from injection pump bleed screw

Fig. 122–View showing location of bleed screw on C.A.V. fuel injection pump assembly.

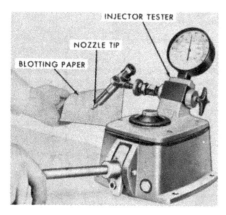

Fig. 123–A fuel injector tester such as the one shown is necessary for checking and adjusting fuel injector assemblies. Nozzle seat leakage check is illustrated; refer to paragraph 147.

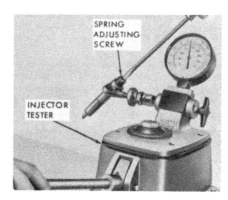

Fig. 124–Adjusting nozzle opening pressure; refer to paragraph 145 for specifications and procedure.

and is free of bubbles, then close the bleed screw. Loosen the fuel injection line connections at fuel injectors and crank the engine until fuel flows at all three connections. Then, tighten the fuel line connections and start engine.

INJECTION NOZZLES

CAUTION: Fuel leaves the injection nozzles with sufficient force to penetrate the skin. When testing, keep your person clear of the nozzle spray.

All Diesel Models

132. Diesel engines are fitted with either Simms or C.A.V. fuel injector assemblies. The injector nozzles have four spray holes spaced in a pattern at 90 degrees apart around the tip of the nozzle.

All 2000 models are fitted with a C.A.V. distributor type fuel injection pump and C.A.V. injectors.

Model 3000 is fitted with either a C.A.V. distributor type fuel injection pump or a Simms multiple plunger type pump. On models with C.A.V. pump, C.A.V. fuel injector assemblies are used; nozzle (C.A.V. part No. BDLL150S6443) has four 0.0114-0.0122 diameter spray holes. On models with Simms pump, Simms fuel injector assemblies are used; nozzle (Simms part No. NL413) has four 0.0102-0.0110 inch diameter spray holes.

All 4000 models are fitted with a C.A.V. distributor type fuel injection pump; however, either C.A.V. or Simms injector assemblies may be installed in production. Some early production 4000 engines may have C.A.V. injectors with nozzles (C.A.V. part No. BDLL150S6443) having 0.0114-0.0122 inch diameter spray holes; later engines will have either C.A.V. fuel injector assemblies with nozzles (C.A.V. part No. BDLL150S6476)

having four 0.0122-0.0130 inch diameter spray holes or Simms fuel injector assemblies with nozzles (Simms part No. NL461) having four 0.0122-0.0130 inch diameter spray holes. The Simms nozzles (part No. NL461) are not available for service; however, the C.A.V. nozzles are completely interchangeable and may be installed in the Simms injector holder. When servicing early units with C.A.V. part No. BDLL150S6443 nozzles, be sure to use the same part to service individual units; however, if all three nozzles are being installed, the C.A.V. BDLL150S6476 nozzles with larger spray hole diameter may be installed.

The C.A.V. or Simms part number is etched on the larger diameter of the fuel injector nozzle. Corresponding Ford part numbers and model usage are as follows:

Tractor Model	C.A.V. or Simms Part Number	Ford Part Number
2000, 3000	BDLL150S6443	C5NE-9E527-A
4000	BDLL150S6476	C5NE-9E527-C
3000 W/Simms pump	NL413	C5NE-9E527-A

133. **TESTING AND LOCATING A FAULTY NOZZLE.** If engine does not run properly and a faulty injection nozzle is indicated, such a faulty nozzle can be located as follows: With engine

running, loosen the high pressure line fitting on each nozzle holder in turn, thereby allowing fuel to escape at the union rather than enter the injector. As in checking for faulty spark plugs in a spark ignition engine, the faulty unit is the one which, when its line is loosened, least affects the running of the engine.

144. **NOZZLE TESTER.** A complete job of testing and adjusting the fuel injection nozzle requires use of a special tester such as shown in Fig. 123. The nozzle should be tested for opening pressure, spray pattern, seat leakage and leak back.

Operate the tester until oil flows and then connect injector nozzle to tester. Close the tester valve to shut off passage to tester gage and operate tester level to be sure nozzle is in operating condition and not plugged. If oil does not spray from all four spray holes in nozzle, if tester lever is hard to operate or other obvious defects are noted, remove nozzle from tester and service as outlined in paragraph 150. If nozzle operates without undue pressure on tester lever and fuel is sprayed from all four spray holes, proceed with following tests:

145. **OPENING PRESSURE.** While slowly operating tester lever with valve to tester gage open, note gage pressure at which nozzle spray occurs. This gage pressure should be 2720-2794 psi. If gage pressure is not within these limits, remove cap nut and turn adjusting screw (See Fig. 124) as required to bring opening pressure within specified limits. If opening pressure is erratic or cannot be properly adjusted, remove nozzle from tester and overhaul nozzle as outlined in paragraph 150. If opening pressure is within limits, check spray pattern as outlined in following paragraph.

146. **SPRAY PATTERN.** Operate the tester lever slowly and observe nozzle spray pattern. All four (4) sprays must be similar and spaced at approximately 90° to each other in a nearly horizontal plane. Each spray must be well atomized and should spread to a 3 inch diameter cone at approximately 8 inches from nozzle tip. If spray pattern does

Fig. 125–Left; fuel injector leak off line connections. Right view; Removing the fuel injector assembly. Refer also to paragraph 149.

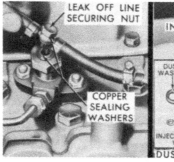
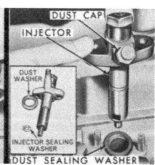

not meet these conditions, remove nozzle from tester and overhaul nozzle as outlined in paragraph 150. If nozzle spray is satisfactory, proceed with seat leakage test as outlined in following paragraph:

147. SEAT LEAKAGE. Close valve to tester gage and operate tester lever quickly for several strokes. Then, wipe nozzle tip dry with clean blotting paper, open valve to tester gage and push tester lever down slowly to bring gage pressure to 200 psi. below nozzle opening pressure and hold this pressure for one minute. Apply a piece of clean blotting paper (see Fig. 123) to tip of nozzle; the resulting oil blot should not be greater than one-half inch in diameter. If nozzle tip drips oil or blot is excessively large, remove nozzle from tester and overhaul nozzle as outlined in paragraph 150. If nozzle seat leakage is not excessive, proceed with nozzle leak back test as outlined in following paragraph.

148. NOZZLE LEAK BACK. Operate tester lever to bring gage pressure to approximately 2300 psi., release lever and note time required for gage pressure to drop from 2200 psi. to 1500 psi. Time required should be from 10 to 40 seconds. If time required is less than ten seconds, nozzle is worn or there are dirt particles between mating surfaces of nozzle and holder. If time required is greater than forty seconds, needle may be too tight a fit in nozzle bore. Refer to paragraph 150 for disassembly, cleaning and overhaul information.

NOTE: A leaking tester connection, check valve or pressure gage will show up in this test as excessively fast leak back. If, in testing a number of injectors, all show excessively fast leak back, the tester should be suspected as faulty rather than the injectors.

149. **REMOVE AND REINSTALL INJECTORS.** Before removing injectors, carefully clean all dirt and other foreign material from lines, injectors and cylinder head area around the injectors. Disconnect the injector leak-off line (Fig. 125) at each injector and at the fuel return line. Disconnect the injector line at the pump and at the injector. Cap all lines and openings. Remove the two retaining nuts and carefully remove the injector from cylinder head.

Prior to reinstalling injectors, check the injector seats in cylinder head to see that they are clean and free from any carbon deposit. Install a new copper washer in the seat and a new cork dust sealing washer around the body of the injector. Insert the injector in cylinder head bore, install retaining

washers and nuts and tighten the nuts evenly and alternately to a torque of 10-15 Ft.-Lbs. Position new leak-off fitting gaskets below and above each banjo fitting and install the banjo fitting bolts to a torque of 8-10 Ft.-Lbs. Reconnect leak-off line to return line. Check the fuel injector line connections to be sure they are clean and reinstall lines tightening connections at pump end only. Crank engine until a stream of fuel is pumped out of each line at injector connection, then tighten the connections. Start and run engine to be sure that injector is properly sealed and that injector line and leak-off line connections are not leaking.

150. **OVERHAUL INJECTORS.** Unless complete and proper equipment is available, do not attempt to overhaul diesel nozzles. Equipment recommended by Ford is Kent-Moore J8666 Injector Nozzle Tester and J8537 Injector Nozzle Service Tool Set. This equipment is available from the Kent-Moore Organization, Inc., 28635 Mound Road, Warren, Michigan.

Refer to Fig. 126 and proceed as follows: Secure injector holding fixture (J8537-11) in a vise and mount injector assembly in fixture. Never clamp the injector body in vise. Remove the cap nut and back-off adjusting screw, then lift off the upper spring disc, injector spring and spindle. Remove the nozzle retaining nut using nozzle nut socket (J8537-14), or equivalent, and remove the nozzle and valve. Nozzles and valves are a lapped fit and must never be interchanged. Place all parts in clean fuel oil or calibrating fluid as they are disassembled. Clean injector assembly exterior as follows: Soften hard carbon deposits formed in the spray holes and on needle tip by soaking in a suitable carbon solvent, then use a soft wire (brass) brush to remove carbon from needle and nozzle exterior. Rinse the nozzle and needle immediately after cleaning to prevent the carbon solvent from corroding the highly finished surfaces. Clean the pressure chamber of the nozzle with a 0.043 reamer (J8537-4) as shown in Fig. 127. Clean the spray holes with the proper size wire probe held in a pin vise (J4298-1) as shown in Fig. 128. To prevent breakage of wire probe, the wire should protrude from pin vise only far enough to pass through the pin holes. Rotate pin vise without applying undue pressure. Use a 0.011 wire probe (Kent-Moore part No. J8537-2) with C.A.V. nozzle having part No. BDLL150S6443 etched on large diameter, a 0.012 inch diameter wire probe (Kent-Moore No. J8537-3) with a C.A.V. nozzle having part No.

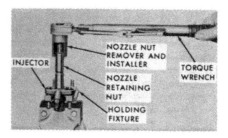

Fig. 126–View showing injector assembly mounted on holding fixture; nozzle retaining nut is being tightened with torque wrench.

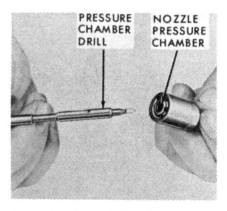

Fig. 127–Cleaning nozzle tip cavity with pressure chamber drill.

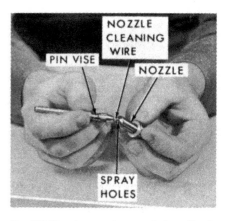

Fig. 128–Cleaning nozzle spray holes with wire probe held in pin vise.

BDLL150S6476 or Simms nozzle having part No. NL461, or a 0.009 inch diameter wire probe (Kent-Moore No. J8537-1) with Simms nozzle having part No. NL413.

The valve seats in nozzle are cleaned by inserting the valve seat scraper (J8537-18) into the nozzle and rotating scraper. Refer to Fig. 129. The annular groove in top of nozzle and the pressure chamber are cleaned by using (rotating) the pressure chamber carbon remover tool (J-8537-15) as shown in Fig. 130.

When above cleaning is accomplished, back flush nozzle and needle by installing reverse flushing adapter (J8537-6) on the nozzle tester and inserting nozzle and needle assembly tip

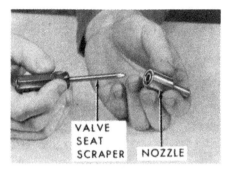

Fig. 129—Use scraper to clean carbon from valve seat in nozzle body.

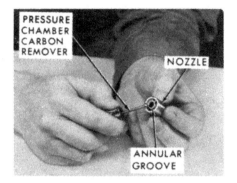

Fig. 130—Pressure chamber carbon remover is used to clean annular groove as well as clean carbon from pressure chamber in nozzle.

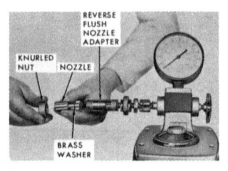

Fig. 131—A back flush attachment is installed on nozzle tester to clean nozzle by reverse flow of fluid; note proper installation of nozzle in the adapter unit.

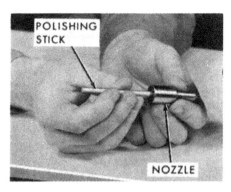

Fig. 132—Polishing needle valve seat with tallow and polishing stick.

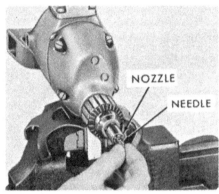

Fig. 133—Chuck small diameter of nozzle in slow speed electric drill to lap needle to nozzle if leak back time is excessive or needle sticks in nozzle. Hold pin end of needle with vise grip pliers.

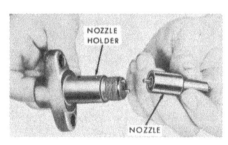

Fig. 134—Be sure dowel pins in nozzle holder body enter mating holes in nozzle.

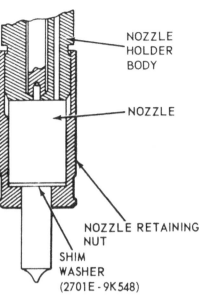

Fig. 135—Cross-sectional view showing shim washer installed between nozzle and nozzle retaining nut.

end first into the adapter and secure with knurled nut. Rotate the needle in nozzle while operating tester lever. After nozzle is back flushed, the seat can be polished by using a small amount of tallow (J8537-28) on the end of a polishing stick (J8537-21) and rotating stick in nozzle as shown in Fig. 132.

If the leak-back test time was greater than 40 seconds (refer to paragraph 148), or if needle is sticking in bore of nozzle, correction can be made by lapping the needle and nozzle assembly. This is accomplished by using a polishing compound (Bacharach No. 66-0655 is suggested) as follows: Place small diameter of nozzle in a chuck of a drill having a maximum speed of less than 450 RPM. Apply a small amount

of polishing compound on barrel of needle taking care not to allow any compound on tip or beveled seat portion, and insert needle in rotating nozzle body. Refer to Fig. 133. Note: It is usually necessary to hold upper pin end of needle with vise-grip pliers to keep the needle from turning with the nozzle. Work the needle in and out a few times taking care not to put any pressure against seat, then withdraw the needle, remove nozzle from chuck and thoroughly clean the nozzle and needle assembly using back flush adapter and tester pump.

Prior to reassembly, rinse all parts in clean fuel oil or calibrating fluid and assemble while still wet. The injector inlet adapter (Simms only) normally

does not need to be removed. However, if the adapter is removed, use a new copper sealing washer when reinstalling. Position the nozzle and needle valve on injector body and be sure dowel pins in body are correctly located in nozzle as shown in Fig. 134. Install the ⅜-inch shim washer (see Fig. 135) and nozzle retaining nut and tighten nut to a torque of 50 Ft.-Lbs. Note: Place injector in holding fixture (J8537-11) and tighten nut with socket (J8537-14). Install the spindle, spring, upper spring disc and spring adjusting screw. Connect the injector to tester and adjust opening pressure as in paragraph 145. Use a new copper washer and install cap nut. Recheck nozzle opening pressure to be sure that installing nut did not change adjustment. Retest injector as outlined in paragraphs 146 through 148; renew nozzle and needle if still faulty. If the injectors are to be stored after overhaul, it is recommended that they be thoroughly flushed with calibrating fluid prior to storage.

**FUEL INJECTION PUMP
Models 2000, 3000 and
4000 With C.A.V. Pump**

151. **PUMP TIMING.** Refer to Fig. 136. The C.A.V. fuel injection pump is correctly timed when installed with scribe line on pump body aligned with the "O" mark on engine front plate.

NOTE: Some mechanics may prefer to set injection pump timing to obtain maximum engine horsepower on a dynamometer. If following this procedure, do not loosen pump mounting bolts when engine is running. Stop engine to make change in timing, then restart and recheck horsepower output on dynamometer.

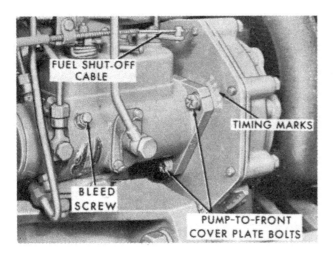

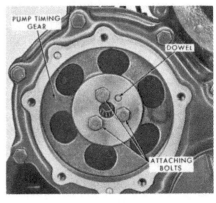

Fig. 136–View showing timing marks on C.A.V. fuel injection pump body and engine front plate. Timing is correct when scribe line on pump is aligned with "O" mark on engine front plate.

Fig. 137–View with pump drive gear cover removed from timing gear cover showing C.A.V. fuel injection pump drive gear, attaching bolts and dowel pin.

152. FUEL INJECTION PRESSURIZATION VALVES. Fuel injection pressurization (delivery) valves are available for installation on models equipped with C.A.V. fuel injection pumps. These valves will eliminate diesel engine flutter problem where experienced. To install the valves, proceed as follows:

Thoroughly clean the fuel injection pump and injector lines. Remove the No. 1 cylinder fuel injection line from injection pump, then remove and discard the line connector. Install the straight pressurization valve (Ford part No. C5NN-9N022-A) using a new gasket, tighten valve to torque of 22.7 Ft.-Lbs. (262 inch-pounds) and reconnect the injector line. Remove the No. 2 and No. 3 injector lines from pump, remove and discard line connectors, then install the remaining two pressurization valves (Ford part No. C5NN-9N022-B) and tighten to a torque of 22.7 Ft.-Lbs. Reconnect injector lines to pressurization valves and bleed the lines as outlined in paragraph 131.

153. R&R FUEL INJECTION PUMP. Thoroughly clean the pump, lines and connections and area around pump. Remove the pump to injector lines, disconnect the fuel inlet and outlet (return) lines from injection pump and immediately cap all openings. Disconnect the throttle control rod and fuel shut-off cable. Remove cover plate from engine timing gear cover and remove the three cap screws retaining pump drive gear to pump drive hub; refer to Fig. 137. Unbolt the injection pump from engine front plate and remove the pump assembly. The pump drive gear will remain in the engine timing cover and cannot become out of time; however, the engine should not be turned when the pump is removed.

To reinstall fuel injection pump, reverse removal procedures. Align scribe mark on pump body with "O" mark on engine front plate as shown in Fig. 136

and tighten pump retaining bolts to a torque of 26-30 Ft.-Lbs. Tighten injection pump drive gear retaining bolts to a torque of 20-25 Ft.-Lbs. Bleed the diesel fuel system as outlined in paragraph 131.

Model 3000 With Simms Pump

154. LUBRICATION. The Simms fuel injection pump is lubricated by oil sump in the pump cambox. After each 300 hours of operation, the pump should be drained, the cambox breather cleaned and the cambox refilled to proper level with new, clean engine oil. Use same weight and type oil as for engine crankcase; refer to Fig. 120 for location of breather (B), drain plug (D), oil level plug (L), and filler plug (F).

Whenever installing a new or rebuilt fuel injection pump, be sure the cambox is filled with engine oil to level of the oil level plug (L) before attempting to start engine. There will be some oil dilution with diesel fuel during engine operation and after engine is stopped, some of the fuel-oil mixture may run from outlet tube (O).

155. PUMP TIMING. To check and adjust pump timing, proceed as follows:

Drain coolant from radiator and remove lower radiator hose. Open the hood, swing battery out and remove the rocker arm cover. Remove cover plate from front side of engine timing gear cover and remove the flywheel timing hole cover plate from right rear side of engine. Turn engine until intake valve on No. 1 cylinder closes, then continue to turn engine slowly until the 19 degree BTDC timing mark on flywheel is aligned with arrow at edge of timing hole. The timing marks on fuel injection pump should then be exactly aligned as shown in Fig. 138, if not, loosen the three cap screws that retain gear to fuel injection pump drive hub and rotate the pump camshaft until marks are aligned. Tighten the cap screws to a torque of 20-25 Ft.-Lbs. Recheck timing marks and if aligned, reinstall the engine timing gear cover, lower radiator hose and timing hole cover. Refill cooling system.

156. R&R FUEL INJECTION PUMP. Thoroughly clean the pump, lines and connections and the area around the pump. Proceed as outlined in paragraph 155 to bring the flywheel and pump timing marks into alignment and shut-off the fuel. Remove the

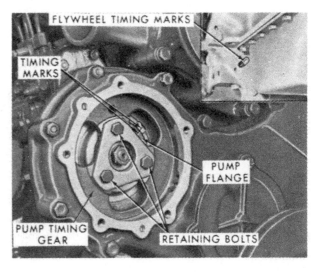

Fig. 138–View with pump drive gear cover removed from timing gear cover to show Simms fuel injection pump drive gear, retaining bolts and timing marks. Inset, top right, shows location of timing hole for flywheel timing marks.

Fig. 139–View showing location of high idle (maximum) speed stop screw and (slow) idle speed stop screw on C.A.V. fuel injection pump.

pump to injector lines, disconnect the fuel inlet and outlet lines from fuel lift pump and the filter to injection pump line and immediately cap all openings. Disconnect the throttle and fuel shut-off controls. Remove the three cap screws retaining gear to injection pump drive hub and remove the gear clamping plate. Remove the cap screws retaining pump to engine front plate and remove the pump assembly. The pump drive gear will remain in the engine timing gear cover and cannot become out-of-time; however, the engine should not be turned with pump removed. Remove excess fuel leak-off line from pump and plug the opening.

To reinstall the fuel injection pump, reverse the removal procedures and time the pump as outlined in paragraph 155. Tighten the fuel injection pump retaining cap screws to a torque of 26-30 Ft.-Lbs. and the gear to drive hub retaining cap screws to a torque of 20-25 Ft.-Lbs. Bleed fuel injection pump as outlined in paragraph 129.

INJECTION PUMP DRIVE GEAR
All Diesel Models
157. To remove the fuel injection pump drive gear, first remove the engine timing gear cover as outlined in paragraph 77 or 80. Then, remove the three cap screws, retainer plate (models with Simms injection pump only) and the fuel injection pump drive gear.

Prior to installing gear, turn engine crankshaft so that timing marks on crankshaft gear and camshaft gear point towards center of idler (camshaft drive) gear. Then, remove the self-locking cap screw retaining idler gear to front face of cylinder block, remove the idler gear and reinstall with timing marks aligned with marks on crankshaft gear and camshaft gear. Tighten idler gear cap screw to a torque of 100-105 Ft.-Lbs.

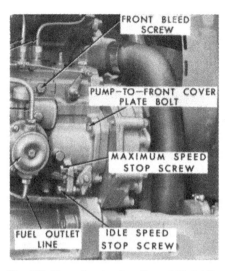

Fig. 140–View showing location of high idle (maximum) speed stop screw and (slow) idle speed stop screw on Simms fuel injection pump.

On models with C.A.V. pump, turn pump shaft so that pump drive gear can be installed on dowel pin in pump drive adapter with timing marks on pump drive gear aligned with timing mark on idler (camshaft drive) gear. Install and tighten the three pump drive gear retaining cap screws to a torque of 20-25 Ft.-Lbs.

On models with Simms pump, place the pump drive gear on adapter hub with timing mark aligned with timing mark on idler gear. Place retainer plate on gear, install socket wrench on nut on front end of pump camshaft and turn pump camshaft until timing mark on adapter is aligned with timing mark (pointer) on pump front plate. Then, turn the pump slowly in a counter-clockwise direction (as viewed from front of engine), if necessary, so that the cap screws can be installed through retainer plate and drive gear into the pump drive adapter. Turn the engine until No. 1 piston is at 19 degrees

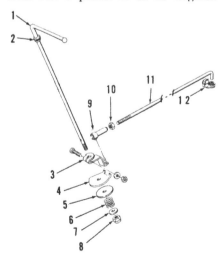

Fig. 141–Exploded view of typical hand throttle lever and linkage; some models have spring washer instead of coil spring (6).

BTDC on compression stroke. With pump drive gear retaining cap screws loose, turn the pump camshaft with socket wrench so that pump timing marks are aligned (see Fig. 138), then tighten the drive gear retaining cap screws to a torque of 20-25 Ft.-Lbs.

Reinstall timing gear cover as per appropriate paragraph.

DIESEL GOVERNOR ADJUSTMENTS
Models With C.A.V. Pump
158. To check idle speed adjustment, proceed as follows: Start engine and bring to a normal operating temperature. Disconnect throttle linkage from governor arm on fuel injection pump. Hold the governor arm against the slow idle speed stop screw; slow idle speed should then be 650 RPM. If it is not, loosen the lock nut on the slow idle speed stop screw (Fig. 139) and turn the screw in or out to obtain proper slow idle speed, then tighten the lock nut. Hold the injection pump governor arm against the high idle (maximum) speed stop screw; engine speed should then be 2175-2225 RPM on models 2000 and 3000 or 2395-2445 on model 4000. If not within specified speed range, the high idle speed stop screw should be adjusted. CAUTION: The high idle speed stop screw is sealed at the factory with a sealing wire and cover tube; this seal should not be broken on tractors within factory warranty by other than Ford authorized diesel service personnel. To adjust high idle speed, break the wire seal, remove the cover tube and loosen the lock nut on adjusting screw; then, turn screw in or out to obtain specified high idle speed, tighten lock nut and reseal the screw. When reconnecting throttle linkage, refer to paragraph 160 or 161 and adjust linkage if necessary.

Models With Simms Injection Pump
159. Start engine and bring to normal operating temperature. Disconnect throttle linkage from fuel injection pump governor arm and hold the arm so that stop lever contacts the slow idle speed stop screw (see Fig. 140). If engine speed is not then approximately 650 RPM, loosen the lock nut on stop screw and turn screw in or out until proper slow idle speed is obtained and tighten lock nut. Hold the arm so that stop lever is against the high idle (maximum) speed stop screw; engine high idle speed should then be 2175-2225 RPM. If high idle speed is not within the specified range, the stop screw should be adjusted. CAUTION: The high idle (maximum) speed stop screw adjustment is sealed with a sealing

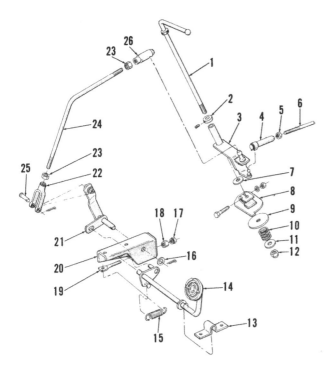

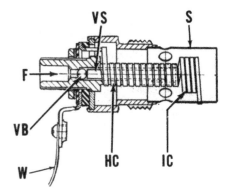

Fig. 142–Exploded view of foot operated throttle assembly typical of those available on some models. Bell crank (3) pivots on throttle lever shaft (1) allowing foot pedal to control throttle link (6) without moving hand lever. Refer also to Fig. 153.

1. Hand lever
2. Lock collar
3. Bell crank
4. End assembly
5. Lock nut
6. Throttle link
7. Washer
8. Friction plate
9. Friction disc
10. Spring
11. Washer
12. Adjusting nut
13. Support bracket
14. Foot pedal assy.
15. Tension spring
16. Washer
17. Jam nut
18. Adjusting nut
19. Adjusting bolt
20. Pivot bracket
21. Bell crank
22. Link end
23. Lock nuts
24. Link
25. Clevin pin
26. End assembly

Fig. 143–Cross-sectional view of diesel Thermostart unit (optional) that is located in intake manifold. Refer to Figs. 144 and 145 for fuel reservoir and fuel line connections.

F. Fuel inlet
HC. Heater coil
IC. Ignition coil
S. Shield

VB. Valve ball
VS. Valve stem
W. Wire to key switch

wire at the factory; this seal should not be broken by anyone other than Ford authorized diesel service personnel if tractor is within factory warranty. To adjust high idle speed, break wire seal, loosen lock nut and turn adjusting screw in or out until proper speed is obtained; then, tighten lock nut and reseal adjusting screw. When reconnecting throttle linkage, refer to paragraph 160 or 161 and adjust linkage if necessary.

THROTTLE LINKAGE ADJUSTMENTS
Models With Hand Operated Throttle Only

160. Refer to exploded view of typical hand operated throttle linkage in Fig. 141. With the throttle link (11) disconnected from injection pump governor arm and with engine governed speed adjusted as outlined in paragraph 158 or 159, proceed as follows: Move hand lever (1) to rear against slow idle stop and hold pump governor arm against slow idle speed stop screw; forward end of link (11) should then enter hole in governor arm without binding. If not, loosen lock nut (10) and turn link in or out of end assembly (9) as required. With link length properly adjusted, reconnect link to pump governor arm with retainer (12) and tighten lock nut (10). With engine running, check operation of throttle; there should be sufficient tension on spring (6) (some models have spring washer instead of coil spring) so that hand lever will remain in any desired position, yet move without excessive binding.

Tighten or loosen spring adjusting nut (8) as required. Recheck engine governed speed (paragraph 158 or 159) with throttle linkage connected.

Models With Foot Operated Throttle

161. Refer to Fig. 142 for exploded view of typical foot operated throttle linkage. With the throttle link (6) disconnected from fuel injection pump governor arm and engine governed speed adjusted as outlined in paragraph 158 or 159, proceed as follows: Disconnect link (24) by removing clevis pin (25) and move hand throttle lever (1) to rear against slow idle stop. Hold fuel injection pump governor lever against slow idle speed stop screw and push link (6) to rear; forward end of link should then enter hole in pump governor arm without binding. If not, loosen lock nut (5) and turn link in or out of end assembly (4) so that correct link length is obtained, connect link to pump governor arm with retainer and tighten lock nut. With hand throttle lever to rear against slow idle stop, pull link (24) to rear. It should then be necessary to depress foot pedal slightly so that clevis pin (25) can be inserted through link end (22) and bell crank (21). If not, loosen lock nuts (23) and lengthen the link (24). With clevis pin installed, slowly depress foot throttle pedal; pedal should contact downward stop at same time fuel injection pump governor lever contacts high idle (maximum) speed stop screw. If not, shorten or lengthen the link (24) as required and tighten the lock nuts (23). When hand lever is against rear (slow idle) stop, spring (15) should return linkage

to slow idle position. If not, loosen jam nut (17) and tighten adjusting nut (18) to increase spring tension. With engine running, check operation of hand throttle lever; there should be sufficient tension on spring (10) so that hand lever will remain in any desired position, yet move without excessive binding. Tighten or loosen spring adjusting nut (12) as required. Recheck engine governed speed (paragraph 158 or 159) with all throttle linkage connected.

THERMOSTART COLD WEATHER STARTING AID
All Diesel Models So Equipped

162. Refer to Fig. 143 for cross-sectional view of the Thermostart unit which is located in the engine intake manifold. The fuel inlet (F) is connected to a small reservoir (see Figs. 144 and 145). In operation, turning the key switch to either "HEAT" or "HEAT-START" position connects Thermostart wire (W—Fig. 143) to battery heating the heater coil (HC) and ignition coil (IC). The hot coils react to pull the valve stem (VS) away from valve ball (VB) allowing ball to unseat and fuel to flow through the unit, where it is heated and ignited, and into the intake manifold.

To operate the unit in cold weather, turn the key-starter switch to "HEAT" position for 15 seconds, then turn switch to "HEAT-START" position to start engine. If engine does not start after 10 seconds, return switch to "HEAT" position for 10 seconds, then back to "HEAT-START". After engine is started, turn switch to "ON" position.

CAUTION: DO NOT attempt to use ether starting fluid and Thermostart unit at same time; use ether starting

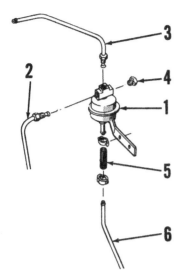

Fig. 144–View showing Thermostart reservoir and fuel line connections for models with C.A.V. fuel injection pump.

1. Reservoir assembly
2. Injector excess fuel return line
3. Return line to fuel tank
4. Plug (replaces fuel pump to reservoir line used on early models
5. Flexible connector tube
6. Line to Thermostart fuel inlet

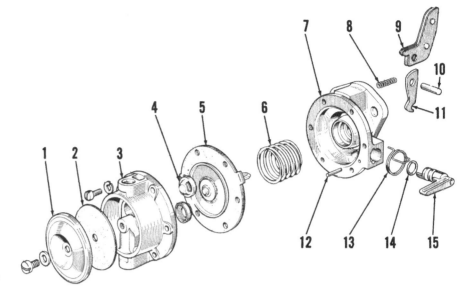

Fig. 146–Exploded view of fuel lift pump assembly that is mounted on Simms multiple plunger fuel injection pump. Fuel lift (transfer) pump is not used on models with C.A.V. fuel injection pump as the C.A.V. pump has an internal vane type transfer pump.

1. Cover	6. Diaphragm spring	11. Diaphragm lever
2. Pulsator diaphragm	7. Inner body	12. Retaining pin
3. Outer body	8. Cam lever spring	13. Return spring
4. Valves	9. Cam lever	14. "O" ring
5. Pump diaphragm	10. Pivot pin	15. Primer lever

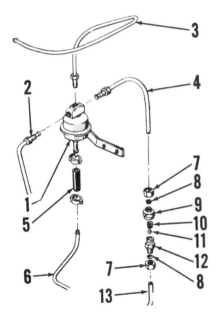

Fig. 145–View showing Thermostart reservoir and fuel line connections for models with Simms fuel injection pump.

1. Reservoir assembly
2. Injector excess fuel return line
3. Return line to fuel tank
4. Line to pressure valve
5. Flexible connector tube
6. Line to Thermostart fuel inlet
7. Compression nut
8. Compression sleeve
9. Connector, tube to valve assembly
10. Valve spring
11. Valve ball
12. Valve body
13. Line from pressure outlet of fuel filter assembly

fluid only if Thermostart unit and intake manifold are cold.

For connecting fuel lines to Thermostart unit and reservoir, refer to Fig. 144 for models with C.A.V. fuel injection pump or to Fig. 145 for models with Simms fuel injection pump. NOTE: When diesel fuel lines have been disconnected, always make certain that fuel is present at the Thermostart fuel inlet (F—Fig. 143) before tightening the connection at that point. Any attempt to operate the Thermostart unit without fuel available may cause failure of the unit allowing fuel oil to leak into the intake manifold.

Service of the Thermostart unit and/or fuel reservoir consists of renewing the defective unit.

FUEL LIFT PUMP
Models With Simms
Fuel Injection Pump

163. The fuel lift pump is mounted on the outside of the fuel injection pump and is driven from a cam on the injection pump camshaft. The fuel pump is of the diaphragm type, and component parts are available for service. Refer to exploded view of the fuel pump assembly in Fig. 146. The inlet and outlet valves (4) are staked into the outer body (3); when renewing valves, insert in body as shown and carefully stake in position. The primer lever retaining pin (12) must be installed with outer end below flush with the machined surface of inner body (7). After inserting

pivot pin (10) in inner body, securely stake pin in place.

To test pump, operate the primer lever with lines disconnected and with fingers closing the inlet port; pump should hold vacuum after releasing primer lever. With finger closing outlet port, there should be a well defined surge of pressure when operating primer lever. Note: On models with Thermostart, check the pressure valve in filter to Thermostart reservoir line if loss of fuel pump pressure is suspected, but pump appears in good condition. Refer to Fig. 146.

Fig. 147–The governor driver (flyball) assembly is accessible after removing governor housing and outer race from front of timing gear cover; refer also to Fig. 148. Retaining nut has left hand thread.

GOVERNOR (Non-Diesel)

Non-diesel engines are equipped with a variable speed centrifugal flyball type governor that is mounted on front end of the ignition distributor drive shaft; refer to Fig. 147.

SPEED ADJUSTMENT
Early Series 2000-3000-4000

164. Refer to Fig. 149 for assembled view of governor linkage and to Fig. 150 for exploded view. Before attempting any adjustments, first check to be sure that linkage moves freely without binding or excessive looseness.

With engine not running, disconnect governor rod (27) at forward end. Move hand throttle lever to wide open position and carburetor throttle arm against high speed stop. Loosen jam nut (28) if necessary and adjust governor rod until front end easily enters hole in governor arm. Lengthen governor rod one more turn to assure complete opening of carburetor throttle, reconnect governor rod and tighten jam nut (28).

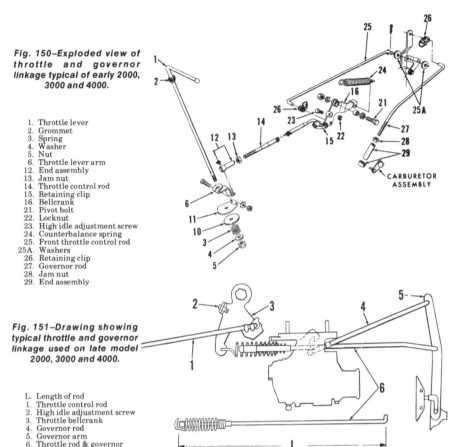

Fig. 150—Exploded view of throttle and governor linkage typical of early 2000, 3000 and 4000.

1. Throttle lever
2. Grommet
3. Spring
4. Washer
5. Nut
6. Throttle lever arm
12. End assembly
13. Jam nut
14. Throttle control rod
15. Retaining clip
16. Bellcrank
21. Pivot bolt
22. Locknut
23. High idle adjustment screw
24. Counterbalance spring
25. Front throttle control rod
25A. Washers
26. Retaining clip
27. Governor rod
28. Jam nut
29. End assembly

Fig. 151—Drawing showing typical throttle and governor linkage used on late model 2000, 3000 and 4000.

L. Length of rod
1. Throttle control rod
2. High idle adjustment screw
3. Throttle bellcrank
4. Governor rod
5. Governor arm
6. Throttle rod & governor spring assembly

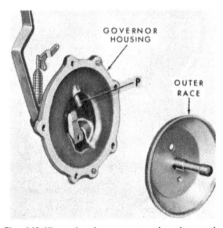

Fig. 148—View showing governor housing and outer race removed from front of timing gear cover. Pin (P) in housing keeps outer race from turning.

Move hand throttle lever to slow idle position and check to be sure carburetor throttle arm is firmly held against slow speed stop screw. Remove right hand steering gear sheet metal cover, if necessary; disconnect end assembly (12) from throttle lever arm (6) and adjust length of control rod (14) as required.

Start and run engine until normal operating temperature is attained. Turn idle speed adjusting screw on carburetor throttle arm to obtain the recommended 600-650 RPM slow idle speed.

Move hand throttle lever to wide open position and adjust high idle screw (23), if necessary, to the following:

Tractor Model	Maximum No-Load Speed
2000	2200-2300
3000	2285-2335
4000	2395-2445

Move the hand throttle control lever to slow idle position and, if necessary, readjust slow idle speed to 600-650 RPM. Stop engine and reinstall steering gear cover.

Late 2000-3000-4000

165. Refer to Fig. 151 for schematic view of linkage and to Fig. 152 for alternate bellcranks which may be used on 3 cylinder units. First make sure linkage is free without excessive looseness. Move hand throttle to high speed position and adjust governor rod (4) if necessary, to fully open carburetor throttle. Adjust length of throttle rod (L) to 11 25/32 inches if bellcrank is

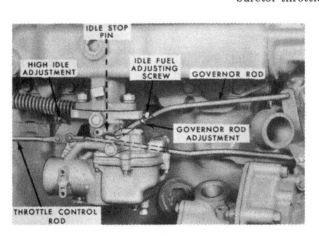

Fig. 149—View showing governor and throttle linkage typical of early 2000, 3000 and 4000.

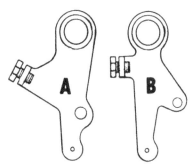

Fig. 152–Two different shape bellcranks have been used on late three cylinder engines. Adjust throttle rod to 11 25/32 inches if bellcrank (A) is used; to 12 25/32 inches for bellcrank (B).

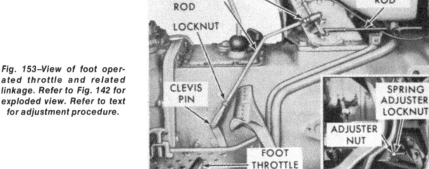

Fig. 153–View of foot operated throttle and related linkage. Refer to Fig. 142 for exploded view. Refer to text for adjustment procedure.

shaped like that show in (A—Fig. 152); or 12 25/32 if bellcrank (B) is used.

Start and warm engine and adjust slow idle speed to 650-700 RPM. Adjust high idle speed as follows, by turning adjustment screw (2).

Tractor Model	High Idle Speed
2000	2200-2300
3000	2285-2385
4000	2395-2495

166. ACCELERATOR (FOOT) PEDAL ADJUSTMENT. To adjust the foot operated throttle linkage on models so equipped, first disconnect the accelerator rod (see Fig. 153) from hand operated throttle lever arm and adjust the throttle linkage as outlined in paragraph 164 or 165. Then, adjust the length of the accelerator rod as follows: Move the hand throttle lever to wide open position and hold the foot pedal down against the step plate (foot rest). Adjust the length of the accelerator rod so that it can be reconnected with hand and foot operated throttle controls in this position. Tighten the accelerator rod adjustment lock nuts after reconnecting rod, then recheck both high no-load and slow idle speeds. Readjust linkage if necessary, then reinstall the cover on right side of steering gear. Note: If foot pedal will not return throttle to slow idle position when released, tighten the return spring (see inset in Fig. 153).

R&R AND OVERHAUL

167. Refer to Fig. 154 for an exploded view of engine governor; plus governor drive which also includes distributor drive shaft and housing.

Governor housing (20 or 26) can be removed after disconnecting linkage and removing the retaining cap screws. Outer race (15) can be withdrawn after removing governor housing. Remove drive shaft nut (14) if indicated, and

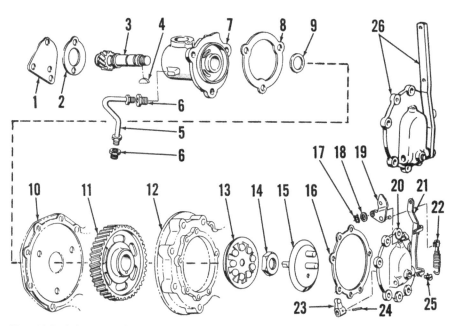

Fig. 154–Exploded view of non-diesel governor and distributor drive assembly. Governor housing and lever assembly (26) is used on late 3 cylinder units. On early 3 cylinder, governor spring (22) is connected between throttle arm (21) and governor arm (19).

1. Rear cover plate	7. Distributor drive housing	14. Nut (L.H. thread)	21. Throttle arm
2. Gasket	8. Gasket	15. Outer race	22. Governor spring
3. Gear, washer & shaft assy.	9. Thrust washer	16. Gasket	23. Governor arm
4. Woodruff key	10. Engine front plate	17. Snap ring	24. Pin
5. Oil line	11. Drive gear	18. Outer washer	25. Plug
6. Fittings	12. Timing gear cover	19. Governor lever	26. Governor housing (late)
	13. Driver assembly	20. Governor housing	

withdraw governor driver assembly (13).

Governor driver (weight) unit is only available as an assembly. Lever (21) and arm (23) are available for early governors but not available for late units.

With governor driver removed, distributor drive housing, shaft and associated parts can be removed after removing ignition coil, distributor, oil feed line and attaching cap screws. Drive gear (11) can only be removed

after removing timing gear cover as outlined in paragraph 77 or 80.

Distributor drive shaft (3) can be removed from housing (7) after removing rear cover (1), Woodruff key (4) and thrust washer (9).

Renew parts which are worn, damaged or questionable, and assemble by reversing the disassembly procedure. Time the ignition as outlined in paragraph 175 and adjust governor linkage if necessary, as in paragraph 164 or 165.

COOLING SYSTEM

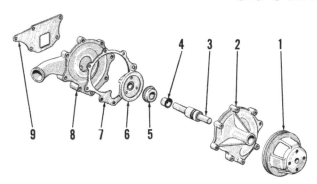

Fig. 155—Exploded view of water pump assembly used on all engines. Seal (5) seats against hub of impeller (6).

1. Pulley
2. Housing
3. Shaft & bearing assy.
4. Water slinger
5. Seal assembly
6. Impeller
7. Gasket
8. Rear cover
9. Mounting gasket

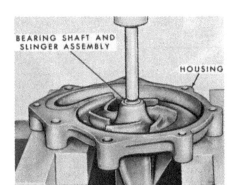

Fig. 156—View showing shaft and bearing assembly being pressed from impeller and housing.

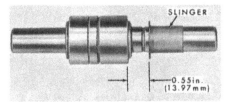

Fig. 157—Flange of water slinger should be 0.55 inch from edge of outer bearing race as shown above.

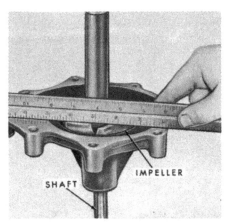

Fig. 158—Press impeller onto shaft with ¾-inch I.D. pipe so that impeller is flush with rear of housing as shown.

Fig. 159—Press pulley onto shaft so that distance from center of belt groove in pulley is 2½ inches from rear face of housing.

RADIATOR PRESSURE CAP AND THERMOSTAT

All Models

168. A 7 psi radiator pressure cap is used on all models.

On non-diesel engines, thermostat is located in front end of engine intake manifold. Thermostat in diesel engines is located in front end of cylinder head. On all models, thermostat is accessible after draining coolant from radiator and removing water outlet connection from intake manifold or cylinder head.

Standard thermostat for all models should start to open at 188° F., and be fully open at 212° F. Optional thermostats with opening temperatures of 160° F., 168° F., and 178° F. are available; however, use of the standard thermostat is recommended for all conditions.

RADIATOR

All Models

169. **R&R RADIATOR.** To remove the radiator, drain the cooling system, disconnect front mounted air cleaner hose and the headlight wires and remove the grille and radiator shell from tractor. Disconnect the hoses and if so equipped, disconnect transmission oil cooler tubes from radiator lower tank. Disconnect the engine breather tube from rocker arm cover and from the fan shroud and remove tube. Then, unbolt

and remove radiator and fan shroud assembly from tractor. Remove shroud from radiator if necessary.

On models with transmission oil cooler, the radiator lower tank assembly contains a heat exchanger and, on some models, lower tank is available separately from the radiator assembly.

To reinstall radiator, reverse removal procedure.

WATER PUMP

All Models

170. Water pump can be removed after removing radiator as outlined in paragraph 169. Refer to exploded view of water pump in Fig. 155 and disassemble pump as follows:

Remove fan from pulley and using standard two-bolt puller, remove pulley from shaft. Remove rear cover (8) and press shaft and bearing assembly (3) out towards front of housing as shown in Fig. 156. Drive the seal (5 —Fig. 155) out towards rear of housing.

Using a length of 1 5/16 inch I.D. pipe, press new seal into housing. Check to see that flange on water slinger (4) is located 0.55 from edge of bearing race as shown in Fig. 157, then press shaft into front of housing until outer bearing race is flush with front end of housing. Using a length of ¾-inch I.D. pipe, press impeller onto shaft as shown in Fig. 158 so that impeller is flush with rear end of housing. Press pulley onto shaft so that center of belt groove in pulley is 2½ inches from rear face of housing as shown in Fig. 159. Install rear cover with new gasket and tighten retaining cap screws to a torque of 18-22 Ft.-Lbs.

Reinstall water pump assembly by reversing removal procedure and tighten retaining cap screws to a torque of 23-28 Ft.-Lbs.

ELECTRICAL SYSTEM

GENERATOR AND REGULATOR

All Models

171. Ford generators and regulators are used. The generator is a two-pole shunt wound type with type "B" circuit; that is, one field coil terminal is grounded to generator frame and the

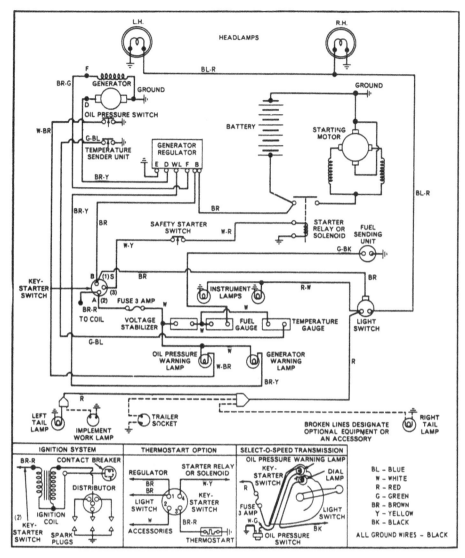

Fig. 160–Wiring diagram for all models with "C" prefix to tractor serial number. Refer to page 2 of this manual for serial number location. For tractors with "A" or "B" prefix to tractor serial number, refer to Fig. 161.

other field coil terminal is connected to armature terminal through the regulator. Specifications are as follows:

Max. output (hot) @ 1350
  RPM and 15 volts 22
Renew brushes if shorter
  than 13/32-inch
Min. brush spring tension
  with new brushes ..... 18 ounces
Field coil current .... 2 amps.
Field coil resistance .... 6 ohms
Max. commutator runout 0.002
Max. armature shaft runout .. 0.002
Cutout Relay:
  Cut-in speed (approximate
    engine RPM) 765
  Cut-in voltage 12.4-13.2
  Cut-out voltage 9.5-11.0
  Cut-out current .... −4 amps.
  Armature-to-core air gap 0.035-0.045
  Contact blade movement 0.010-0.020
Current Regulator:
  On-load setting (amps.) 21-23
  Armature-to-core air gap 0.054

Voltage Regulator:
  Opening voltage:
    50° F. 14.9-15.5
    68° F. 14.7-15.3
    86° F. 14.5-15.1
    104° F. 14.3-14.9
  Armature-to-core air gap 0.053

STARTING MOTOR
Non-Diesel Starting Motor

172. Ford tractors with non-diesel engine are equipped with a starting motor of the type shown in Fig. 162. This starting motor utilizes a series-parallel connected field coil arrangement, an integral positive engagement drive assembly and a moveable pole piece which, together with one of the field coils, acts as a solenoid to engage the drive assembly. When motor is not in use, one of the field coils is grounded through the actuating coil contacts. Closing starter switch completes the circuit resulting in the moveable pole

piece being attracted by the field coil. A lever attached to the moveable pole piece engages the drive gear pinion with the flywheel ring gear. When the moveable pole piece is fully seated, it opens the field coil grounding contacts which applies full field power for normal starting motor operation.

Service specifications are as follows:
C5NF-11001-B—C7NF-11001-B—D1NN-11001-A Starting Motor
Brush spring tension (min.
  with new brushes) .... 40 oz.
Min. brush length .... ¼-inch
Commutator min. diameter 1.46 inches
Max. armature shaft end play .. 0.058
Max. armature shaft runout 0.005
No-load test:
  Volts 12
  Amps 70
  RPM 6000-9500
Loaded test (with warm engine):
  Amps 150-200
  Engine RPM 150-200

C5NF-11001-A, C6NF-11001-A or C7NF-11001-A Starting Motor
Brush spring tension (min.
  with new brushes) .... 40 oz.
Min. brush length .... ¼-inch
Commutator min. diameter 1.46 inches
Max. armature shaft end play .. 0.048
Max. armature shaft runout 0.005
No-load test:
  Volts 12
  Amps 60
  RPM 5220-9440
Loaded test (with warm engine):
  Amps 225-275
  Engine RPM 150-200

Diesel Starting Motor

173. Most diesel engines are equipped with a Ford 5 inch diameter starting motor and relay assembly. Closing the starter switch energizes the solenoid; movement of the solenoid plunger engages the drive pinion and closes a two-stage switch. If the teeth of the drive pinion butt against teeth on flywheel, only the first stage of the switch is closed which will allow current to flow to one field coil. This will provide enough power to turn starter until drive pinion is in position to engage flywheel ring gear teeth; then, full engagement of drive pinion will close second stage of switch energizing all four field coils.

When drive pinion is in engaged position, there should be a clearance of 0.010-0.020 between drive pinion and thrust collar. To check clearance, first energize solenoid with 6-volt power source, then check clearance with feeler gage as shown in Fig. 163. If clearance is not within 0.010 to 0.020, refer to Fig. 164, loosen locknut and turn pivot pin as required to obtain proper clearance. Then, tighten

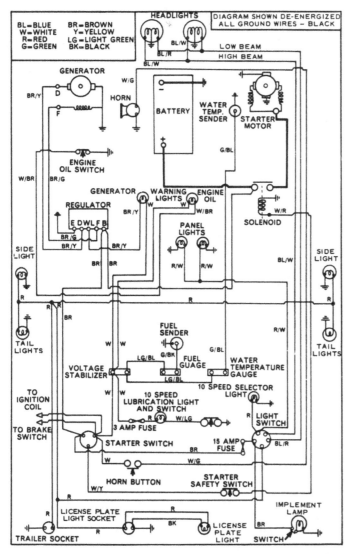

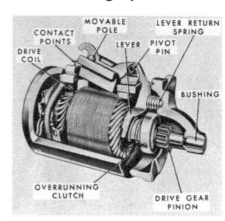

Fig. 162–Cut-away view of starting motor used on models 2000, 3000 and 4000 with non-diesel engine. A similar starting motor is used on some model 3000 with diesel engine.

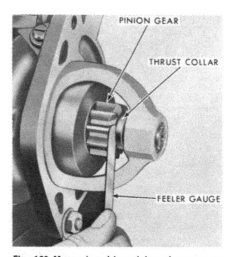

Fig. 163–Measuring drive pinion clearance on diesel engine starting motor. Refer to Fig. 164 for adjustment and to text for measuring and adjusting procedure.

Fig. 161–Wiring diagram for all models with "A" or "B" prefix to tractor serial number. Refer to page 2 of this manual for serial number location. Refer to Fig. 160 for wiring diagram for models with "C" prefix.

Fig. 164–Adjusting drive pinion clearance on diesel engine starting motor.

locknut and recheck clearance.

Service specifications are as follows:

Starting Motor & Relay Assembly
Brush spring tension (min.
  with new brushes) . . . . 42 oz.
Min. brush length 5/16-inch
Commutator min. diameter 1.53 inches
Max. armature shaft end play . . 0.020
Max. armature shaft runout 0.005
Drive pinion clearance
  (engaged) 0.010-0.020
No-load test:
  Volts 12
  Amps 100
  RPM 5500-7500
Loaded test (with warm engine):
  Amps 250-300
  Engine RPM 150-200
NOTE: Beginning with production date 1-73, series 2000 and 3000 tractors are equipped with a 4½ inch diameter starting motor which is basically similar to the 5 inch diameter starting motor. The same service procedure can be used.

IGNITION SYSTEM

174. **SPARK PLUGS.** Autolite AG5 spark plugs are recommended for all gasoline engines. Set electrode gap to 0.023-0.027. Install spark plugs with dry threads and tighten to a torque of 26-30 Ft.-Lbs.

175. **IGNITION TIMING.** Breaker contact gap is 0.024-0.026. Firing order is 1-2-3 on three cylinder engines. To install and time the distributor, proceed as follows:

Remove the No. 1 (front) spark plug and turn engine slowly until air is forced out spark plug hole, then continue turning engine slowly until 0° (TDC) flywheel mark is aligned with arrow in inspection opening of engine rear cover plate, then, place the distributor with dust cover and rotor installed, in drive housing with rotor pointing towards No. 1 cylinder distributor cap terminal. This should properly mesh the distributor gear with the

drive shaft gear. Loosen the bolt clamping the timing arm to the distributor base and rotate distributor until breaker points just start to open. Hold distributor housing in this position, center the timing arm slot on bolt hole in drive housing and tighten the timing arm clamp bolt. Reinstall spark

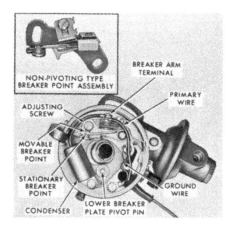

Fig. 165–View of 3-cylinder ignition distributor with cap and rotor removed. Two types of breaker point assemblies are used; refer to inset for non-pivoting type.

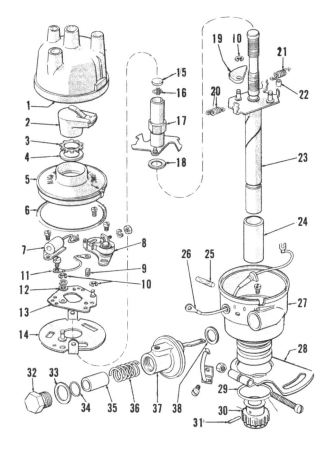

Fig. 166–Exploded view of Ford 3-cylinder ignition distributor assembly.

1. Distributor cap
2. Rotor
3. Retainer
4. Seal
5. Dust cover
6. "O" ring
7. Condenser
8. Breaker points
9. Cam lubricant felt
10. Retainer rings
11. Ground wire
12. Spring washer
13. Upper breaker plate
14. Lower breaker plate
15. Wick
16. Cam retainer ring
17. Distributor cam
18. Thrust washer
19. Advance weights
20. Secondary advance spring
21. Primary advance spring
22. Rubber sleeve
23. Distributor shaft
24. Bushing
25. Wick
26. Primary ignition wire
27. Distributor base
28. Timing arm assembly
29. "O" ring
30. Drive gear
31. Spring pin
32. Plug
33. Gasket
34. Shim washer
35. Stop
36. Spring
37. Diaphragm assembly
38. "O" ring

plug, distributor cap and spark plug wire, start engine and set timing with timing light as follows:

176. IGNITION TIMING WITH TIMING LIGHT. A Power Timing Light can be used to check distributor advance mechanism. Stamped timing marks run from 0° (TDC) to 30° (BTDC) which is not sufficient to register combined advance. A suggested procedure for checking the distributor and adjusting timing is as follows:

Connect timing light to No. 1 spark plug and open timing hole cover on right, front side of engine rear plate, then start the engine and run at high idle rpm. Retard ignition timing until 30° timing mark aligns with timing pointer, tighten clamp screw and shut off engine. Disconnect and plug vacuum advance line. Restart engine and, with engine running at high idle speed, recheck timing which should now be 18° BTDC. Adjust the vacuum advance, if necessary, as outlined in paragraph 179.

With engine still running at high idle speed and with vacuum advance line disconnected, reset ignition timing as follows:

Model 2000 .... 22° BTDC
Model 3000
W/C5NF-12127A Dist. ..... 18° BTDC
Model 3000
W/C7NF-12127A Dist. ..... 22° BTDC
Model 4000 Before
June 1968 .... 24° BTDC
Model 4000 After
June 1968 .... 22° BTDC

Tighten clamp screw and reduce engine speed to slow idle; timing should retard to 0°-6° BTDC. If is does not, overhaul distributor as in paragraph 180 or adjust centrifugal advance mechanism as in paragraph 178.

177. DISTRIBUTOR TEST AND OVERHAUL. The breaker contact gap for all models is 0.024-0.026. Cam dwell angle for all models is 35-38 de-

grees; however, most dwell meters will not have a 3-lobe position for 3-cylinder engines. Therefore, the cam dwell for 3-lobe distributor cam should test 17½-19 degrees when dwell meter is set for 6-lobe (6-cylinder) position.

Breaker contact points may be either the conventional pivoted type or pivotless. On pivoted type points, breaker arm spring tension should be 17-21 ounces when measured at end of breaker points contact, or 15-18 ounces when measured at center of contact points. To adjust spring tension, loosen nut holding breaker arm spring and move slotted end of spring towards pivot point to decrease tension, or away from pivot point to increase tension. Tension on pivotless point set is non-adjustable. Refer to Fig. 165 for views showing both types of breaker points.

For distributor test stands (synchroscopes), refer to the following test data for both centrifugal advance test data and vacuum advance test data: (All data is in distributor RPM and distributor degrees).

C5NF-12127-C—C7NF-12127-B Distributor for Model 2000 Tractors

Centrifugal Advance Data:

Distributor RPM	Degrees Advance
200-350	−0.5 to +0.5
400	0 to 1
600	4 to 5
850	9 to 10
900	9 to 11

Vacuum Advance at 1000 Distributor RPM:

Inches Mercury	Degrees Advance
1	0
3	0 to 0.5
5	0 to 1
6	0 to 3.5
7	2.2 to 5.2
8	3.5 to 6.5
9	4.5 to 7.5
15	4.5 to 7.5

C5NF-12127-A Distributor for Model 3000 Tractors

Centrifugal Advance Data:

200-550	−0.5 to +0.5
600	0 to 1
700	2 to 3
850	5.2 to 6.2
1000	5.5 to 6.5
1200	5.75 to 7.0
2000	7.2 to 8.8

Vacuum Advance at 1000 Distributor RPM:

Inches Mercury	Degrees Advance
1	0
3	0 to 0.5
5	0 to 1
7	0 to 1.5
8	0 to 3
9	1 to 4
10	2 to 5.2
12	3.8 to 6.8
13	4.5 to 7.5
15	4.5 to 7.5

C5NF-12127-B Distributor for Model 4000 Tractors

Centrifugal Advance Data:

Distributor RPM	Degrees Advance
200-525	−0.5 to +0.5
550	0 to 1
800	4 to 5
1000	7 to 8.5
1100	9 to 10
1200	9 to 11

Vacuum Advance at 1000 Distributor RPM:

Inches Mercury	Degrees Advance
1	0
3	0 to 0.5
5	0 to 1
6	0 to 3
7	2 to 5
9	4.5 to 7.5
15	4.5 to 7.5

C7NF-12127-A Distributor for Model 3000 Tractors

Centrifugal Advance Data

Distributor RPM	Degrees Advance
200-525	−0.5 to +0.5
900	5.5 to 6.8
1150	9 to 11

Vacuum advance at 1000 Distributor RPM

Inches Mercury	Degrees Advance
1	0
5	0 to 1
8	4 to 7
9	4.5 to 7.5

C7NF-12127-D Distributor for Model 4000 Tractors

Centrifugal Advance Data

Distributor RPM	Degrees Advance
200-475	−0.5 to +0.5
900	5 to 7
1150	9 to 11

Vacuum Advance At 1000 Distributor RPM

Inches Mercury	Degrees Advance
1	0
5	0 to 1
8	2 to 5
11	5 to 7.5

178. ADJUST CENTRIFUGAL ADVANCE. If the distributor centrifugal advance did not fall within specifications given in paragraph 177, proceed as follows to adjust or correct centrifugal advance mechanism:

Refer to Fig. 167 and check to be sure sleeve (22) is in place on tang of distributor shaft plate. Note: Top view of distributor in Fig. 167 is with breaker plate (14—Fig. 166) removed; however, distributor shaft can be rotated and sleeve (22—Fig. 167) and spring adjustment tabs (T) observed through hole in breaker plate.

If low RPM centrifugal advance is not within specified limits, turn dis-

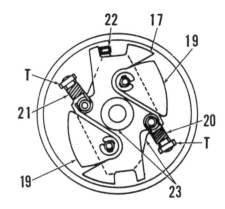

Fig. 167–Drawing showing top of distributor assembly with breaker plates removed. Refer to Fig. 166 for parts identification. Adjusting tabs for advance springs are (T).

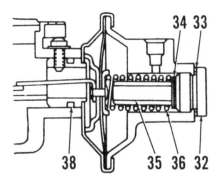

Fig. 168–Cross-sectional view of vacuum advance mechanism showing proper location of vacuum advance stop (35) and adjusting shim washers (34). Refer to Fig. 166 for parts identification.

tributor shaft so that primary spring (21) adjusting tab (T) is in view through hole in breaker plate and bend tab in to increase the advance or out (away from distributor shaft) to decrease the advance.

If high RPM centrifugal advance is not within specified limits, turn distributor shaft so that secondary spring (20) adjusting tab (T) is in view through hole in breaker plate and bend tab in to increase the advance or out (away from distributor shaft) to decrease the advance. NOTE: Secondary advance spring should be loose on tang when the distributor shaft is stationary.

After adjusting centrifugal advance springs, recheck centrifugal advance throughout low, intermediate and high distributor RPM ranges given in test data. Renew advance springs if centrifugal advance is not within specified limits throughout the test RPM range and cannot be adjusted. Advance springs for model 2000 have 7½ coils and are color coded blue; advance springs for all other models have six coils and are color coded purple.

NOTE: 'Primary" and "Secondary" advance springs are identical, only the adjustment is different.

179. ADJUST VACUUM ADVANCE. If the vacuum advance is not within the specified limits as outlined in paragraph 177, remove plug (32—Fig. 168) and add shims (34) between plug and spring to decrease the advance, or remove shims to increase the advance. Be sure gasket (33) is in good condition or renew gasket when reinstalling plug. Check vacuum unit for leaks after reinstalling plug. Shims are available in four thickness ranges: 0.008-0.010, 0.020-0.022, 0.040-0.042 and 0.080-0.082.

180. OVERHAUL DISTRIBUTOR. Refer to exploded view of the ignition distributor for 3 cylinder engines in Fig. 166.

It is important to properly lubricate distributor whenever servicing the unit. Felts (4, 9, 15, and 25) should be lightly saturated with SAE 10W motor oil. When advance unit is disassembled, fill grooves in top of distributor shaft and lubricate pivot pins with multi-purpose lithium base grease.

When installing new breaker points or adjusting breaker point gap, make certain that after the point retaining screws are tightened, the ground wire (11) is properly positioned as shown in Fig. 165.

To renew distributor shaft and/or shaft bushing, proceed as follows: Remove the vacuum advance assembly and breaker plates. Remove felt wick from top of distributor cam (17—Fig. 166) and extract the retainer (16) with needle nose pliers. Drive the gear retaining pin (31) from gear and press shaft (23) from gear and housing. Press old bushing (24) out towards top of housing with bushing driver (Nuday tool No. SW 503 or equivalent). Lubricate outside of new bushing with motor oil, place flat steel washer against shoulder on driver and press new bushing in with driver until washer seats against top inside surface of housing. Ream the bushing to inside diameter of 0.468-0.469. Lightly oil shaft, insert shaft into housing and press gear onto lower end of shaft so that shaft end play is 0.029-0.042. Using pin hole in gear as a guide, drill retaining pin hole through new shaft, then install pin.

After reassembling distributor, check and adjust advance mechanism as outlined in paragraphs 177, 178 and 179.

CLUTCH

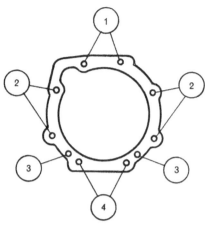

Fig. 169–When reinstalling engine to transmission housing bolts, tighten bolts at locations shown to the following tightening torque specifications.

1. 220-300 Ft.-Lbs.
2. 125-140 Ft.-Lbs.
3. 35-50 Ft.-Lbs.
   (engines with stamped steel oil pan)

4. 220-300 Ft.-Lbs.
   (engines with cast iron oil pan)

Tractors with transmission type PTO or Independent PTO will be equipped with a 11 inch diameter single plate dry type clutch. Tractors with "live" type PTO will be equipped with a dual plate spring loaded disc type clutch having a 11 inch diameter transmission clutch disc and an 8½ inch diameter PTO disc. Tractors with a 10 speed (Select-O-Speed) transmission will have a disc type torque limiting clutch in the flywheel.

TRACTOR SPLIT
All Models Except 4200

181. To split the tractor between engine and transmission, first drain cooling system, disconnect battery ground cable and proceed as follows: Remove the vertical exhaust muffler if so equipped. Disconnect wiring harness from support clips under engine hood, then unbolt and remove the hood. Disconnect proofmeter drive cable at rear end of generator. Remove steering gear side covers from under fuel tank. On 4000 models, disconnect steering drag link from steering gear arm.

On models 2000 and 3000, disconnect the steering drag links from steering gear arms and the radius rods from transmission housing on manual steering models. On models with power steering, disconnect drag links from front axle spindle arms and radius rods from front axle, and disconnect power steering fluid return tube from reservoir and the power steering pump.

On models with "C" prefix to serial number, unplug the wiring connector located at front of fuel tank. On all

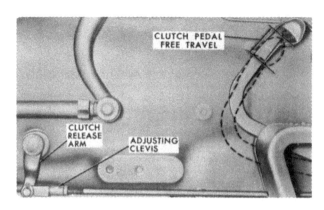

Fig. 170–View showing clutch pedal free travel adjustment on all single clutch and late double clutch models. Refer to text for adjustment specifications.

other models, disconnect wiring from starter relay terminals, generator, voltage regulator, front lights, oil pressure switch and temperature gage sending unit.

Remove the shield from above starting motor and reinstall the fuel filter retaining bolts. Unbolt and remove starting motor and the flywheel access cover from plate between engine and transmission. On models so equipped, unbolt and remove the engine driven hydraulic (piston type) pump assembly. Disconnect the rear throttle control rod under fuel tank and disconnect diesel shut-off cable or gasoline carburetor choke cable. Shut off fuel supply valve at tank and disconnect fuel supply line and diesel excess fuel return line. Unbolt fuel tank and rear hood (fuel tank cover) from support brackets on rear end of engine. On models equipped with horizontal exhaust, disconnect exhaust pipe from muffler under the left step plate. Disconnect transmission oil cooler lines on models so equipped and on model 4000 (except 4110 L.C.G.) remove steering gear arm to provide clearance for the tubes.

Insert wood wedges between front axle and front support. Place supports under front end of transmission housing and support engine with hoist or rolling floor jack, unbolt engine from transmission and roll the front unit away.

To reconnect tractor between engine and transmission, reverse the procedure used to split tractor. Refer to paragraph 129 or 131 for bleeding the diesel fuel system and to paragraph 361 for bleeding the engine driven hydraulic pump assembly. Refer to Fig. 169 for engine to transmission bolt tightening torque.

Model 4200

182. On model 4200, follow same procedures as outlined for other models in paragraph 181; however, disconnect the four power steering lines at connec-

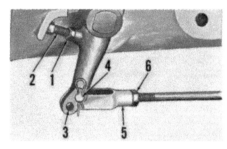

Fig. 171–To reduce clutch pedal height (early models) when PTO operation is not required, change clevis pin from position (4) to (3). Pedal height and free pedal adjustment points are shown; refer to text.

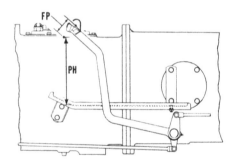

Fig. 172–Drawing showing free pedal (FP) and pedal height (PH) measurements for double clutch equipped tractors.

tions under front of fuel tank and ignore reference to other steering components. Be sure to adequately support tricycle front end when separating the tractor.

LINKAGE ADJUSTMENT
All Models

183. The recommended clutch pedal free travel is 1½ inches for early models or 1¼ inches for late models. To adjust the linkage on all single clutch models and the late double clutch models which have clutch release shaft arm with only one hole, refer to Fig. 170. Disconnect adjusting clevis from clutch release shaft arm, loosen locknut and turn clevis in or out as required.

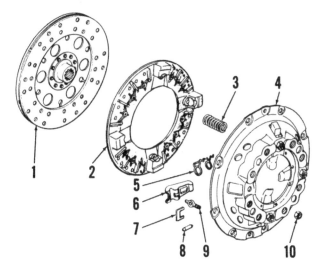

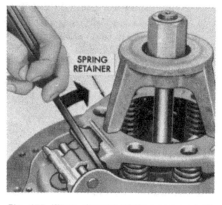

Fig. 173–Exploded view of early type single disc clutch used on three cylinder models so equipped.

1. Clutch disc
2. Pressure plate
3. Clutch spring
4. Clutch cover
5. Anti-rattle spring
6. Release lever
7. Strut
8. Pin
9. Eye bolt
10. Adjusting nut

Fig. 175–When disassembling cerametallic clutch, pry spring retainer to side as shown by heavy arrow, until lever link can be unhooked.

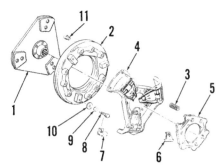

Fig. 174–Exploded view of the (Cerametallic) single disc clutch.

1. Clutch disc
2. Pressure plate
3. Clutch spring
4. Cover
5. Retainer
6. Link
7. Clip
8. Adjusting screw
9. Locknut
10. Washer
11. Dowel

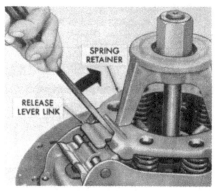

Fig. 176–Unhooking lever link.

On early double clutch models which have clutch release shaft arm with two holes, both pedal height and free play must be adjusted. Refer to Fig. 171 and proceed as follows: Loosen locknut (1) and carefully depress clutch pedal by hand until free play is just removed. Hold pedal position and back out stop screw (2) until it contacts housing stop. Refer to Fig. 172 and measure pedal height (PH) from footrest. Pedal height should be 9¼ inches for early models or 8½ inches for late models. If it is not, loosen clevis locknut (6—Fig. 171), remove clevis pin (4) and turn clevis as required until specified pedal height is obtained. After adjusting pedal height, turn stop screw (2) until free play, measured at pedal pad, is 1½ inches on early models or 1¼ inches on late models.

After completing adjustment on double clutch models, make an operational check as follows: Remove pto cap, engage pto lever and start engine. Push pedal down against footrest (or stop) and check to see if pto shaft stops turning. Make further adjustments as needed until pto clutch fully releases

when pedal is fully depressed. If proper adjustment cannot be accomplished, overhaul and/or adjust clutch unit.

R&R AND OVERHAUL
Single Clutch Models

Early type clutches used a full cover with four release fingers, and clutch disc facings were a full circle fibrous type as shown in Fig. 173. Beginning with October 1969 production, and continuing through August 1971 production, a ceramic button ("cerametallic") clutch was installed which uses a skeleton type cover with three release fingers as shown in Fig. 174. The "cerametallic" clutch units are not interchangeable with the earlier (prior 10/69) units and both the flywheel and clutch shaft must be renewed if the "cerametallic" type clutch is installed in early models. Beginning with September 1971 production, the "cerametallic" clutch was discontinued and a clutch similar to the early full circle (prior 10/69) clutch was installed. However, clutch disc of the latest full circle type is not interchangeable with the early type as it has square cut splines which fit the current clutch shaft that was introduced with the "cerametallic" clutch.

184. To remove the clutch unit after tractor is split, remove the retaining cap screws and lift off the unit.

When reinstalling, use a clutch pilot tool and position clutch disc as follows: On all models using full circle disc, except model 4000 with independent PTO, position disc with long hub toward pressure plate (rearward). On model 4000 with independent PTO, position disc with long hub toward flywheel (forward). On all models using "cerametallic" disc, position disc with long hub toward flywheel (forward).

Install cover assembly on flywheel and tighten cover retaining cap screws to 23-30 Ft.-Lbs. of torque.

Place clutch cover on bed of a press or

on clutch disassembly tool. On models with full cover, refer to Fig. 173 and remove the four finger adjusting nuts (10).

On (Cerametallic) units carefully depress spring retainer (5—Fig. 174) until it clears clutch cover (4) by approximately ¼-inch at nearest point. Using a screwdriver or similar tool, pry spring retainer sideways away from release lever pivot (Fig. 175), and unhook each lever link (Fig. 176) in turn, until all three are disconnected.

NOTE: It may be necessary to thread adjusting screws (8—Fig. 174) into pressure plate for additional finger clearance.

On all models, slowly release the pressure and disassemble cover unit. On late models, DO NOT remove release levers or pivot pins from clutch cover (4—Fig. 174); unit is available only as an assembly. Other parts of all clutches are available individually.

Various color-coded clutch springs have been used. Renew any springs which are rusted, distorted, heat discolored, cracked, or which fail to meet the test specifications which follow: Note: All springs are tested at a length of 1 11/16 inches except where noted.

Given free lengths and test loads are recommended minimum.

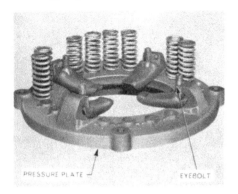

Fig. 177–Partially assembled view of single clutch pressure plate assembly.

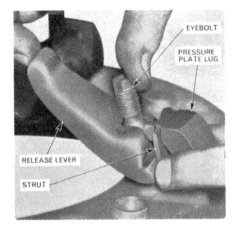

Fig. 178–Installing clutch release lever, strut and eyebolt.

Part No.	Color Code	Free Length	(In.) Lbs. Test
C5NN-7572A	Dark Green/Cream	3 3/64	107
C5NN-7572C	Buff/Lt. Blue	2 45/64	150
C5NN-7572F	Yellow/Lt. Blue	2 5/8	140
C9NN-7572A	Lt. Blue Double Stripe	2 55/64	53
C9NN-7572B	White Dbl. Stripe	2 13/16	63

Assemble by reversing the disassembly procedure. Use new self-locking adjusting nuts when assembling early models. Finger height should be as outlined in the following table using a new clutch disc and measuring from friction surface of flywheel:

Early Model 2000,
    3000, 4000 . . . . 1.945-1.965
All Models With Cerametallic
    Clutch 1.970-2.030

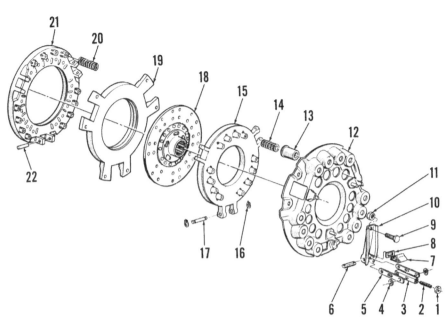

Fig. 179–Exploded view of double disc clutch assembly.

1. Jam nut	7. Retaining clips	13. Spring cups
2. Adjusting screw	8. Shims	14. PTO pressure springs
3. Center link	9. Adjusting screws	15. PTO pressure plate
4. Lever spacers	10. Release levers	16. Snap rings
5. Lever links	11. Jam nuts	17. Pins
6. Struts	12. Cover assembly	18. PTO friction disc
19. Intermediate pressure plate	20. Transmission springs	21. Transmission pressure plate
22. Pins		

Double Clutch Models

185. To remove the clutch after tractor split, first depress release levers and secure to clutch cover with wire; then unbolt and remove cover assembly. Transmission clutch disc can be renewed after cover is off, to renew pto clutch disc, it will be necessary to disassemble cover. Use a suitable pilot when reinstalling and leave pilot in place until levers are unwired. Position discs with full circle lining with long hub rearward. Position discs with ceramic buttons with long hub forward. Tighten clutch cover retaining cap screws to a torque of 13-15 Ft.-Lbs.

To disassemble the removed double clutch, mount the unit on a suitable disassembly tool or on the bed of a press and use a tripod fixture to apply pressure to cover (12—Fig. 179). With pressure applied, remove struts (6) and cotter pins retaining release levers (10). Do not lose or intermix clip shims (8) as levers are removed. Lift off levers and remove pto pressure plate pins (17), then slowly release the pressure and disassemble the clutch unit.

PTO clutch springs (14) on all models are color coded Violet. The springs have an approximate free length of 3 17/64 inches and should test 110-120 lbs. when compressed to a height of 1.95 inches.

Three different transmission clutch springs (20) have been used. The original production spring was color coded

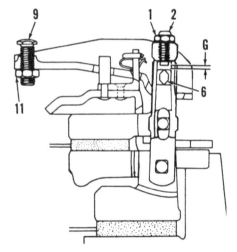

Fig. 180—Cross sectional view of release lever used on dual clutch. Gap (G) controls release of pto clutch. Gap must be adjusted after clutch is installed.

Light Blue. This spring had an approximate free length of 3 7/16 inches and tested 120 lbs. when compressed to a height of 2 17/64 inches. The service spring for early models is color coded Yellow. It has a free length of 3 7/16 inches and should test 92-102 lbs. when compressed to a height of 2 17/64 inches. On late models with Cerametallic button clutch, the transmission clutch spring is color coded Dark Blue. This spring has a free length of approximately 3½ inches and should test 76 lbs. when compressed to a height of 2 17/64 inches. Be sure springs meet test specification and that correct springs are used.

Assemble by reversing the disassembly procedure. Install clip shims (8) of sufficient thickness to apply a slight preload to retaining clip (7) when cotter pin is installed. Shims (8) are available in thicknesses of 0.005, 0.010, 0.015 and 0.020.

Set release lever height to 2.110 inches, measured from hub of pto friction disc (18—Fig. 179) to top of adjusting screw (9), after clutch is installed on flywheel. Adjust release lever height by loosening locknut (11) and turning adjusting screw (9) as necessary. All levers must be adjusted to the same height within 0.010 inch.

Adjust pto clutch release linkage after clutch is installed by loosening locknuts (1—Fig. 180) and backing out turning screws (2). The gap (G) between adjusting screw and release lever strut should be set at 0.050-0.054 inch for tractors with either an external clutch pedal return spring or an internal coil type return spring. Gap (G) should be set at 0.070-0.074 inch for tractors with internal leaf type pedal return spring. Turn adjusting screw (2) until correct

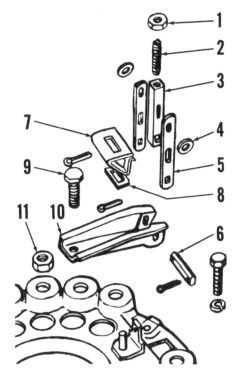

Fig. 181–Exploded view of double clutch finger showing correct parts relationship. Refer to Fig. 179 for parts identification.

gap is obtained, then tighten locknut while holding adjusting screw. All levers must be adjusted to same height.

AUXILIARY TRANSMISSION

Ford 2000 and 3000 models prior to 1971 production that are equipped with a four speed transmission have available an Auxiliary Transmission providing an overdrive and underdrive for each of the four gear ranges making possible a selection of twelve forward and three reverse speeds. This three-range (over, under and direct) transmission is interposed between the engine clutch and main transmission.

REMOVE AND REINSTALL

All Models So Equipped

186. Separate (split) tractor between engine and transmission housing as outlined in paragraph 181. Disconnect the clutch release bearing springs (40 —Fig. 182) and remove the bearing. Drain the auxiliary transmission and main transmission. Remove the socket head screw (1) from auxiliary transmission shift lever (2) and remove the lever. Then, unbolt and remove the auxiliary transmission from mounting flange (49) on front of main transmission housing.

The auxiliary transmission mounting flange can be removed at

this time by removing the four cap screws (50) and pulling flange and shims (48) from tractor transmission housing.

Reinstall by reversing removal procedure and check bearing preload as outlined in paragraph 191.

OVERHAUL

187. **SHIFTER RAILS AND FORKS.** To remove shifter rails and forks, the auxiliary transmission must be removed as outlined in paragraph 186. Refer to Fig. 182, remove the retaining cap screws and lift shifter cap assembly (9) from gear case. Remove the two detent caps (6), detent springs (7) and detent balls (8) from shifter cap. Remove lock wires and the drilled head set screws (35) from shifter fingers (forks) (33). Remove the Welch plugs (38) from the shifter rod bores at the rear of the shifter cap, then drive the shifter rods (39) out the front of shifter cap. Remove the set screw (4) from the interlock bore and remove the interlock (5). Remove the snap ring (13) from the groove on the selector shaft (12) and slide the selector fork (34) and snap ring toward the end of the selector

shaft which is opposite from the shoulder stop. Remove the Woodruff key (11), then withdraw selector shaft from the shifter cap. Remove selector shaft seal (36).

Wash all parts in a suitable solvent and position the shifter rods (39) and selector shaft (12) in their respective bores and check for freedom of movement; if binding occurs due to rods or shaft being bent, renew the bent parts. If binding occurs due to rods or shaft being scored, it may be possible to recondition the scored part by using Crocus cloth. Inspect the shifter fingers (forks) (33) and selector fork (34) and renew them if they show signs of contact at points other than the contact pads. Inspect the interlock (5) for flat spots or signs of scoring and renew if necessary.

188. When reassembling, use new seal and Welch plugs and proceed as follows: Place the selector shaft oil seal (36) in its bore and drift into position using a socket of proper size. Start the selector shaft in its bore and place the selector fork (34) and snap ring (13) on shaft. Install Woodruff key in its slot, then slide selector shaft into place and position snap ring in its groove. Start a shifter rod (39) in the bore farthest (opposite) from the interlock plug and be sure the end with the two grooves is toward the detent end of shifter cap. Place a shifter finger (fork) (33) over shifter rod so that the recess in shifter finger will engage selector fork (34). Note: The shifter fingers (forks) and shifter rods are identical and can be interchanged. Position parts and align the center groove in the shifter rod (39) with the tapped hole in the shifter finger (33), then secure shifter finger with the drilled head cap screw (35). Install the interlock (5) and tighten socket head cap screw (4). Install the other shifter rod (39) and shifter finger (33), then install lock wires in the drilled head set screws (35) and the holes provided in the shifter fingers. Place the detent balls (8) and springs (7) in their bores and install the detent caps (6). Install new Welch plugs (10 and 38).

Use a new gasket and install shifter cap assembly to gear case making sure the shift fingers engage the shift collars.

189. **CLUTCH SHAFT.** With tractor separated (split) as in paragraph 181 and the unit drained, the clutch shaft (21—Fig. 182) can be removed by unbolting the support (17) and pulling the support and clutch shaft assembly from the gear case. Remove the roller bearing (22) and the thrust washer (23) from the pilot end of

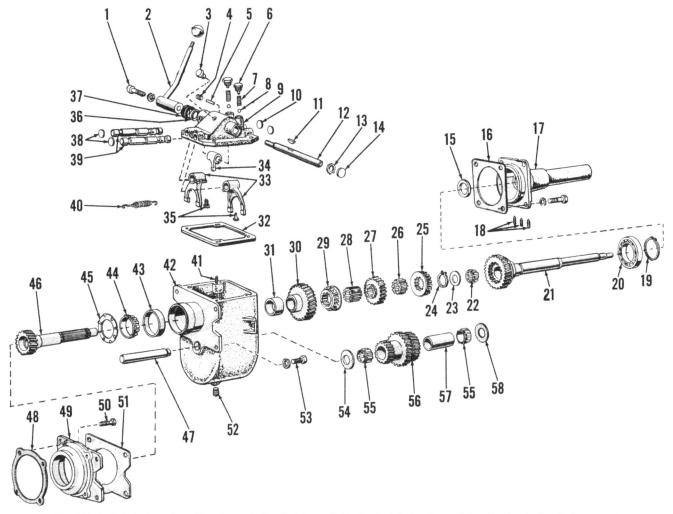

Fig. 182–Exploded view of auxiliary transmission that is available for installation in models with standard shift four-speed transmission.

1. Cap screw	11. Woodruff key	20. Ball bearing	30. Low range gear
2. Shift lever	12. Selector shaft	21. Input (clutch) shaft	31. Spacer
3. Breather	13. Snap ring	22. Roller bearing	32. Gasket
4. Socket head plug	14. Welch plug	23. Washer	33. Shift forks
5. Interlock	15. Oil seal	24. Snap ring	34. Shift finger
6. Detent caps	16. Gasket	25. Shift collar	35. Set screws
7. Detent springs	17. Release bearing	26. Coupling	36. Seal
8. Detent balls	support	27. High range gear	37. Grommet
9. Shift cover	18. Set screws	28. Coupling	38. Welch plugs
10. Welch plugs	19. Snap ring	29. Shift collar	39. Shift rails

40. Clutch release	49. Mounting flange
springs	50. Cap screws
41. Roll pin	51. Gasket
42. Transmission case	52. Drain plug
43. Bearing cup	53. Cap screws
44. Bearing cone & roller	54. Thrust washer
45. Oil slinger	55. Roller bearings
46. Drive shaft	56. Cluster gear
47. Cluster gear shaft	57. Spacer
48. Shims	58. Thrust washer

drive shaft (46). Remove the three socket head screws (18) from front support (17) and withdraw the clutch shaft. Remove the snap ring (19) and press ball bearing (20) from clutch shaft. Remove oil seal (15) using OTC No. 956 bearing puller and slide hammer.

Wash all parts in a suitable solvent, inspect them for damage and renew as necessary.

When reassembling, install new oil seal (15), press bearing (20) onto clutch shaft and secure with snap ring (19). Install the clutch shaft and bearing in the front support and secure with the three socket head screws (18). Note: Use caution when inserting shaft into support to avoid damaging oil seal. Fill unit with proper lubricant. Place the

thrust washer (23), then the roller bearing (22) on the pilot end of drive shaft (46) and using new gasket, install the front support and clutch shaft assembly after making certain that the oil return hole in front support is on the bottom.

190. DRIVE SHAFT. With the shifter cap assembly removed as in paragraph 187 and the front support and clutch shaft assembly removed as in paragraph 189, refer to Fig. 182 and remove snap ring (24) from drive shaft (46). Then remove front shift collar (25), short spline sleeve (26), step-up gear (27), rear shift collar (29), long spline sleeve (28), step-down gear (30) and spacer (31) from drive shaft. NOTE. Identify shift collars (25 and 29) so they can be reinstalled in their proper positions. Pull the drive shaft,

oil slinger (45) and taper bearing (44) from rear of gear case; then, using a suitable press or puller, remove the taper bearing and oil slinger from the drive shaft. Remove bearing cup (43) from gear case if necessary.

Clean all parts in a suitable solvent, inspect and renew as necessary.

To reinstall, place oil slinger (45) on drive shaft and press taper bearing onto shaft. Place shaft and bearing in gear case and install spacer (31) and step-down gear (30) with hub of gear facing rearward. Install long spline sleeve (28), then position the rear shift collar (29) with teeth toward rear. Install step-up gear (27) with the shift collar engaging teeth rearward, and the short spline sleeve (26), then position the front shift collar (25) with engaging teeth forward. Install the retaining snap ring (24). Then, install shifter cap assembly.

NOTE: With the drive shaft assembly and the shifter cap assembly installed, a bearing pre-load must be established as follows:

191. Remove mounting flange (49) and install approximately 0.050 thickness of shims (48) behind the mounting flange. Be sure flat side of mounting flange is at top and reinstall the flange. Shims (48) are available in thicknesses of 0.003, 0.005 and 0.012. Position new mounting flange gasket (51), install the auxiliary transmission and measure drive shaft end play. Now, remove shims from behind the mounting flange so that drive shaft has zero end play plus removing an additional 0.005 thick shim to preload the drive shaft bearing.

Fill unit with specified lubricant and install the clutch shaft as outlined in paragraph 189.

192. **CLUSTER GEAR AND SHAFT.** Remove the drive shaft assembly as outlined in paragraph 190. Refer to Fig. 182 and drift the cluster shaft (47) out toward rear of gear case and catch cluster gear (56) and thrust washers (54 and 58) as shaft is driven out. Remove the roller bearings (55) and spacer (57) from cluster gear.

Check cluster gear, bearings and shaft for wear. If wear or damage is found on any of the parts, it is recommended that the cluster shaft, gear and bearings be renewed as a unit.

Reassembly procedure is evident; however, keep in mind that the larger of the two end gears of the cluster gear is toward front of gear case. Reassemble balance of parts and reinstall in tractor.

FOUR-SPEED TRANSMISSION

Tractors having a four speed transmission as covered in this section may also be equipped with an auxiliary transmission interposed between engine clutch and the four speed transmission input shaft. The auxiliary transmission provides an underdrive, direct drive and overdrive ratio for each four speed transmission ratio thus providing twelve forward and three reverse speeds. For service information on the auxiliary transmission refer to paragraph 186.

LUBRICATION

193. Transmission lubricant capacity is 6 quarts. Oil level plug is located on right rear side of transmission housing. Recommended lubricant is SAE 80-EP lubricant (Ford specification No. M2C 53-A). Recommended lubricant change interval is after each 1200 hours of service. Filler plug is located in transmission shift cover; drain plug is located in rear center of transmission housing.

REMOVE AND REINSTALL

194. To remove the transmission, first split tractor between engine and transmission as outlined in paragraph 181, then, remove the steering gear and fuel tank as a unit from top of transmission. Following procedures outlined in paragraph 284, remove the transmission from rear axle center housing.

Reinstall transmission by reversing removal procedures. Refill transmission with proper lubricant as outlined in paragraph 193.

OVERHAUL

195. **SHIFTER RAILS AND FORKS.** Shifter rails and forks can be removed after removing transmission top (shift) cover and splitting tractor between transmission and rear axle center housing as outlined in paragraph 284.

Disconnect starter safety switch wire from switch, unbolt and remove shift lever and transmission top cover assembly (see Fig. 183) and remove the 1st-3rd shift rail detent spring (9—Fig. 184) and detent ball (10). Loosen locknut (12) and remove the 1st-3rd shifter fork lock screw (11), then remove shift rail (4) and fork (13). Remove shift plate pivot screws (27) from both sides of transmission housing, then remove the shift plates (19 and 20). Remove the detent ball caps (23), springs (25) and detent balls (26) from both sides of transmission housing. Remove the cup plugs (1) from rear face of housing at shift rail (2 and 3) bores, loosen the lock nuts (7 and 15), remove the lock screws (8 and 14), then remove the reverse shift rail (3) and fork (6) and the 2nd-4th shift rail (2) and fork (16).

Renew any worn or damaged parts and reinstall by reversing removal procedure. The square cornered slots in the reverse and 2nd-4th shift rails must face inward for engagement with shift plates.

196. **MAIN DRIVE GEAR (CLUTCH SHAFT).** (Note: On models equipped with auxiliary transmission, refer to paragraphs beginning with 186). The main drive (input) gear (clutch shaft) can be removed after splitting engine from transmission as outlined in paragraph 181. However, if bearing adjustment is required, the transmission assembly must be completely removed from tractor; refer to paragraph 194.

Disconnect the clutch release bearing springs and remove the release bearing and hub from bearing support (1—Fig. 185). Unbolt the bearing support from transmission housing and remove the support and main drive gear (6) as an assembly. Remove main drive gear, bearing cone (5), bearing cup (4) and oil seal (3).

Renew any worn or damaged parts and reassemble as follows: Install new seal in support with lip of seal to rear (towards bearing.) Drive new bearing cup into support and new bearing cone

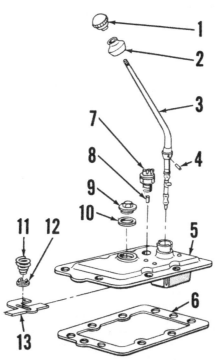

Fig. 183–Exploded view of four-speed transmission shift cover. Refer to Fig. 184 for remainder of shift mechanism.

1. Shift knob	
2. Dust seal	
3. Shift lever	
4. Pin	8. Switch plunger
5. Cover assembly	9. Filler plug
6. Gasket	10. Gasket
7. Starter safety switch	11. Spring
	12. Spring seat
	13. Neutral latch

and attach a pull scale to cord; a steady pull of 22-44 pounds should be required to rotate shaft with shift forks in neutral position. Note: A stub shaft must be installed to check bearing preload if equipped with power take-off. Add or remove shim between bearing retainer (or PTO support) and housing to obtain proper preload. Shims are available in thicknesses of 0.003, 0.005 and 0.012.

Reinstall main shaft and gear assembly by reversing removal procedure; however, before installing 1st-3rd shift fork, check bearing preload as follows: Wrap cord around rear end of shaft and attach pull scale; a steady pull of 27-47 pounds should be required to rotate shaft with lower shift forks in neutral position. Adjust preload by varying number of shims between bearing retainer and transmission housing; shims are available in thicknesses of 0.003, 0.005 and 0.012. Note: Preload on main shaft bearings should be in comparable range of adjustment to countershaft bearing preload.

198. **COUNTERSHAFT.** After removing the main shaft (paragraph 197), the countershaft and gear assembly can be removed as follows: Remove the 2nd-4th shifter rail and fork and remove the bearing retainer (35—Fig. 185) or the PTO bearing and shifter support (paragraph 319). Then, remove the countershaft and gears as an assembly.

With bearing puller, remove the rear bearing cone (31) and oil seal sleeve (32). With gear puller, remove gear (25) and bearing cone (24). Remove remaining gears and shift collars from shaft. If necessary to renew front bearing cup, it can be removed without removing the expansion plug from bore in front of bearing; however, if the plug is damaged or removed, a new 2.56 diameter expansion plug should be installed. Renew the oil seal (36) in bearing retainer or PTO support, and if necessary, renew bearing cup (34). Renew all other excessively worn or damaged parts and reassemble countershaft using Fig. 185 as a guide. Reinstall countershaft by reversing removal procedure and check bearing preload on countershaft as in paragraph 197.

199. **REVERSE IDLER GEAR.** After removing countershaft as in paragraph 198, remove reverse idler gear as follows: Remove the reverse shift rail (if not already removed) and the reverse shift fork. Pull idler shaft (42—Fig. 185) out towards rear of housing and remove the gears (37 and 38). Bushing (39) in gear (38) is renewable. Renew all other excessively worn or damaged parts and install new "O" ring (43).

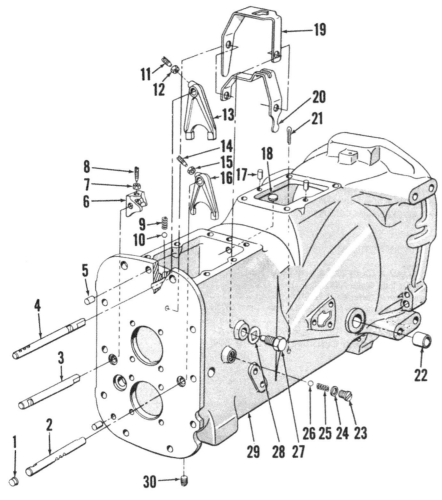

Fig. 184—Exploded view of four-speed transmission case and shifter mechanism. Refer to Fig. 183 for shift cover assembly.

1. Plug	9. Spring	17. Dowel pins
2. 2nd-4th shift rail	10. Detent ball	18. Welch plug
3. Reverse shift rail	11. Set screw	19. Reverse shift plate
4. 1st-3rd shift rail	12. Jam nut	20. 2nd-4th shift plate
5. Dowel pin	13. 1st-3rd shift fork	21. Drain hole pin
6. Reverse shift fork	14. Set screw	22. Clutch shaft
7. Jam nut	15. Jam nut	bushings
8. Set screw	16. 2nd-4th shift fork	23. Detent cap

24. Gasket	
25. Spring	
26. Detent ball	
27. Pivot bolts	
28. Gasket	
29. Transmission case	
30. Drain plug	

onto main drive gear until seated. Lubricate shaft of main drive gear and insert through support. Install support and gear as a unit using new gasket (2) and tighten retaining cap screws to a torque of 40-50 Ft.-Lbs. Transmission bearings should be preloaded and if any end play of main drive gear is noted, refer to paragraph 197 and adjust bearing preload.

197. **MAIN (OUTPUT) SHAFT.** To remove the transmission main (output) shaft and gear assembly, first remove transmission from tractor as in paragraph 194, then proceed as follows:

Following the procedure outlined in paragraph 195, remove the upper (1st-3rd) shift rail and fork and the shift plates. Remove the main drive gear as outlined in paragraph 196, or the auxiliary transmission as in paragraph 186. Unbolt and remove the main shaft rear bearing retainer (21—Fig. 185)

and any shims (19) located between the retainer and transmission housing. Slide the reverse shifter fork and the 2nd-4th shifter fork forward to provide clearance. Note: It may be necessary to remove the reverse shifter rail, but shift fork can remain in housing. Then, remove the main shaft and gear assembly. Using gear pullers, remove the end gears and bearing cones from shaft, then slide remaining gears and shifter collars from shaft. Renew any excessively worn or damaged parts. Install new oil seal (22) in retainer (21) with lip of seal forward (towards bearing). Refer to the exploded view in Fig. 185 (items 8 through 22) for reassembly guide.

Before reinstalling transmission main shaft and gear assembly, it is advisable to check bearing preload on transmission countershaft. Wrap a cord around rear end of countershaft

1. Bearing support
2. Gasket
3. Oil seal
4. Bearing cup
5. Bearing cone & roller
6. Input (clutch) shaft
7. Bearing cone & roller
8. Thrust washer
9. 4th gear
10. 2nd gear
11. Output shaft
12. 3rd gear
13. Shift-collar
14. Connector
15. 1st gear
16. Thrust washer
17. Bearing cone & roller
18. Sleeve
19. Shims
20. Bearing cup
21. Bearing retainer
22. Oil seal
23. Bearing cup
24. Bearing cone & roller
25. PTO drive gear
26. 4th gear
27. Connector
28. Shift collar
29. 2nd gear
30. Countershaft
31. Bearing cone & roller
32. Sleeve
33. Shims
34. Bearing cup
35. Bearing retainer
    (without PTO)
36. Oil seal
37. Reverse idler
38. Idler driven gear
39. Bushings
40. Thrust washer
41. Thrust washer
42. Reverse idler shaft
43. "O" ring
44. Snap ring

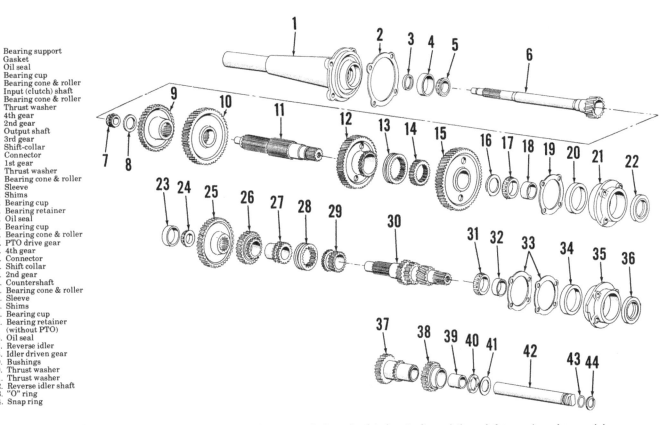

Fig. 185–Exploded view of four-speed transmission gears, shafts and related parts. Items 1 through 6 are not used on models equipped with auxiliary transmission; refer to Fig. 182.

Reinstall reverse idler unit by reversing removal procedure and make sure that grooves in face of washer (40) are towards washer (41) and that tang on washer (41) is placed in notch in transmission housing reverse idler boss.

200. PTO SHIFTER AND/OR

COUNTERSHAFT BEARING SUPPORT. For information on overhaul of the PTO shifter and bearing support on models so equipped, refer to paragraph 319.

The PTO shifter mechanism and countershaft bearing support can be unbolted and removed from rear of transmission after splitting tractor between rear axle center housing and transmission as outlined in paragraph 284. Be careful not to lose or damage countershaft bearing preload adjustment shims located between shifter support and transmission housing.

SIX-SPEED TRANSMISSION

Starting in Mid 1970, the Four-Speed transmission was phased out and a Six-Speed Transmission was substituted as a product option. The transmission consists of a three-speed plus reverse gear unit with combined dual range output gears. The transmission is similar in construction to the Eight Speed transmission unit and many of the parts are interchangeable.

LUBRICATION

201. Transmission lubricant capacity is 15 quarts. Maintain lubricant at level of check plug opening located on right side of transmission housing just behind brake pedals. Recommended lubricant is SAE 80 EP gear oil (Ford Specification No. M2C53-A.) NOTE:

An additional pint is required on models with transmission pto to maintain proper fluid level. Recommended lubricant change period is 1200 hours.

REMOVE AND REINSTALL

202. To remove the transmission unit, first split tractor between engine and transmission housing as outlined in paragraph 181, then remove steering gear assembly and fuel tank from top of transmission housing as a unit. Support transmission housing from an overhead hoist and rear axle center housing separately; then separate transmission from center housing as outlined in paragraph 284.

Reinstall transmission by reversing the removal procedure.

OVERHAUL

203. **SHIFT COVER.** To remove the shift cover (9—Fig. 186) remove attaching cap screws and lift the cover slightly. Disconnect safety switch wiring from switch unit and withdraw wire from cover unit; then lift off the cover.

Shift levers and safety switch can be serviced at this time. The safety switch threads into a retainer (20) which slides onto high-low shift rail (21).

The safety switch is actuated by plunger (16) which rides in a cross bore in high-low shift rail as shown, and by a corresponding shaft detent for the steel ball (17). The switch is closed only when high-low shift lever is in neutral position. The retainer is positioned by a

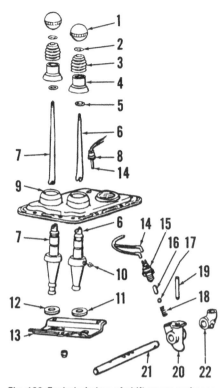

Fig. 186–Exploded view of shift cover and associated parts used on six speed transmission.

1. Shift knob	12. Washer
2. Snap ring	13. Stop plate
3. Spring	14. Switch wire
4. Retainer	15. Safety switch
5. O-ring	16. Plunger
6. Range shift lever	17. Ball
7. Main shift lever	18. Spring
8. Grommet	19. Dowel
9. Cover	20. Retainer
10. Pin	21. Shift rail
11. Washer	22. Gate

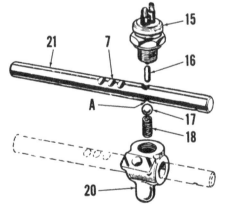

Fig. 187–Plunger (16) lies in cross bore of range shift rail as shown. Refer to Fig. 186 for parts identification and to Fig. 188 for installed view.

locating pin (19) which is pressed into a bore in shift cover (9) and enters the positioning hole in top of retainer. Refer also to Figs. 187 and 188.

To disassemble shift levers, unbolt and remove stop plate (13—Fig. 186). Remove knobs (1). Slightly depress lever spring (3) and unseat and remove snap ring (2); then withdraw lever downward out of cover unit. Assemble

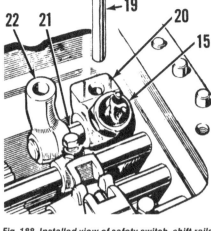

Fig. 188–Installed view of safety switch, shift rails and associated parts. Positioning dowel (19) is pressed into top cover and enters locating hole in retainer (20). Refer to Fig. 186 for parts identification and to Fig. 189 for assembled view of cover.

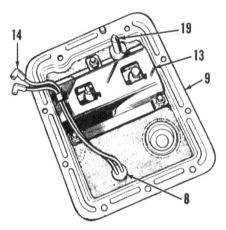

Fig. 189–Assembled view of transmission top cover as viewed from bottom. Refer to Fig. 186 for parts identification.

1. Transmission housing
2. High/low shift rail
3. Low/intermediate rail
4. High/reverse rail
5. Plug
6. Interlock pin
7. Plug
8. Spring
9. Detent ball
10. Connector
11. Low/intermediate fork

by reversing the disassembly procedure.

When reinstalling cover, be sure to reconnect the safety switch wires and check to be sure locating pin (19—Fig. 188) in cover enters positioning hole in retainer (20). Tighten cover retaining cap screws to a torque of 25-30 Ft.-Lbs. Check to be sure safety switch operates properly before releasing tractor for service.

204. **SHIFT RAILS AND FORKS.** Refer to Fig. 190 for exploded view. Upper shift rails, forks and associated parts can be removed for service after removing shift cover as outlined in paragraph 203 and transmission rear support plate as in paragraph 206. High speed shift fork (13) slides on an extension of reverse idler shaft (17) and removal is outlined in paragraph 208.

Be sure to recover detent ball (9), spring (8) and plunger (7) for each rail as unit is disassembled. Interlock plunger (6) lies in a bore in transmission housing wall between high/reverse rail (4) and low/intermediate rail (3). Plunger can be removed with a suitable magnet after removing plug (5) and rail (4). The transmission housing is interchangeable with the one used on 8-speed models, and the left-hand bore for shift rails remains vacant on 6-speed units.

Install by reversing the removal procedure. Be sure the tang on reverse fork (12) properly engages the notch in high speed fork (13) during assembly. Also be sure that interlock pin (6) is properly positioned between the rails (3 & 4). Tighten set screws and lock

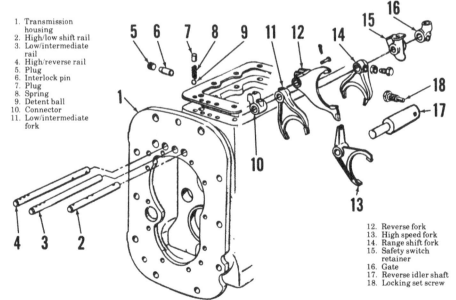

12. Reverse fork
13. High speed fork
14. Range shift fork
15. Safety switch retainer
16. Gate
17. Reverse idler shaft
18. Locking set screw

Fig. 190–Exploded view of shifter rails and forks used in six speed transmission. Reverse idler shaft (17) serves as rail for high speed fork (13). Interlock pin (6) lies between rails (3 & 4) and prevents accidental shifting into two gears at once.

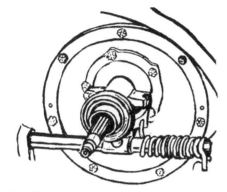

Fig. 191—Assembled view of transmission input shaft and clutch release mechanism used on single clutch models.

nuts to a torque of 20-25 Ft.-Lbs.

205. INPUT SHAFT, CLUTCH RELEASE BEARING SUPPORT AND FRONT SUPPORT PLATE. On models with single disc clutch, transmission input shaft can be removed after splitting tractor between engine and transmission housing as outlined in paragraph 181. On dual clutch models, clutch release mechanism, bearing support and pto input shaft can be removed for service after front split, but removal of transmission input shaft will require transmission removal as outlined in paragraph 202.

On all models, drain transmission and make clutch split. Disconnect pedal linkage and on models so equipped, unhook clutch return spring from stop on inside of clutch housing. Refer to Fig. 191. Remove the bolt securing clutch release fork to shaft and withdraw the shaft; then lift out clutch release bearing, hub and spring. Remove the inner circle of cap screws securing shaft sleeve to front plate and withdraw the sleeve. On dual clutch models, pto input shaft will be removed with sleeve. Remove outer bolt circle retaining support plate and withdraw the plate; noting, on dual clutch models, that plate must come straight forward until free of pto countershaft front bearing.

On single clutch models, unseat snap ring (10—Fig. 193) and withdraw input

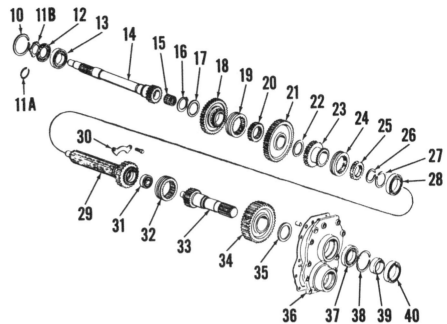

Fig. 193—Transmission upper shafts, gears and associated parts used in six speed models.

10. Snap ring	17. Thrust washer	25. Connector	33. Output shaft
11A. Snap ring (Trans. pto)	18. Intermediate gear	26. Snap ring	34. Range gear
11B. Snap ring (Live pto)	19. Shift collar	27. Snap ring	35. Thrust washer
12. Plate (Live pto)	20. Connector	28. Bearing	36. Rear support plate
13. Bearing	21. Low gear	29. Mainshaft	37. Bearing
14. Input shaft	22. Thrust washer	30. Bearing plate	38. Snap ring
15. Pilot bearing	23. Reverse gear	31. Pilot bearing	39. Seal sleeve
16. Snap ring	24. Shift collar	32. Shift collar	40. Oil seal

shaft (14) and bearings. On dual clutch models (with transmission removed), remove bearing (41—Fig. 194) using a suitable puller, and unseat and remove snap ring (42). Remove snap ring (64) coupling (63) and snap ring (62) from rear of pto countershaft; then drift countershaft and rear bearing (61) rearward until drive gear (44) can be removed. Transmission input shaft can now be removed by unseating snap ring (10—Fig. 193) and withdrawing shaft and bearing (13). Refer to Fig. 195 for differences between single clutch and dual clutch models.

Assemble by reversing the removal procedure. Tighten cover retaining cap screws to a torque of 23-30 Ft.-Lbs.

206. REAR SUPPORT PLATE, PTO DRIVE SHAFT, OUTPUT SHAFT AND SECONDARY COUNTERSHAFT. On single clutch models, rear support plate (36—Fig. 193) and associated parts can be removed after splitting tractor between transmission and rear axle center housing as outlined in paragraph 284. On dual clutch models it will be necessary to remove transmission as outlined in paragraph 202 and front support plate as in paragraph 205.

Remove locking wire and rear cover cap screws. On single clutch models, pry support plate from its doweled position on transmission housing. On dual clutch models, remove pto countershaft front bearing (41—Fig. 194) using a suitable puller; remove snap ring (42), then bump pto countershaft (57) rearward until rear plate (36) is free of dowels. PTO output shaft will be removed with rear plate; remove snap ring (64), coupling (63) and snap ring (62), then bump the pto output shaft rearward out of plate.

Transmission output shaft (33—Fig. 193) and secondary countershaft (54—Fig. 194) will remain in transmission housing. Refer to Fig. 196. Partially withdraw secondary countershaft until front bearing clears housing bore, then lower countershaft to bottom of com-

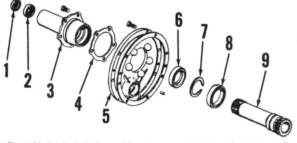

1.	Oil seal
2.	PTO shaft bearing
3.	Shaft sleeve
4.	Gasket
5.	Front support plate
6.	PTO shaft oil seal
7.	Snap ring
8.	PTO shaft bearing
9.	PTO input shaft

Fig. 192—Exploded view of front support plate and associated parts used on dual clutch models. Transmission input shaft (14—Fig. 193) will remain with transmission and is not removed with front support plate.

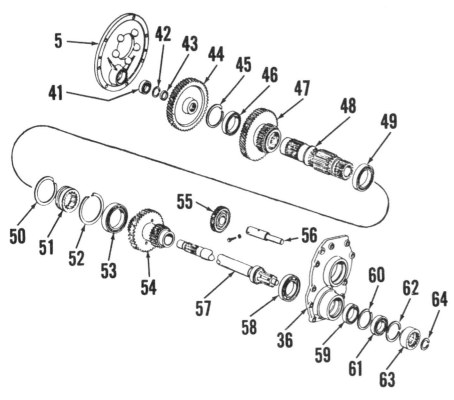

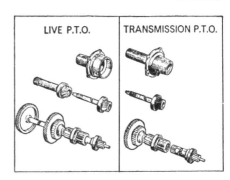

Fig. 195–View showing differences in pto drive train in Live PTO models (left) and Transmission PTO models (right). On Transmission pto models, PTO drive shaft (57–Fig. 194) splines into transmission countershaft (48) as shown.

Fig. 194–Transmission lower shafts, gears and associated parts used on six speed models with live pto.

5. Front support plate	46. Bearing	52. Snap ring	58. Bearing
36. Rear support plate	47. Countershaft main	53. Bearing	59. Oil seal
41. Bearing	gear	54. Secondary	60. Snap ring
42. Snap ring	48. Countershaft	countershaft	61. PTO bearing
43. Thrust washer	49. Bearing	55. Reverse idler	62. Snap ring
44. PTO drive gear	50. Snap ring	56. Reverse idler shaft	63. PTO coupling
45. Snap ring	51. Shift collar	57. PTO drive shaft	64. Snap ring

Fig. 196–With secondary countershaft partially withdrawn and lowered to floor of housing, output shaft assembly can be removed.

partment. Output shaft can now be withdrawn and secondary countershaft can be lifted out after output shaft is removed.

Secondary countershaft front bearing (53—Fig. 194) can be removed from shaft using a suitable press or puller; and two 3/16-2 inch steel rods or bolts inserted in holes provided, to apply pressure to inside of inner race of bearing.

Assemble by reversing the disassembly procedure. Tighten rear support plate attaching cap screws to a torque of 23-30 Ft.-Lbs.

207. MAINSHAFT ASSEMBLY. To remove the transmission mainshaft assembly (29—Fig. 193), first remove transmission from tractor and remove front support plate, rear support plate, input and output shafts and the shift mechanism as previously outlined.

Remove snap ring (16) from front of shaft and withdraw thrust washer (17). Remove the two cap screws retaining bearing plate (30) and lift out the plate; then slide shaft (29) and bearing (28) rearward while lifting gears, connectors, thrust washer and couplings out top opening. Keep sliding couplings (19 and 24) with their respective couplings (20 and 25). Units are interchangeable but matched, and should not be inter-

mixed.

Use Figs. 193 and 197 as guides if necessary, and assemble by reversing the disassembly procedure. Tighten bearing retaining plate cap screws to a torque of 72-96 inch-pounds.

208. REVERSE IDLER. To remove the reverse idler shaft and associated parts, first remove transmission mainshaft as outlined in paragraph 207.

Remove the idler shaft set screw from outside left of transmission housing. Move shaft forward and lift out high speed fork (13—Fig. 190) and its accompanying sliding coupling; then slide shaft rearward while lifting out reverse idler gear (55—Fig. 194).

Be sure to install the gear with long hub forward, and make sure high speed fork is installed and properly engaging the groove of high speed sliding coupling. Tighten idler shaft retaining set screw to a torque of 15-18 Ft.-Lbs.

209. MAIN COUNTERSHAFT. With transmission mainshaft removed as outlined in paragraph 207 and reverse idler removed as in paragraph 208, remove main countershaft (48—Fig. 194) as follows:

Remove snap ring (52) from front end of secondary countershaft bearing bore and snap ring (50) from rear of main

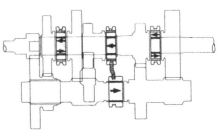

Fig. 197–Cross sectional view of transmission gears showing shift collars.

countershaft rear bearing (49). Using a step plate of proper size in front bore of hollow shaft, a suitable drift and heavy hammer, bump the shaft rearward until both bearings are free of bores and shaft is loose in transmission housing. Move the shaft rearward until cluster gear (47) is adjacent to housing center wall and insert wooden protector blocks between gear and wall; then continue to bump countershaft rearward until front bearing (46) is free and cluster gear can be removed.

Install shaft and components by reversing the removal procedure.

EIGHT-SPEED TRANSMISSION

(Models 2000, 3000 and 4000)

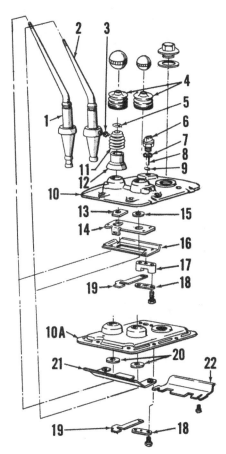

Fig. 198—Exploded view of 8-speed shift cover assembly used on early models 2000, 3000 and 4000. Shift cover may be either cast iron (10) or pressed steel (10A). Refer to Fig. 199 or 200 for late production shift cover assemblies.

1. Main shift lever	12. Retainers
2. High-low shift lever	13. Sliding plate
3. Pins	14. Lever stop plate
4. Dust covers	15. Washer
5. Snap rings	16. Baffle plate
6. Starter safety switch	17. Switch plunger
7. "O" ring	retainer
8. Switch plunger	18. Support plate
9. Roller	19. Switch connector
10. Cast iron cover	20. Washers
10A. Steel cover	21. Baffle plate
11. Springs	22. Baffle plate

LUBRICATION

210. Transmission lubricant capacity is 17 quarts; maintain lubricant to level of plug located on right side of transmission housing just behind brake pedals. Recommended lubricant is SAE 80-EP gear lubricant (Ford specification No. M-2C53-A). Recommended lubricant change interval is after each 1200 hours of service. Filler plug is located in transmission shift cover; drain plug is at rear center of transmission housing.

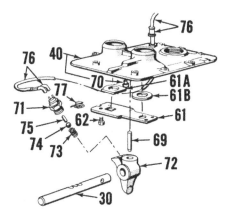

Fig. 199—Exploded view of late production cast iron shift cover assembly. Starter safety switch (71) and retainer (72) are mounted on high-low shift rail (30) instead of in cover.

30. High-low shift rail	70. Dowel
40. Shift cover	71. Safety switch
61. Shift lever stop plate	72. Switch retainer
61A. Sliding plate	73. Spring
61B. Washer	74. Steel ball
62. Cap screws	75. Switch plunger
69. Locating pin	76. Starter wiring

REMOVE AND REINSTALL

211. To remove transmission, first split tractor between engine and transmission as outlined in paragraph 181 or 182, then remove steering gear assembly (steering support on model 4200) and fuel tank from top of transmission as a unit. Following procedures outlined in paragraph 284 or 299, remove the transmission assembly

Fig. 201—Exploded view of 8-speed transmission shift mechanism as used on models 2000, 3000 and 4000. Refer to Fig. 198 for early shift cover assembly and to Fig. 199 or 200 for late production shift covers.

1. Retainer
2. Spring
3. Plunger
4. Pin
5. Selector arm
6. 4th, low & 8th shift fork
7. Interlock pins
8. Plug
9. 4th & 8th shift rail
10. 2nd, 6th & reverse shift rail
11. 7-3, 5-1 shift rail
12. High-low shift rail
13. Detent balls (4)
14. Springs (4)
15. Plungers (4)
16. 2nd, 6th & reverse fork
17. Connector
18. 7-3, 5-1 shift fork
19. Gate
20. High-low shift fork
21. Oil level plug

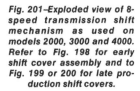

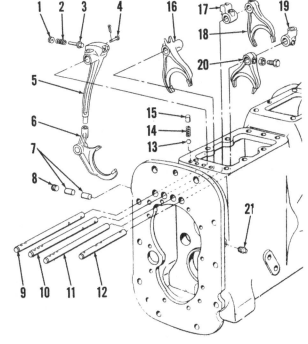

from rear axle center housing.

Reinstall transmission by reversing removal procedure.

OVERHAUL

212. **SHIFT COVER.** On model 4200 rowcrop, remove the operator's platform. Then, on all models, proceed as follows:

On early production transmissions with starter safety switch on top of

Fig. 200—Exploded view of late production pressed steel shift cover assembly. Refer to Fig. 199 for cast iron cover.

30. High-low shift rail	70. Dowel
40A. Shift cover	71. Safety switch
61F. Washers	72. Switch retainer
66. Lever stop plate	73. Spring
67. Self-locking nuts	74. Steel ball
68. Locating pin	75. Switch plunger
	76. Wiring

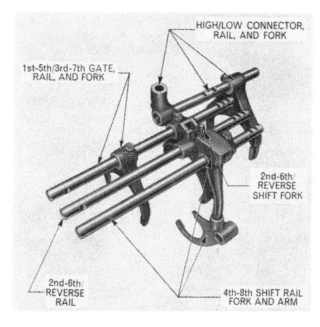

1st-5th/3rd-7th GATE, RAIL, AND FORK

HIGH/LOW CONNECTOR, RAIL, AND FORK

2nd-6th/ REVERSE SHIFT FORK

2nd-6th/ REVERSE RAIL

4th-8th SHIFT RAIL FORK AND ARM

Fig. 202–Assembled view of shift mechanism for 8-speed transmission on models 2000, 3000 & 4000. Refer to Fig. 201 for exploded view. Late production starter safety switch is not shown on high-low shift rail; refer to Figs. 199 and 200.

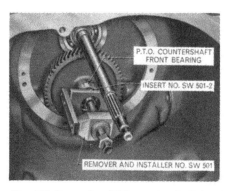

Fig. 205–Removing PTO countershaft front bearing with special puller (Nuday tool No. SW-501).

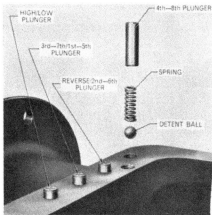

Fig. 203–View showing location of shift rail detent plungers, springs and balls. Plungers are retained by transmission shift cover.

Fig. 204–View of front end of transmission assembly.

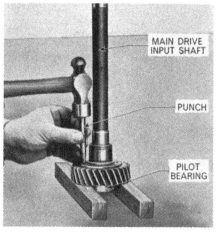

Fig. 206–Driving pilot bearing from rear end of main drive input (clutch) shaft.

1.320 on models with cast iron cover (10). Tighten safety switch retainer, roller support and shift lever stop plate cap screws to a torque of 14-17 Ft.-Lbs. When reinstalling cover, install dowel cap screw without a washer in left rear corner of cover and install cap screws with lockwashers in remaining holes. Tighten dowel cap screw first, then tighten remaining cap screws to a torque of 23-29 Ft.-Lbs. Be sure that engine starter can be engaged only when main gear change lever is in neutral position.

On late production transmissions, unbolt and lift shift cover assembly, then disconnect wiring (76—Fig. 199 or 200) from starter safety switch (71), then remove shift cover from transmission. The starter safety switch can now be serviced at this time. Other than components shown in Figs. 199 or 200, shift lever assemblies are same as shown in exploded view in Fig. 198. When reinstalling cover, be sure that dowel pin (70—Fig. 199 or 200) and switch retainer locating pin 69—Fig. 199 or 68—Fig. 200) are in place and reconnect wiring to starter safety switch. Be sure dowel pin and retainer locating pin properly engage shift cover as it is being lowered into place. Tighten cover retaining cap screws to a torque of 23-29 Ft.-Lbs. Be sure that starter will engage only when high-low shift lever is in neutral position.

213. SHIFT RAILS AND FORKS. To remove the shift rails and forks, it is necessary to first remove the transmission from tractor (refer to paragraph 211), remove top cover (paragraph 212), remove input shaft and front support plate (paragraph 214) and the rear support plate, PTO driveshaft, output shaft assembly and secondary counter-

shaft (paragraph 215). Then, refer to exploded view of the shift mechanism in Fig. 201, to the assembled view in Fig. 202 and proceed as follows:

Remove the four detent plungers and springs (see Fig. 203) from their bores. Remove a detent ball as each corresponding shift rail is removed. Move the high-low shift rail rearward until sliding coupling (see Fig. 212) can be removed from mainshaft and high-low shift fork. Note: On late production transmission, refer to Fig. 199 or 200 and remove starter safety switch (71) and retainer (72) as high-low shift rail (30) is being removed. Turn the rail so that locknut can be loosened and the set screw removed from high-low shift fork and remove the fork out rear opening. Then, turn the rail so that locknut can be loosened and the set screw removed from connector on front end of shaft, withdraw the shaft from rear of housing and remove the connector and detent ball. Loosen the locknuts and remove the set screws from remaining shift forks and arms, withdraw the rails one at a time from rear of housing and remove the forks,

shift cover, disconnect wiring, then unbolt and remove shift cover assembly. Refer to Fig. 198 for guide to disassembly and reassembly of cover. Starter safety switch plunger (8—Fig. 198) should measure 0.990-0.995 on models with steel cover (10A) or 1.315-

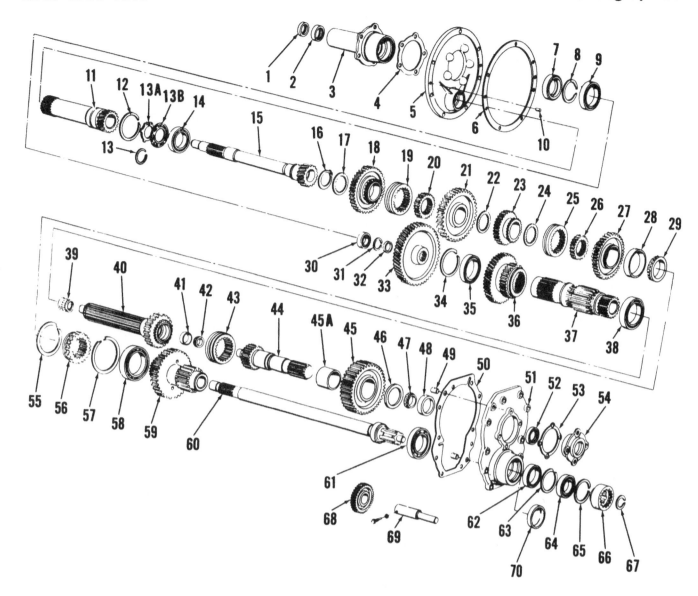

Fig. 207–Exploded view of eight-speed ("live" PTO) transmission gears, shafts and related parts as used on models 2000, 3000 and 4110; refer to Fig. 208 for model 4000 transmission. Refer to Fig. 216 for drawing showing relative gear location when installed in transmission and for number of teeth per gear for identification. Gear ratios of early and late production differ. On models with transmission type PTO (single disc clutch), PTO shaft (60) is splined into countershaft (37) and PTO drive gears (11 and 33) are not used.

1. Oil Seal	13A. Bearing plate	23. Reverse gear	35. Ball bearing
2. Needle bearing	retaining ring	24. Thrust washer	36. Countershaft main
3. Clutch release	13B. Bearing plate	25. Shift collar	gear
bearing support	14. Bearing	26. Connector	37. Countershaft
4. Gasket	15. Transmission input	27. 2nd gear	38. Ball bearing
5. Front support plate	(clutch) shaft	28. Bearing cup	39. Roller bearing
6. Gasket	16. Snap ring	29. Bearing cone & roller	40. Main shaft
7. Oil seal	17. Thrust washer	30. Ball bearing	41. Bearing cup
8. Snap ring	18. 3rd gear	31. Snap ring	42. Bearing cone & roller
9. Bearing	19. Shift collar	32. Thrust washer	43. Shift collar
10. Dowel pins	20. Connector	33. PTO countershaft	44. Output shaft
11. PTO input shaft	21. 1st gear	gear	45. Output shaft gear
12. Snap ring	22. Thrust washer	34. Snap ring	45A. Bushing
13. Snap ring			

46. Thrust washer	59. Secondary
47. Bearing cone & roller	countershaft
48. Bearing cup	60. PTO countershaft
49. Dowel pins	61. Ball bearing
50. Gasket	62. Oil seal
51. Rear support plate	63. Snap ring
52. Gasket	64. Ball bearing
53. Shims	65. Snap ring
54. Bearing support	66. PTO coupling sleeve
55. Snap ring	67. Snap ring
56. Sliding gear	68. Reverse idler
57. Snap ring	69. Idler shaft
58. Ball bearing	70. Plug (w/o PTO)

arms and detent balls. Note: The 4th-8th shift fork is located on the reverse idler shaft; refer to paragraph 217 for shaft and fork removal procedure. Remove the interlock plunger bore plug (see Fig. 215) and the two interlock plungers (7—Fig. 201).

To reassemble, proceed as follows: Insert the 4th-8th shift arm in socket of shift fork and install upper 4-8th shift rail through housing and arm so that set screw holes are aligned, install and tighten the set screw and locknut to a torque of 20-25 Ft.-Lbs. and move rail to neutral position. Next, install the 2nd-6th/reverse shift rail and fork, tighten set screw and locknut to a torque of 20-25 Ft.-Lbs. and move rail to neutral position. Insert second interlock plunger against the shift rail, then install remaining shift rails, forks and

connector, tightening set screws and locknuts to a torque of 20-25 Ft.-Lbs. and install interlock plunger bore plug. Insert detent balls, springs and plungers in their bores on top of shift rails and move all shift rails to neutral position.

214. **INPUT SHAFT, CLUTCH RELEASE BEARING SUPPORT AND FRONT SUPPORT PLATE.** On all models without a PTO and on

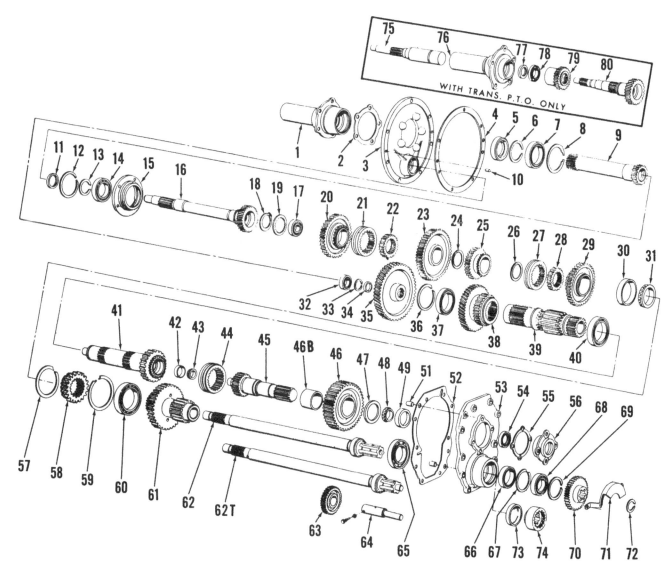

Fig. 208–Exploded view of model 4000 eight-speed transmission gears, shafts and related parts; refer to Fig. 207 for models 2000, 3000 and 4110. On models with transmission type PTO, items 75 through 80 are used instead of 1, 5, 6, 7, 8, 9, and 11. Refer to Fig. 217 for diagram showing gear location and for gear identification by number of teeth. Early and late production transmissions have different gear ratios.

1. Clutch release bearing support	16. Transmission input (clutch) shaft	31. Bearing cone & roller	45. Output shaft
2. Gasket	17. Ball bearing	32. Ball bearing	46. Output shaft gear
3. Front support plate	18. Snap ring	33. Snap ring	46B. Bushing
4. Gasket	19. Thrust washer	34. Thrust washer	47. Thrust washer
5. Oil seal	20. 3rd gear	35. PTO countershaft gear	48. Bearing cone & roller
6. Snap ring	21. Shift collar	36. Snap ring	49. Bearing cup
7. Ball bearing	22. Connector	37. Ball bearing	51. Dowel pins
8. Snap ring	23. 1st gear	38. Countershaft main gear	52. Gasket
9. PTO drive shaft	24. Thrust washer	39. Countershaft	53. Rear support plate
10. Dowel pins	25. Reverse gear	40. Ball bearing	54. Oil seal
11. Oil seal	26. Thrust washer	41. Main shaft	55. Shims
12. Snap ring	27. Shift collar	42. Bearing cup	56. Bearing support
13. Snap ring	28. Connector	43. Bearing cone & roller	57. Snap ring
14. Ball bearing	29. 2nd gear	44. Shift collar	58. Shift collar
15. Retainer	30. Bearing cup		59. Snap ring
			60. Ball bearing

61. Secondary countershaft	71. Shield
62. PTO countershaft (independent PTO)	72. Snap ring
	73. Plug (w/o PTO)
62T. PTO countershaft (transmission PTO)	74. PTO coupling sleeve (transmission PTO)
63. Reverse idler	75. Clutch shaft
64. Idler shaft	76. Clutch release bearing support
65. Ball bearing	77. Oil seal
66. Oil seal	78. Ball bearing
67. Snap ring	79. PTO drive gear
68. Ball bearing	80. Transmission drive (input) shaft
69. Snap ring	
70. Hydraulic pump drive gear	

models 2000 and 3000 with "transmission type" PTO, transmission input shaft can be removed after splitting tractor between engine and transmission housing as outlined in paragraph 181 or 182. On model 4000 with PTO and on models 2000 and 3000 with "live" PTO (double clutch), it is necessary to remove the transmission as outlined in paragraph 211. With transmission removed or with tractor split be-

tween engine and transmission, as required, proceed as follows:

Refer to Fig. 204 and on models so equipped, disengage the clutch release fork return spring from stop inside transmission housing. Remove the bolt from release fork, slide release shaft from housing and remove fork. Remove clutch release bearing and hub assembly from support, then unbolt and remove bearing support from front

support plate. On models 2000 and 3000 with "live" PTO and on model 4000 with independent PTO, PTO input shaft will be removed with bearing support. Then, unbolt and remove front support plate from transmission housing. NOTE: On models 2000 and 3000 with "live" PTO and on all 4000 with PTO, the PTO countershaft front bearing is located in front support plate and plate must be removed squarely.

Fig. 209–Removing the rear support plate and PTO countershaft assembly.

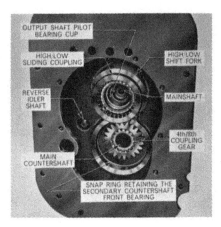

Fig. 212–View showing rear transmission with output shaft assembly and secondary countershaft assembly removed. Refer to text for further disassembly procedure.

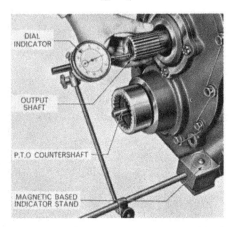

Fig. 213–Checking transmission output shaft end play with dial indicator.

Fig. 210–Transmission output shaft assembly must be removed and reinstalled in conjunction with the secondary countershaft assembly.

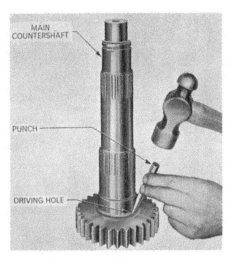

Fig. 214–Removing bearing from rear end of main countershaft. Model 4000 shaft is shown; however, bearing removal on models 2000, 3000 and 4110 is similar.

Fig. 211–Secondary countershaft front bearing can be removed by pressing against steel pins inserted through shaft against inner bearing race as shown.

On model 4000 with PTO, refer to Fig. 208 and unbolt shield (71) on models equipped with independent PTO. Then, on all model 4000, remove snap ring (72) from rear end of PTO countershaft, gear (70) or coupling (74) and remove the snap ring (69) from transmission rear support plate.

On models 2000 and 3000 with live PTO, refer to Fig. 207 and remove snap ring (67) and coupling (66) from rear end of PTO countershaft, then remove snap ring (65) from transmission rear support plate.

On models 2000 and 3000 with live PTO and all 4000 with PTO, refer to Fig. 205 and remove the PTO countershaft front bearing as shown. Then, remove the snap ring and thrust washer from front end of PTO countershaft, bump the countershaft rearward out of transmission and lift the PTO countershaft gear out of front opening in transmission housing.

On all model 4000, unbolt the transmission input shaft bearing retainer (15—Fig. 208) and remove input shaft, bearing and retainer assembly. The pilot bearing (17) should be removed with the input shaft and can be removed from shaft as shown in Fig. 206.

On all models 2000 and 3000, remove the snap ring (12—Fig. 207) and pull the input shaft and bearing (14) assembly from transmission. The pilot bearing (39) for transmission mainshaft can also be removed at this time if not removed with input shaft.

To reassemble transmission, reverse removal procedures. On model 4000 with independent PTO, tighten the shield (71—Fig. 208) retaining cap screws to a torque of 23-30 Ft.-Lbs. and install new locking wire through rear support retaining cap screw heads. On model 4000, tighten input shaft bearing retainer cap screws to a torque

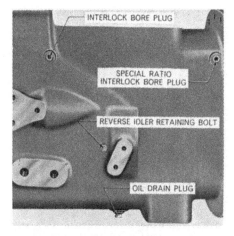

Fig. 215–View showing interlock bore plug, reverse idler shaft retaining bolt and oil drain plug. Disregard special ratio interlock bore plug as interlock is not used at this location on U.S. production tractors.

of 23-30 Ft.-Lbs. On all models, tighten front support plate and clutch release bearing support retaining cap screws to a torque of 23-30 Ft.-Lbs.

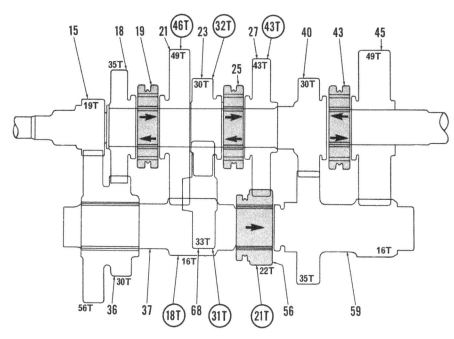

215. REAR SUPPORT PLATE, PTO DRIVE SHAFT, OUTPUT SHAFT AND SECONDARY COUNTERSHAFT. On all models not equipped with power take-off and on models 2000 and 3000 with transmission type PTO, the rear support plate, PTO drive shaft (if so equipped), output shaft and secondary countershaft can be removed after splitting tractor between transmission housing and rear axle center housing as outlined in paragraph 284 or 299. On other models, it will be necessary to remove the transmission as outlined in paragraph 211. With the transmission removed or tractor split at transmission and rear axle center housing, proceed as follows:

On models 2000 and 3000 with "live" PTO and all 4000 equipped with PTO, refer to paragraph 214 and remove the clutch release shaft, bearing support, front support plate and the PTO driven gear from front compartment of transmission housing.

Fig. 216–Diagram showing gear location on models 2000, 3000 and 4110; refer to Fig. 217 for model 4000. Number of teeth on gear is indicated by number with "T" suffix. Early and late production transmissions have different gear ratios; number of teeth on late production gears are circled and number of teeth on early production gears are on or immediately adjacent to gear.

15. Transmission input (clutch) shaft	23. Reverse gear	37. Countershaft	56. Sliding gear
18. 3rd gear	25. Shift collar	40. Main shaft	59. Secondary countershaft
19. Shift collar	27. 2nd gear	43. Shift collar	68. Reverse idler
21. 1st gear	36. Countershaft main gear	45. Output shaft gear	

On all models, cut the locking wire and remove the cap screws retaining rear support plate to housing. Pry the support plate loose on models not equipped with PTO or on models 2000 and 3000 with transmission type PTO. On 4000 with PTO and 2000 and 3000 with live PTO, bump the PTO countershaft rearward to loosen rear support plate. Then, on all models, refer to Fig. 209 and remove the rear support plate and, if so equipped, the PTO shaft as an assembly. With rear support plate removed, pull the secondary countershaft rearward until front bearing is free of bore in housing, then withdraw the transmission output shaft assembly and the secondary countershaft as shown in Fig. 210.

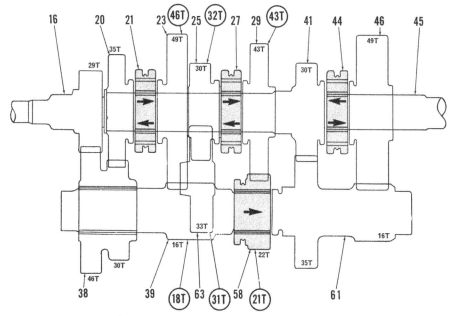

Fig. 217–Diagram showing gear location on model 4000; refer to Fig. 216 for models 2000, 3000 and 4110. Number of teeth on each gear is indicated by numbers with "T" suffix. Early and late production transmissions have different gear ratios; number of teeth of late gears are circled and number of teeth on early gears are on or immediately adjacent to gears.

16. Transmission input (clutch) shaft	25. Reverse gear	39. Countershaft	46. Output shaft gear
20. 3rd gear	27. Shift collar	41. Main shaft	58. Sliding gear
21. Shift collar	38. Countershaft main gear	44. Shift collar	61. Secondary countershaft
23. 1st gear		45. Output shaft	63. Reverse idler

Disassemble components as follows: Remove the snap ring retaining PTO coupling or hydraulic pump drive gear to rear end of PTO shaft and remove the coupling or gear. Remove snap ring at rear side of PTO shaft bearing in rear support and drive or press shaft and bearing out to rear of support plate. Remove PTO shaft seal from rear support plate. Drive the oil seal, then the output shaft rear bearing cup, from output shaft bearing retainer. Remove bearings from secondary countershaft if necessary; refer to Fig. 211. For front bearing removal procedure; press the output shaft from gear, thrust washers and bearing cone, then remove bearing cone from front end of shaft. Carefully inspect all parts and renew any that are excessively worn or damaged.

Reassemble components and reinstall in transmission by reversing removal and disassembly procedure and by observing the following: Install oil seals with lips to inside of transmission (forward). Be sure that the high-low coupling is engaged on teeth on rear end of mainshaft and that snap ring is in groove in lower bearing bore of transmission housing; refer to Fig. 212. Place the secondary countershaft and the output shaft assembly into rear compartment of transmission as shown in Fig. 210. Then, align front bearing of secondary countershaft with bore in housing and bump the shaft and bearing assembly forward until front bearing seats against snap ring. Install rear support plate with new gasket, but without output shaft bearing retainer or PTO shaft. Tighten rear support plate retaining cap screws to a torque of 24-30 Ft.-Lbs. Insert PTO shaft and bearing into rear of transmission, place PTO gear in front compartment and slide the shaft through the gear. Bump the shaft forward until thrust washer and snap ring can be installed on shaft at front side of gear and snap ring can be installed in rear support plate at rear side of the PTO shaft bearing. Install the front PTO shaft bearing, front support plate, bearing support and clutch release shaft as outlined in paragraph 214. Install the transmission output shaft bearing retainer with at least 0.060 thickness of shims between retainer and rear support plate and tighten retaining cap screws to a torque of 24-30 Ft.-Lbs. Measure output shaft end float with dial indicator as shown in Fig. 213, then remove the retainer and remove shim thickness equal to shaft end float plus an additional zero to 0.002 thickness to provide proper bearing preload. Reinstall retainer and tighten cap screws to a torque of 24-30 Ft.-Lbs. Shims are available in thicknesses of 0.003, 0.005

and 0.012. Install locking wire through rear support retaining cap screw heads and reinstall PTO coupling or hydraulic pump drive gear.

216. MAINSHAFT ASSEMBLY. To remove the mainshaft assembly, first remove transmission from tractor and remove front support plate, rear support plate, input and output shafts and the shift mechanism as outlined in previous paragraphs, then proceed as follows:

Remove the snap ring at front end of transmission mainshaft and withdraw the mainshaft from rear end of transmission. Remove the gears, sliding couplings, connectors and thrust washers as the mainshaft is withdrawn. Remove ball bearing from rear end of shaft. Remove bearing cup from inside rear bore of shaft as shown in Fig. 214. Carefully inspect all parts and renew any excessively worn or damaged parts. Bushings in inside diameter of gears are not serviced separately; renew bushing and gear assembly if bushing is worn beyond further use.

Reinstall mainshaft assembly by reversing removal procedure. For reassembly guides, refer to Figs. 207 and 216 for models 2000 and 3000 and to Figs. 208 and 217 for model 4000.

217. REVERSE IDLER. With the transmission mainshaft assembly removed as outlined in paragraph 216, proceed as follows:

Refer to Fig. 215 and remove reverse idler shaft retaining cap screw from outside of transmission housing. Push the reverse idler shaft forward and remove the 4th-8th shift fork. The 4th-8th coupling-gear can now be removed from rear end of main countershaft.

Push the idler shaft rearward out of housing and remove the reverse idler gear.

To reinstall reverse idler, proceed as follows: Place reverse idler gear in housing with long hub of gear forward and insert idler shaft through housing and gear with small diameter of shaft to rear. Place the 4th-8th sliding coupling on teeth of main countershaft with shifter fork groove to front and insert shifter fork in groove of coupling. Slide the idler shaft forward far enough to install fork, then slide shaft back through fork and install the retaining cap screw with a new rubber coated sealing washer. Tighten the cap screw to a torque of 15-18 Ft.-Lbs.

218. MAIN COUNTERSHAFT. With the transmission mainshaft removed as outlined in paragraph 216 and the reverse idler removed as in paragraph 217, proceed as follows:

Remove the snap ring from secondary countershaft front bearing bore and the snap ring from bore at rear side of main countershaft rear bearing. Then, using suitable drift, drive the main countershaft rearward until both the front and rear bearings are free from their bores and the assembly is resting in bottom of transmission housing. Place suitable wood or soft metal blocks between the cluster gear and rear supporting wall of transmission main compartment, then continue to drive countershaft rearward forcing the cluster gear and front bearing from shaft. Remove the shaft from rear of transmission and remove the cluster gear and bearing from main compartment.

To reinstall main countershaft, reverse the removal procedure.

"SELECT-O-SPEED" (10-SPEED) TRANSMISSION

(All Models)

The "Select-O-Speed" transmission is a planetary gear drive unit providing ten forward and two reverse speeds. Desired gear ratio is selected by moving a control lever and starting, stopping or changing gear ratios is accomplished without operation of a conventional clutch. A foot operated feathering valve is provided for interrupting the gear train in case of emergency or for close maneuvering such as the hitching or unhitching of implements.

On models 2000, 3000 and 4000, the

"Select-O-Speed" transmission is available with three PTO options; it is available without a PTO, with a single speed 540 RPM PTO or with a two speed 540-1000 RPM PTO which also incorporates a ground drive speed PTO.

For a better understanding of the operation of the "Select-O-Speed" transmission, fundamental operating principles of a planetary gear system are outlined in following paragraphs 219 through 226.

PLANETARY GEAR POWER FLOW

All Models

219. Refer to Fig. 218; the three "elements" of a planetary system are the sun gear, pinion carrier and the ring gear. When any element is rotated, the other two elements will also turn unless one is held by an external force. Depending upon which element is held, power can be applied or taken out at

the sun gear, pinion carrier or ring gear. The possible means of obtaining different gear ratios from a planetary gear set are as follows:

220. Turning the sun gear and holding the ring gear forces the pinions to turn within the ring gear moving the pinion carrier with them. Thus, the carrier turns in the same direction as the sun gear but at a slower speed (Underdrive gear ratio).

221. Turning the ring gear and holding the sun gear forces the pinions to turn within the ring gear moving the carrier with them; the pinion carrier turns in the same direction as the ring gear, but at a slower speed (Underdrive gear ratio).

222. Turning the pinion carrier and holding the ring gear forces the pinions to turn within the ring gear causing the sun gear to turn in the same direction as the carrier, but at a higher speed (Overdrive gear ratio).

223. Turning the pinion carrier and holding the sun gear forces the pinions to turn around the sun gear causing the ring gear to turn in the same direction as the carrier, but at a higher speed (Overdrive gear ratio).

224. Turning the sun gear and holding the pinion carrier forces the pinions to act as idlers turning the ring gear in opposite direction from the sun gear and at a slower speed (Underdrive reverse gear ratio).

225. Turning the ring gear and holding the pinion carrier forces the pinions to act as idlers turning the sun gear in the opposite direction from ring gear and at a higher speed (Overdrive reverse gear ratio).

226. Locking any two units of a planetary system together results in a solid drive unit and if any element is turned, all three elements turn in the same direction at the same speed (Direct drive gear ratio).

TRANSMISSION OPERATION
All Models

227. **PLANETARY SYSTEMS.** The "Select-O-Speed" transmission utilizes three planetary systems designated, from front of transmission to rear, as "A", "B" and "C" planetary

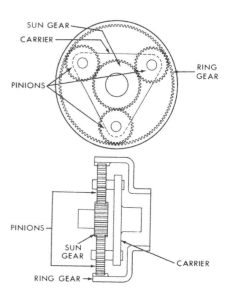

Fig. 218–Drawing showing elements of planetary gear system. Refer to paragraphs 219 through 226 for different gear ratios obtainable from such a gear set.

units. Elements of each unit are designated as "A" sun gear, "A" carrier, "A" ring gear, etc.

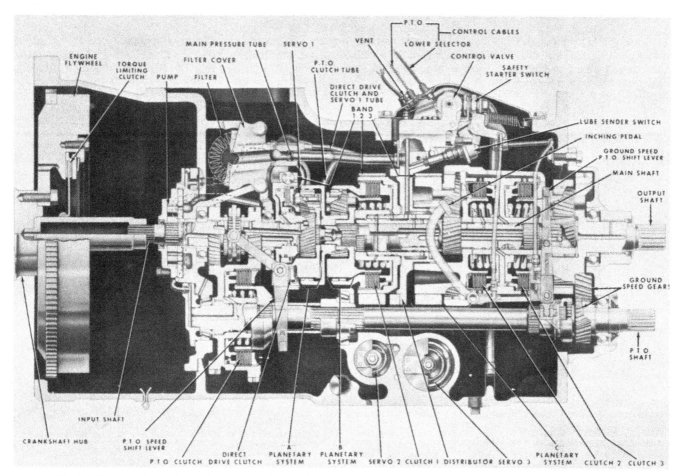

Fig. 219–Cross-sectional view of Select-O-Speed transmission used in models 2000 and 3000. Transmission for model 4000 is similar, but does not have planetary reduction unit at rear of transmission.

On models 2000 and 3000, a fourth "D" (rear) planetary system is used as a final reduction unit only. Power enters at the "D" sun gear, the "D" ring gear is held stationary and power is taken from the "D" carrier which is integral with the transmission output shaft.

Six basic speed ratios, five forward and one reverse, are obtained by various combinations of holding or applying power to the elements of the "B" and "C" planetary units. The "A" (front) planetary unit is used as a direct drive-overdrive unit to double the basic speed ratios providing ten forward and two reverse speeds.

228. PLANETARY CONTROLLING UNITS. Three brake bands and four multiple disc clutches are used to control the "A", and "B", and "C" planetary units to provide the different speed ratios.

The three brake bands are designated, from front to rear, as Band 1, Band 2 and Band 3. The bands are actuated by hydraulic servos designated Servo 1, Servo 2 and Servo 3 to correspond with the band numbers. The servos contain springs that work in the opposite direction from hydraulic pressure. The servos apply the bands with spring pressure and release the bands utilizing hydraulic pressure. The bands, when applied, hold planetary elements stationary as follows:

Band Applied	Planetary Element Held
Band 1	"A" Sun Gear
Band 2	"B" Ring Gear and "C" Sun-Gear
Band 3	"C" Carrier

The multiple disc clutch packs are designated, from front to rear, as the Direct Drive Clutch, Clutch 1, Clutch 2 and Clutch 3. (Clutch 2 and Clutch 3 are contained in the same housing.) The clutches are engaged by hydraulic pressure against a piston within the clutch unit and are disengaged by spring pressure returning the piston to disengaged position. The clutches, when engaged, lock planetary elements together as follows:

Clutch Engaged	Planetary Elements Locked
Direct Drive	"A" Sun Gear to "A" Carrier
Clutch 1	"B" Carrier to "B" Ring Gear
Clutch 2	"B" Carrier to "C" Carrier
Clutch 3	"B" Carrier to "D" Sun Gear (2000 & 3000) or to Output Shaft (4000)

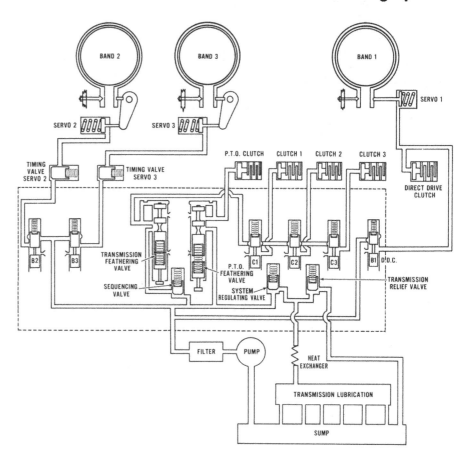

Fig. 220–Schematic diagram of Select-O-Speed transmission hydraulic system. Refer to Fig. 221 for diagram showing operation of feathering valve and to Figs. 222 and 223 for operating principles of the servo 2 and 3 timing valves.

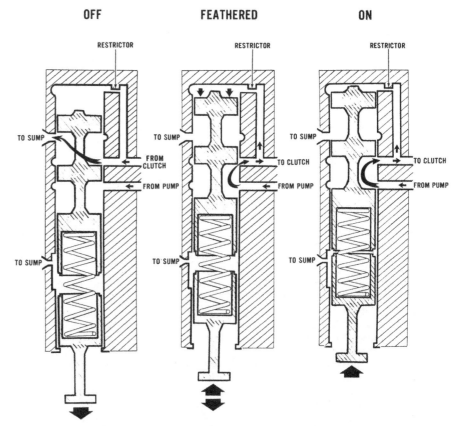

Fig. 221–Cross-sectional views showing transmission and PTO feathering valve assemblies in "OFF", "FEATHERED" and "ON" positions.

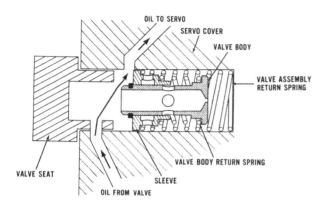

OIL TO SERVO
SERVO COVER
VALVE BODY
VALVE ASSEMBLY RETURN SPRING
VALVE BODY RETURN SPRING
SLEEVE
OIL FROM VALVE
VALVE SEAT

Fig. 222—Cross-section view of Servo 2, Servo 3 timing valve. Valve position during time when hydraulic pressure is being applied to servo is shown. Refer to sequence views in Fig. 223 for valve positions during time servo hydraulic pressure is being released.

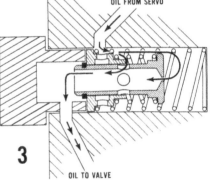

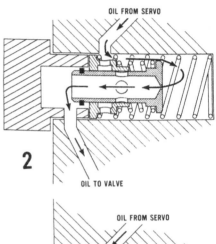

OIL FROM SERVO
OIL TO VALVE
1

OIL FROM SERVO
OIL TO VALVE
2

OIL FROM SERVO
OIL TO VALVE
3

Fig. 223—Sequence views of Servo 2-Servo 3 timing valve during time servo hydraulic pressure is being released, and servo spring pressure is applying band. In view 1, valve position allows a fast initial pressure drop in servo to begin band application. As servo pressure decreases, timing valve spring moves valve to permit a relatively slow continued servo pressure drop as shown in view 2 to permit gradual band application. In view 3, servo hydraulic pressure has decreased far enough to allow timing valve spring to move valve so that a final dumping of oil from servo can take place, completely applying the band by servo spring pressure.

"SELECT-O-SPEED" HYDRAULIC SYSTEM

All Models

229. HYDRAULIC CIRCUITS. Refer to Fig. 220 for diagram of the "Select-O-Speed" hydraulic circuits.

Pressure is supplied directly to the Band 1, Band 2 and Band 3 control valves and PTO feathering valve (direct transmission circuit); indirectly to the Clutch 1, Clutch 2 and Clutch 3 control valves via the transmission sequencing valve and the transmission feathering valve (indirect transmission circuit) and indirectly to the transmission cooling and lubrication circuit via the transmission regulating valve.

Refer to Fig. 221 for diagrams showing operation of the transmission and PTO feathering valves. With the inching pedal (transmission feathering valve control pedal) up or the PTO control handle pulled out, pressure is directed to the indirect transmission circuit or to the PTO clutch. With the inching pedal fully depressed or the PTO control handle pushed in, the feathering valves block pressure from the indirect transmission circuit or from the PTO clutch. With the inching pedal partially depressed or the PTO control handle partly pulled out, pressure in the indirect transmission circuit or PTO clutch can be "feathered" to provide smooth starting of tractor or smooth engagement of the PTO clutch. This is similar to slipping a conventional clutch and the transmission or PTO clutch should not be operated for any period of time with the control in "feathering" position.

The transmission **sequencing valve** separates the direct and indirect hydraulic circuits so that oil is supplied only to the direct circuit when system oil pressure drops below valve setting of 125 psi. This prevents Bands 2 and 3 from locking up the transmission when large quantities of oil are required such as during the 4-5 and 5-4 shifts.

The system **regulating valve** controls pressure within the transmission hydraulic system to 200-210 psi (at 800 engine RPM and oil temperature of 80-120°F.). When pressure in system exceeds the regulating valve pressure setting, the valve opens and oil passes into the transmission cooling (on some models) and lubrication circuit. When system pressure is below regulating valve setting, oil pressure in lubrication circuit drops to zero and the oil pressure warning light will come on. Note: warning light circuit was discontinued in March 1968.

The transmission **relief valve** limits maximum pressure in the hydraulic system to 220-230 psi on models with oil cooling system (heat exchanger) and to 210-220 psi on models without heat exchanger. Oil by-passing the relief valve is returned to the sump via the return tube attached to control valve assembly.

Timing valves are incorporated in the Servo 2 and Servo 3 circuits to permit gradual application of bands and to prevent both Band 2 and Band 3 from being engaged at the same time during the 4-5 and 5-4 shifts. During these shifts, one servo is pressurized to release its corresponding band and pressure is released from other servo to apply its corresponding band. Oil flow to the servo (See Fig. 222) to release the band is not restricted as the valve assembly return spring will be compressed and oil will by-pass the valve. On the other servo return oil will first compress the valve body return spring (Step 1—Fig. 223) and oil will flow through both the valve orifice and the four annular ports. As servo oil pressure begins to drop, the valve body return spring expands moving the valve to close off the four annular ports (Step 2) which slows down the flow of oil from servo and allows band to gradually apply. When servo pressure drops further, the valve body spring expands to full length again opening the four annular ports (Step 3) to provide a final dumping of oil to permit the band to be completely applied.

When the Band 1—Direct Drive Clutch spool in control valve assembly is closed to pressurize the circuit, oil pressure will apply the Direct Drive Clutch as the Servo 1 spring is compressed to release Band 1. Conversely, as the spool is opened to release circuit pressure, the spring in Servo 1 will apply Band 1 as the Direct Drive Clutch is released.

POSITION OR GEAR	DIRECT DRIVE CLUTCH	BAND 1 (B1)	BAND 2 (B2)	BAND 3 (B3)	CLUTCH 1 (C1)	CLUTCH 2 (C2)	CLUTCH 3 (C3)
PARK (P)		A	A	A			
R2		A		A	A		
R1	A			A	A		
NEUTRAL (N)	A						
1ST	A			A			A
2ND	A			A		A	
3RD		A		A			A
4TH		A		A		A	
5TH	A		A				A
6TH	A		A			A	
7TH		A	A				A
8TH		A	A			A	
9TH	A				A	A	
10TH		A			A	A	

Fig. 224–Chart showing transmission units applied at each gear selector position. Bands are applied by servo spring pressure and are released by hydraulic pressure. Clutches are applied by hydraulic pressure and are released by spring pressure. Hydraulic pressure is indicated by shaded area.

230. **CONTROL POSITIONS (GEAR SELECTION).** Fourteen control positions are provided on the gear selector. Units of the transmission that are applied in each position of the gear selector are shown in the Application of Bands and Clutches Chart (Fig. 224).

LUBRICATION AND FILTERS
All Models

231. As the Select-O-Speed transmission incorporates a hydraulic control system, use of correct lubricating oil and keeping the oil clean is of utmost importance.

Recommended fluid is Ford Specification No. M2C41-A transmission and hydraulic oil. Capacity is 13¼ quarts for model 4000 and model 3000 with heat exchanger, 12 quarts for models 2000 and 3000 without heat exchanger. Fluid should be maintained to bottom of oil level plug (See Fig. 225). Add oil through oil level plug openings.

To remove the pump intake screen, first drain oil from transmission, then remove filter screen as shown in Fig. 226.

The pleated paper filter element should be renewed each 600 hours of service and the fluid changed every 2400 hours to coincide with each 4th oil filter change. To remove old filter, unbolt the filter cover (Fig. 227) and unscrew filter from cover. When reinstalling filter cover, tighten retaining bolts to torque of 35-40 Ft.-Lbs.

SYSTEM ADJUSTMENTS
All Models

Malfunction of the "Select-O-Speed" transmission could be from a number of causes, the most common of which is maladjustment of one or more units of the transmission. Therefore, the first step in correcting troubles would be a

Fig. 225–View showing location of transmission oil level–filler plug on models 2000, 3000 and 4000.

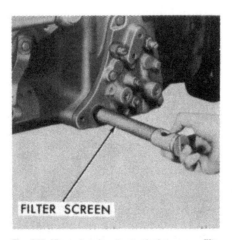

Fig. 226–View showing transmission pump filter screen being removed from model 2000, 3000 and 4000 transmission.

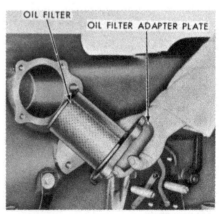

Fig. 227–Removing model 2000, 3000 and 4000 Select-O-Speed transmission oil filter element and adapter plate (cover). Unscrew element from cover.

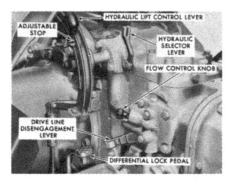

Fig. 228–View showing location of drive line (shaft) disengagement lever on model 4000; lever on models 2000 and 3000 is in similar location.

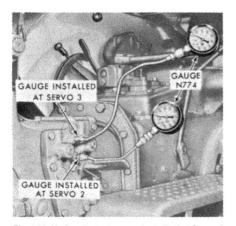

Fig. 229–Hydraulic test gages installed at Servo 2 and Servo 3 pressure check ports in servo cover. Test port for Servo 1 is located in servo cover on right side of transmission case. Model 5000 is shown for illustrative purposes.

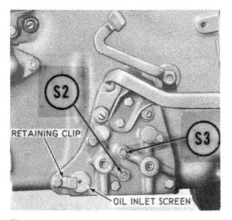

Fig. 230–View showing location of Servo 2 (S2) and Servo 3 (S3) hydraulic pressure test ports in servo cover for models 2000, 3000 and 4000. Refer to Fig. 229 for gages installed at Servo 2 and 3 locations.

complete operational adjustment of the three transmission bands and the pressure control valves within the control valve assembly and correcting any non-alignment of the transmission or PTO controls.

As some of the adjustments are made with the engine running and the gear ratio selector lever in an operational

Fig. 231–View showing method of attaching transmission oil cooler tubes to transmission housing.

1. Inlet tube
1A. Inlet tube (4200)
2. Outlet tube
2A. Outlet tube (4200)
3. Adapter bolts
4. Unions

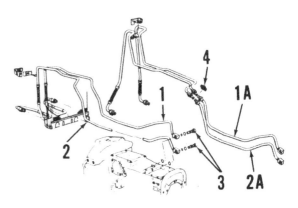

position, the first step is to disengage the traction coupling as follows:

232. **TRACTION COUPLING.** All tractors equipped with a "Select-O-Speed" transmission have a traction coupling sleeve which can be shifted to disengage the transmission output shaft from the differential pinion. The traction coupling shift lever is located as shown in Fig. 228 for 2000, 3000 and 4000 models. Move lever up on models 2000, 3000 and 4000 to disengage the traction coupling.

233. **FLUID LEVEL CHECK.** Before an attempt is made to start or service the tractor, first check the transmission fluid level. To check fluid level, remove pipe plug (See Fig. 225) located on right side of transmission housing. Fluid level should be even with bottom of plug opening with tractor standing level. If fluid level is below bottom of plug opening, add fluid as necessary through the opening. Recommended fluid is Ford M2C41-A transmission and hydraulic oil.

234. **PRESSURE CHECK.** Provisions are made for installation of pressure gages in Servos 1, 2 and 3 hydraulic circuits as shown in Figs. 229 and 230. However, pressure adjustments and system diagnosis can usually be made with gage installed at Servo 2 location only. With a 0-300 psi gage installed at Servo 2, proceed as follows:

Start engine and operate transmission until fluid temperature is 80-120°F. Disconnect the traction coupling as outlined in paragraph 232. With engine running at 800 RPM, place gear selector in Neutral (N) position. Gage reading should then be 200-210 psi. A gage reading higher than 210 psi would indicate need of adjusting the transmission regulating valve as outlined in paragraph 235. If pressure gage reading is below 200 psi, proceed with hydraulic circuit checks as outlined in paragraph 244. If, during the checks, the gage reading remained evenly low, readjust transmission regulating valve as outlined in paragraph 235.

If pressure gage reading with engine running at 800 RPM was in the specified range of 200-210 psi, increase engine speed to 2400 RPM. Pressure gage reading should then be 210-220 psi on models without transmission cooling system (heat exchanger) or 220-230 psi on models with heat exchanger. If pressure is not within the range specified at 2400 engine RPM, readjust transmission relief valve as outlined in paragraph 235.

If transmission lubrication light (on early models so equipped) remains on when pressure gage reading at Servo 2 location is within specified range, check lubrication circuit pressure as follows:

On models without heat exchanger, remove the transmission top cover and the lubrication light sending switch (early models) or plug (late models), then install gage in place of sending switch or plug. With engine running at 800 RPM, gage pressure should be above 14¼ psi. If the pressure gage reading is above specified pressure, renew the sending switch.

On models equipped with transmission oil cooler (heat exchanger), obtain a spare oil line adapter bolt (Ford part No. C5NN-7D192-C, drill through head of bolt and thread to connect a hydraulic pressure gage. Install the drilled adapter bolt and gage in top heat exchanger line (1—Fig. 231). With engine running at 800 RPM, gage reading should be above 22 psi. If gage reading is above that specified, renew lubrication light sending switch.

If lubrication pressure is below specified pressure, a leak in the transmission lubrication circuit is indicated.

235. **CONTROL VALVE ADJUSTMENT.** To adjust the control valve, the valve assembly must first be removed from the transmission (refer to paragraph 251) and the adjusting screw retainers (18 and 19—Fig. 247) be removed from upper valve housing. If valve assembly has been disassembled, turn the adjusting screws in flush with upper valve housing. Then, reinstall control valve and return tube with new gasket and proceed as follows:

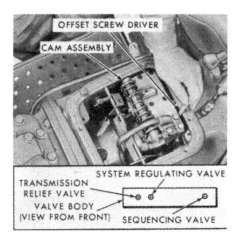

Fig. 232–Adjusting transmission relief valve, regulating valve and sequencing valve on model 2000, 3000 or 4000 with valve mounted in transmission. Refer to text for procedure.

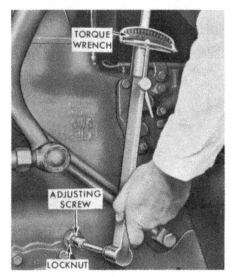

Fig. 233–Adjusting transmission bands with torque wrench. Refer to text for procedure.

With transmission fluid at operating temperature of 80-120° F., a 0-300 psi gage installed at Servo 2 location (Fig. 229 or 230) and with traction coupling disengaged, place valve camshaft in Park position (all control valves out and aligned), ground starter safety switch wire and start engine. Set engine speed to 800 RPM and turn valve camshaft to Neutral position (three detent positions away from Park; Servo valves will be pushed in, clutch valves will be out). With the sequencing valve adjusting screw retainer, turn the transmission regulating valve adjusting screw (See Fig. 232) so that gage reading is 200-210 psi. Set the engine speed at 2400 RPM and with retainer, turn the transmission relief valve adjusting screw so that gage reading is 210-220 psi for models without heat exchanger or 220-230 psi for models with heat exchanger (transmission cooling). Final adjustment for

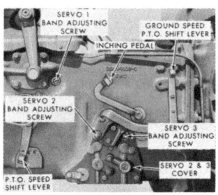

Fig. 234–View showing location of band 1, band 2 and band 3 adjusting screws for models 2000, 3000 and 4000.

the sequencing valve is adjusting screw flush with the upper valve housing.

After final adjustments are made, remove the control valve assembly and reinstall the adjusting screw retainers. If slots in adjusting screws are not aligned for retainer installation, turn the screw in as required (not more than ¼-turn). Reinstall the control valve assembly and the transmission cover as outlined in paragraphs 247 and 251.

236. TRANSMISSION BAND ADJUSTMENT. Before adjusting the transmission bands, operate transmission until fluid temperature is 80-120° F. Disengage the traction coupling as outlined in paragraph 232, refer to Figs. 233 and 234; then, proceed as follows:

BAND NO. 1: Adjust Band 1 with engine stopped. Loosen adjusting screw lock nut at least two full turns while holding adjusting screw from turning. Tighten the adjusting screw to a torque of 19-21 Ft.-Lbs. Check to be sure lock nut did not turn down tight against the sealing washer; if so, back nut off farther and retighten adjusting screw to specified torque. Then, back the adjusting screw out exactly 1¼-turns, hold the adjusting screw stationary and tighten the lock nut to a torque of 20-25 Ft.-Lbs.

BAND NO. 2: Adjust Band 2 with engine running at 800 RPM. Move selector lever to Park position to aid in holding adjusting screw and back off the lock nut at least two full turns while holding screw from turning. Move selector lever to neutral position and tighten the adjusting screw to a torque of 110-130 inch-pounds. Check to be sure lock nut did not turn down tight against sealing washer; if so, back nut off farther and retighten adjusting screw to specified torque. Then, back screw out exactly ¾-turn, move selector lever to Park position to aid in holding adjusting screw stationary and tighten lock nut to a torque of 20-25

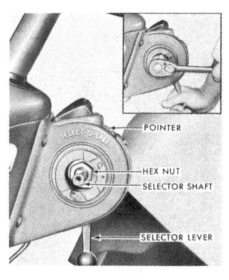

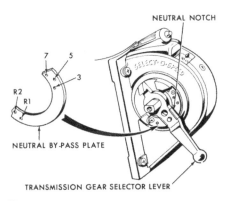

Fig. 235–View showing left end cover removed from gear selector assembly. Inset shows hex nut being loosened to adjust dial pointer.

Fig. 236–Selector lever can be mounted on left side of housing if desired. A neutral bypass plate is available for shuttle work. Numbers indicate possible stop screw locations.

Ft.-Lbs. while holding adjusting screw from turning.

BAND NO. 3: To adjust Band 3, follow the same procedure as outlined for Band 2.

237. SELECTOR DIAL ADJUSTMENT. The individual speed (gear ratio) identifications on the selector dial should be in alignment with the pointer in the selector housing. If not, readjust dial as follows:

Remove the selector shaft end cover from left side of selector unit to expose the shaft and hex nut as shown in Fig. 235. Note: If control lever is mounted on left end of shaft, remove lever and reinstall it on right end of shaft while making adjustment; refer to paragraph 238. Move the selector lever to Neutral (N) position and hold lever firmly in detent notch while loosening hex nut with a deep well socket. The dial can now be moved in either direction to align Neutral (N) identification on dial with pointer. Hold lever to detent and tighten hex nut to torque of 25-35 Ft.-Lbs. Recheck dial alignment; if correct,

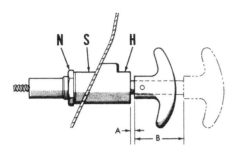

Fig. 237–PTO handle housing (H) is retained to instrument panel by nut (N) and spacer (S). "A" and "B" indicate adjustment measurements; refer to text.

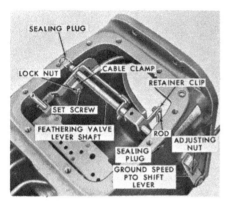

Fig. 238–View showing PTO interlock linkage.

move lever back to left side if desired and reinstall shaft end cover.

238. SELECTOR LEVER POSITION, LEVER STOPS AND NEUTRAL BY-PASS PLATE. The selector lever is normally placed on right side of selector housing for right hand operation; however, if operator prefers, lever can be installed on left side as shown in Fig. 236.

Stop screws can be placed at R2 or R1 and 3rd, 5th or 7th shift positions if desired. When not in use, the stop screws are installed on opposite side of housing from shift lever position.

A neutral by-pass plate can be installed over the neutral notch in housing for shuttle operation. The plate is retained by installing the stop screws through the plate into housing. Install one screw in the R1 or R2 position and the second screw in 3rd, 5th or 7th position as shown in Fig. 236.

239. ADJUST PTO CONTROL. To check adjustment of the PTO control, refer to Fig. 237 and proceed as follows:

With PTO handle pushed in (disengaged position), there should be some clearance (A) between handle and slider housing (H). With a ruler held against hood, measure distance (B) handle moves from disengaged to fully out (engaged) position. If distance is less than 1⅜ inches or more than 1 7/16 inches, the control must be readjusted as follows:

Pull control handle out and disconnect the control cable conduit at both ends. Remove nut (N), lockwasher and spacer (S) from slider housing and pull the handle and slider housing out of hood opening. If there is no clearance between ends of conduit and fittings with handle pulled out, turn handle and slider housing (H) clockwise until there is clearance; then, be sure handle is fully out and turn handle and slider housing counter-clockwise until ends of conduit just contact fittings. Turn handle and slider slightly farther counter-clockwise, if necessary, so that word "PTO" is in vertical position. Reinstall slider housing and reconnect cable conduit. Recheck control handle movement and readjust if necessary.

240. ADJUST PTO INTERLOCK. On models with 2-speed PTO, an interlock mechanism is necessary to prevent

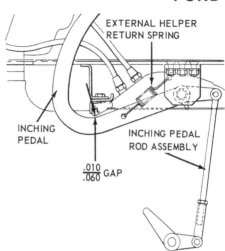

Fig. 239–On model 4200, inching pedal rod must be adjusted to obtain a 0.010-0.060 gap between foot pedal arm and platform at point shown.

GEAR RATIO	DIRECT DRIVE CLUTCH				BAND 1		BAND 2		BAND 3		CLUTCH 1		CLUTCH 2		CLUTCH 3	
	A_h	R_h	A_m	R_m	A	R	A	R	A	R	A	R	A	R	A	R
PARK	P	P	L	P	P	P	P	N	P	N	L	P	L	P	L	P
R2	R1	R2	L	R2	R2	N	L	R2	R2	N	R2	N	L	R2	L	R2
R1	R1	P	R1	N	L	R1	L	R1	R1	N	R1	N	L	R1	L	R1
NEUTRAL	N	P	N	N	L	N	N	N	N	N	N	N	N	N	N	N
1st	1	P	1	N	L	1	L	1	1	N	L	1	L	1	1	N
2nd	2	P	2	N	L	2	L	2	2	N	L	2	2	N	L	2
3rd	1	3	L	3	3	N	L	3	3	N	L	3	L	3	3	N
4th	2	4	L	4	4	N	L	4	4	N	L	4	4	N	L	4
5th	5	P	5	N	L	5	5	N	L	5	L	5	L	5	5	N
6th	6	P	6	N	L	6	6	N	L	6	L	6	6	N	L	6
7th	5	7	L	7	7	N	7	N	L	7	L	7	L	7	7	N
8th	6	8	L	8	8	N	8	N	L	8	L	8	8	N	L	8
9th	9	P	9	N	L	9	L	9	L	9	9	N	9	N	9	9
10th	9	P	L	10	10	N	L	10	L	10	10	N	10	N	10	10

Fig. 240–Operational trouble-shooting chart for Select-O-Speed transmissions for all model 2000, 3000 and 4000 tractors.

Ah—Applied hydraulically and will not release.

Rh—Released hydraulically and will not apply.

Am—Applied mechanically and will not release.

Rm—Released mechanically and will not apply.

A—Applied and will not release.

R—Released and will not apply.

L—Lock-up condition.

possibility of both the engine drive and ground drive PTO from being engaged at the same time. To check interlock adjustment, proceed as follows:

With engine **not** running, push engine drive PTO control knob in and move ground drive PTO lever to "ON" position; it then should not be possible to pull the engine drive PTO control knob out. Move the ground drive PTO lever to "OFF" position and pull engine drive PTO control knob out; it then should not be possible to move the ground drive PTO lever to "ON" position. If both controls can be engaged at the same time, or if neither can be fully engaged even though the other is disengaged, remove top cover from transmission as outlined in paragraph 247, refer to Fig. 238 and proceed as follows: NOTE: Transmission control valve must be installed to adjust PTO interlock although it is not shown installed in Fig. 238.

Loosen lock nut and set screw in cable clamp and move ground drive PTO lever to "ON" position. Then, retighten set screw to clamp cable so that when pulling up on the control cable slider, a 0.005-0.015 clearance remains between end of PTO feathering valve and the cam on feathering valve lever. When adjustment is correct, tighten the lock nut and reinstall transmission top cover. Be sure the engine drive PTO control is properly adjusted as outlined in paragraph 239.

241. **ADJUST INCHING PEDAL** (Model 4200 Only). Refer to Fig. 239; the length of the inching pedal rod assembly must be adjusted to give a 0.010-0.060 gap between the pedal return stop and the pedal shank. Incorrect adjustment may prevent full closing of the transmission feathering valve, thus result in improper operation and/or damage to transmission components.

TROUBLE SHOOTING
All Models

242. **OPERATIONAL (MECHANICAL) CHECK.** If the system adjustments outlined in previous section fail to correct transmission malfunction, the next step in trouble diagnosis would be an operational check. To perform this check, proceed as follows:

Place traction coupling in engaged position, start engine and set engine speed to 800 RPM and depress the inching pedal. Shift the transmission selector to each of the fourteen positions in turn starting with Park (P), gradually release the inching pedal at each position and record the reaction when pedal is released. One of the five following conditions will be encountered.

GEAR RATIO	DIRECT CIRCUIT				INDIRECT CIRCUIT		
	D.D.C.	B1	B2	B3	C1	C2	C3
R2			P		P		
R1	P	P	P		P		
NEUTRAL	P	P	P	P			
1st	P	P	P				P
2nd	P	P	P			P	
3rd			P				P
4th			P			P	
5th	P	P		P			P

Fig. 241–Hydraulic trouble shooting chart for use with gage installed at one or more servo pressure ports. "P" indicates hydraulic pressure applied to unit for different gear ratio selector positions. Refer to text for hydraulic circuit trouble-shooting procedure.

(1) The transmission will operate properly for the control position selected.

(2) The transmission will operate in a different speed ratio than that selected.

(3) The transmission will go into a "neutral" condition in control position other than neutral.

(4) The transmission will go into park condition in control position other than park.

(5) The transmission will lock up and stall the engine.

If transmission seems to operate properly in all fourteen control positions, proceed with TORQUE LIMITING CLUTCH CHECK as in paragraph 243, then the HYDRAULIC CIRCUIT CHECKS as outlined in paragraph 244. If any of the malfunction conditions (2), (3), (4), or (5) are encountered, compare the recorded reactions for each of the control positions with the columns in the operational trouble shooting chart in Fig. 240. The matching column in the trouble shooting chart will indicate the malfunctioning unit and whether the trouble is caused by the unit being continually applied ("A" column) or released ("R" column).

243. **TORQUE LIMITING CLUTCH CHECK.** A defective or worn torque limiting clutch should be suspected if transmission malfunction exists only under heavy loads and in the higher speed ratios, especially if the transmission operates properly in the lower speed ranges.

To check the torque limiting clutch, operate the tractor in 8th, 9th or 10th speed position at wide open throttle and quickly apply both brakes. If tractor forward motion can be halted without pulling the engine down below 1000 RPM or without stalling the engine, renew torque limiting clutch as outlined in paragraph 253.

244. **HYDRAULIC CIRCUIT CHECKS.** Leakage in any of the hydraulic circuits to servos or clutches can be detected by performing the following pressure checks.

With a 0-300 psi hydraulic gage installed at Servo 2 location (see Fig. 229 or 230), transmission fluid temperature at 80-120° F., traction coupling disconnected as outlined in paragraph 232 and with engine running at 800 RPM, record hydraulic gage reading at each selector position shown in the hydraulic trouble shooting chart in Fig. 241 except 5th gear. Normal gage reading for each position is 200-210 psi with a maximum variation of 3 psi. A leak in any one circuit will show up as one of the following conditions; leakage in more than one circuit will show up as a combination of conditions:

A. Low Gage Reading In All Positions—Move gage to Servo 3 location (Fig. 229 or 230) and observe pressure gage reading with selector at 5th gear. A normal gage reading (200-210 psi) would indicate leakage in Servo 2 circuit; a continued low reading would indicate incorrectly adjusted transmis-

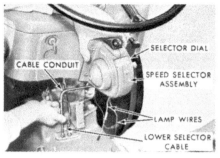

Fig. 242—Removing speed (gear ratio) selector assembly; refer to text.

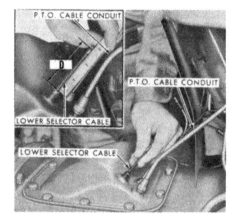

Fig. 243—Measure distance lower selector cable extends from conduit fitting (distance "D"—see inset) before removing cable.

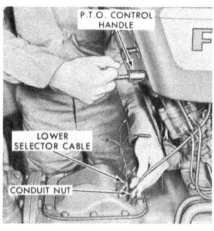

Fig. 244—Removing PTO control cable by turning control handle to unscrew cable from linkage.

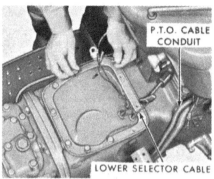

Fig. 245—Screw gear ratio lower selector cable into PTO slider to facilitate top cover installation. Remove cable from PTO slider and screw into control valve cable wheel after cover is installed.

sion regulating valve (refer to paragraph 235), clogged filter and/or worn transmission pump.

B. Low Gage Reading in Positions R2 and R1—Normal reading in other positions and returns to normal in R2 and R1 when inching pedal is fully depressed would indicate leakage in Clutch 1 circuit.

C. Low Gage Reading In Positions R2, N, 1st And 2nd—Normal reading in other positions and depressing inching pedal does not change gage readings would indicate leakage in Servo 1/Direct Drive Clutch circuit.

D. Low Gage Reading In Position N—Normal reading in other positions would indicate leakage in Servo 3 circuit.

E. Low Gage Reading In Positions 1st And 3rd—Normal readings in other positions and gage reading returns to normal in 1st and 3rd when inching pedal is depressed indicates leakage in Clutch 3 circuit.

F. Low Gage Reading In Positions 2 And 4—Normal readings in other positions and gage reading returns to normal in 2 and 4 when inching pedal is depressed indicates leakage in Clutch 2 circuit.

On models 2000, 3000 and 4000 equipped with power take-off, pull PTO

control handle out fully while observing pressure gage; if gage reading falls and remains low, leak is indicated in PTO clutch circuit.

OVERHAUL TRANSMISSION
All Models

CAUTION: The "Select-O-Speed" transmission is a hydraulically controlled unit and merits the same degree of care and cleanliness as for any hydraulic system. Disassembly or service should be attempted only in a clean, dust free shop. Use only lint free paper shop towels to wipe internal transmission parts; lint from cloth shop towels or rags will clog the oil filter screen, etc., and possibly cause serious damage or transmission malfunction.

245. R&R GEAR SELECTOR ASSEMBLY. To remove the gear selector assembly, refer to Fig. 242 and proceed as follows:

Place gear selector in Park (P) position, loosen the conduit retaining nut at top of transmission cover, and remove the four screws retaining selector housing to rear hood panel. Shift the selector to 10th gear position while lifting up on selector housing. Discon-

nect the exposed control cable at lower end of conduit, disconnect the selector lamp and oil pressure warning light wires and remove the selector assembly from tractor.

Prior to turning or removing the lower selector cable, measure the distance (D—Fig. 243) that cable protrudes from conduit fitting and record this measurement for reassembly. The lower cable can now be removed by turning it out of the control valve camshaft wheel, then pulling it out of transmission.

To reinstall selector assembly, first insert lower cable into conduit fitting and thread it into the control valve camshaft wheel until cable protrudes distance (D) measured on disassembly. Note: If measurement was not made on disassembly, or if selector assembly has been disassembled, thread cable into camshaft wheel until measurement (D) is 2¾ inches. Place selector lever in 10th gear position and connect cables at lower end of conduit. Move selector lever to Park position while lowering selector assembly into opening in rear hood panel. Tighten the conduit retaining nut, reconnect light wires and install the four selector housing retaining screws. Check to be sure that selector dial correctly indicates gear ratios; refer to paragraph 237 if adjustment of dial is required.

246. R&R PTO CONTROL ASSEMBLY Refer to Fig. 244 and proceed as follows: Remove the cover from right side of steering gear assembly. Unscrew the conduit nut at each end of control cable conduit and remove the slider housing retaining nut, lockwasher and spacer. Withdraw the slider housing from hood panel and turn the handle and slider assembly clockwise until control cable is unthreaded from linkage in transmission, then pull slider and cable from conduit. Remove the conduit, nut, lockwasher and spacer from under hood panel.

To reinstall, insert cable through hood opening, spacer, lockwasher and nut and slide conduit over the cable. Screw conduit nut onto slider housing loosely and insert lower end of cable into fitting in transmission cover. Turn the slider housing counter-clockwise threading the cable into control linkage. Continue turning slider housing with handle pulled out until the conduit contacts the fitting and letters "PTO" on handle are in vertical readable position. Tighten the conduit nuts, mount slider assembly in hood panel and check adjustment as outlined in paragraph 239. Readjust, if necessary, and reinstall steering gear cover panel.

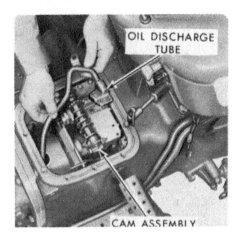

Fig. 246–Removing oil discharge tube on models 2000, 3000 or 4000.

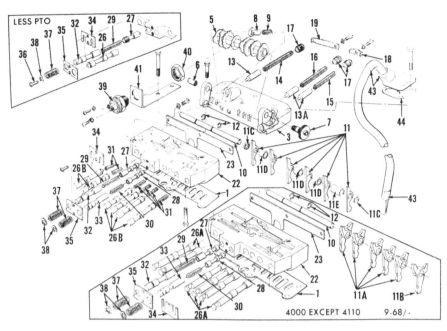

Fig. 247–Exploded view of transmission control valve used on model 2000, 3000 and 4000 tractors. Note different control valve spools used in model 4000 beginning with September 1968 production date. On tractors without PTO, a blocking spool (26) is used to fill PTO feathering valve bore.

1. Gasket	11C. Spacer (.058)	18. Retainer (short)	29. Valve spring
3. Upper valve body	11D. Spacer (.348)	19. Retainer (long)	30. Valve spring
5. Camshaft	11E. Spacer (.878)	22. Lower valve body	31. Valve springs
6. Trunnion	12. Snap rings	23. Valve spring retainer	32. Transmission valve
7. Cable wheel &	13. Sequencing valve	26. Passage blocking	plunger
trunnion	13A. Relief & regulating	spool	33. PTO valve plunger
8. Detent follower	valves	26A. Control valve spools	34. Valve retainer
9. Spring	14. Valve spring	26B. Control valve spools	35. Spring retainer
10. Rocker arm shaft	15. Valve spring	27. Transmission	37. Valve return springs
11. Rocker arms	16. Valve spring	feathering valve	38. Snap rings
11A. Rocker arms	17. Adjusting screws	28. PTO feathering valve	39. Starter safety switch
11B. Rocker arm			

247. R&R TRANSMISSION TOP COVER. First, remove the gear selector assembly as outlined in paragraph 245 and, if equipped with PTO, remove PTO control assembly as outlined in paragraph 246. Unbolt top cover and lift cover up so that starter safety switch wires and oil pressure warning light wire (if so equipped) can be disconnected, then remove cover from transmission.

Before reinstalling cover, be sure the control valve camshaft is in Park position (all valve rocker arms aligned), then proceed as follows: If equipped with PTO, thread the transmission lower selector cable into PTO slider link. Place a new gasket on transmission, position cover so that wires can be threaded under valve camshaft and reconnect starter safety switch wires and oil pressure warning light wire. Insert the cable (see Fig. 245) up through slider guide in cover, position cover on transmission and loosely install cover retaining cap screws. Work the slider up and down with cable and move cover to align slider with guide if necessary. Tighten the cover retaining cap screws to a torque of 20-23 Ft.-Lbs. Remove lower selector cable from PTO slider, insert cable through conduit fitting and thread it into camshaft wheel on control valve. Reinstall selector assembly as in paragraph 245 and the PTO control assembly as in paragraph 246.

249. R&R TRANSMISSION ASSEMBLY. To remove transmission, first split tractor between engine and transmission as outlined in paragraph 181 or 182. Remove gear selector assembly as outlined in paragraph 245 and PTO control assembly as outlined in paragraph 246. Disconnect rear light wires, transmission lube warning light wire and starter safety switch wire, then remove steering gear assembly and fuel tank as a unit from top

of transmission. Refer to paragraph 284 or 299 and remove transmission from rear axle center housing.

Reinstall transmission by reversing removal procedure. Refill with proper lubricant as outlined in paragraph 231.

NOTE: Although complete transmission overhaul requires removal of transmission, most work can be completed after splitting tractor between engine and transmission or between transmission and rear axle center housing.

251. R&R CONTROL VALVE ASSEMBLY. With transmission top cover removed as outlined in paragraph 247, refer to Fig. 246 and proceed as follows: Remove the three cap screws retaining the oil discharge tube manifold and either remove the tube assembly from transmission or move it to rear out of way. Tie the inching pedal down. Remove the two cap screws retaining safety switch bracket, then remove switch and bracket as an assembly. Remove the remaining two valve assembly retaining screws and pry or bump valve assembly loose from gasket. Remove valve assembly from top of transmission. Note: Do not remove the cap screw at left rear corner of valve assembly near the camshaft; this cap screw helps retain valve upper half to lower half only.

To reinstall control valve, place new gasket and valve on oil distributor, then place new oil discharge tube gasket on valve body and install the oil discharge tube retaining cap screws snugly. Note: Lower end of tube clips onto web in bottom of housing; be sure tube is firmly in place and does not contact "C" ring gear. Install the two cap screws in left end of assembly snugly, then turn the camshaft to Park position (all rocker arms aligned and cam for starter switch pointing to rear) and install the safety starter switch assembly so that cam is centered on switch button and the button is fully depressed. Tighten all valve assembly retaining cap screws to a torque of 6-8 Ft.-Lbs. Reinstall transmission cover, selector assembly and, if so equipped, install the PTO control. Refer to paragraphs 247, 246 and 245.

NOTE: If installing control valve assembly for adjustment, be sure the adjusting screw retainers are removed. Also, it is not necessary to install the starter safety switch as the valve must be removed to install the adjusting screw retainers.

252. OVERHAUL CONTROL VALVE. With the control valve assembly removed as outlined in paragraph 251, refer to exploded view of

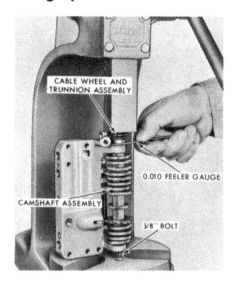

Fig. 249–Press cable wheel and trunnion into camshaft so there is 0.010 clearance between cable wheel and upper valve body. Note ⅜-inch bolt supporting opposite end of camshaft.

Fig. 250–New type camshaft and detent prevents over-shift from Park to 10th gear or vice versa. New camshaft and detent plunger can be installed in place of previous type.

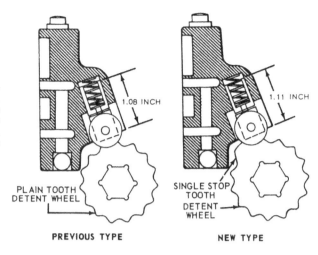

Fig. 251–Exploded view of lower valve body and feathering valve assemblies. Note that PTO plunger (33) has two identification grooves in outer end and PTO feathering valve (29) has shorter land than does the transmission feathering valve (41). PTO feathering valve spring is shorter than transmission feathering valve spring (38).

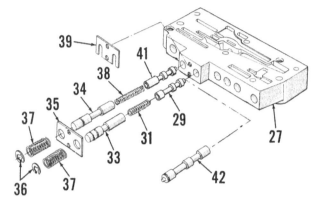

assembly in Fig. 247 and proceed as follows:

NOTE: There are some differences between the valve assembly used in model 4000 beginning with September 1968 production and the assembly used in other models; notably, the late 4000 servo and clutch valves do not have return springs, but are returned by the rocker arms which fit in notches in outer ends of valve spools. However, service procedures and specifications remain essentially the same for all units with any differences noted in the following text:

Remove the two cap screws that hold the halves of control valve assembly together, then separate valve body half (22) from camshaft body half (3). The six control valve spools (26A or 26B) can now be removed. On all models except late 4000, remove the control valve spool return springs (31). Remove the feathering valve retaining plate (34) and withdraw the feathering valve plungers (32 and 33), springs (29 and 30) and feathering valves (27 and 28). NOTE: On models without PTO, a valve spool (26) (same as 26B) is substituted for the PTO feathering valve (28), spring (30) and plunger (33) as these parts are not required. The retainer plate (23) can now be removed from valve body half (22) for better access to clean spool bores. Remove the adjusting screw retainers (18 and 19), adjusting screws (17), springs (14, 15 and 16) and valves (13 and 13A) from camshaft body half (3).

Check the cam followers (rocker arms) (11 or 11A and 11B) and shaft (10) for wear or damage; to disassemble, remove the snap rings (12) and slide shaft from body.

Check the valve camshaft (5), detent

follower (8) and spring (9) and the camshaft cable wheel (7) for wear or damage. To disassemble, thread a ¼-inch diameter, 20 thread, 3 inch long socket head screw (or other hardened steel ¼-inch x 20 bolt with at least 1¾ inches threaded) into the cable wheel (7), hold camshaft with wrench and turn the screw to force cable wheel and trunnion assembly from camshaft. Remove trunnion (6) from opposite end of camshaft using same screw. Note: **Do not** use a common steel or re-threaded bolt for a forcing screw; this type of bolt or screw will twist off in cable wheel or trunnion. The camshaft, detent plunger and spring can now be removed.

Carefully clean all parts in solvent, air dry and check against the following values:

Control Valve Spool Diameter
    Late Model 4000 .... 0.3743-0.3747
    All Other Models 0.3738-0.3742
Feathering Valve Spool Diameter,
    All 0.3743-0.3747
Valve Bore Diameter,
    All 0.3751-0.3758
Spring Free Lengths:
    Camshaft Detent Plunger 0.75
    Control Spool Return 1.24
    Sequencing Valve 2.89
System Regulating Valve 2.47
Transmission Relief Valve 2.06

Transmission Feathering Valve:
    Return 1.22
    Plunger 1.40
PTO Feathering Valve Return .. 1.22
    Plunger 1.12
To reassemble, proceed as follows: Lubricate all parts prior to assembly. Insert detent spring (9) and detent follower (8) in bore of body half (3) and position camshaft (5) so that detent cam is aligned with detent; hold camshaft in place with a ⅝-inch bolt inserted in trunnion (6) end of shaft. Note: The bolt must be long enough to bottom in shaft. Place unit in arbor press as shown in Fig. 249 and press cable wheel into camshaft so there is 0.010 clearance between wheel and valve body. Thread the forcing screw that was used in disassembly into the cable wheel until bottomed; then turn unit over in arbor press and press trunnion (6—Fig. 247), chamfered end first, into camshaft until flush with valve body camshaft support boss.

On late model 4000, insert the shaft (10) through hole in boss of body (3) and install the cam followers (11A and 11B) on shaft as it is pushed into place. Note: Follower (11B) on cable wheel end is constructed with cam arms placed opposite the cam arms of the other five followers (11A). Secure shaft with snap rings (12) inserted in grooves of shaft.

On models other than late 4000, use Fig. 247 as a guide to install shaft (10), cam followers (11) and spacers (11C, 11D and 11E), noting the following: Install followers on shaft with short rounded end seated against cams. Spacers (11C) are 0.058 thick; spacers (11D) are 0.348 thick and spacer (11E) is 0.878 thick. Secure shaft with snap rings (12) inserted in grooves on shaft.

Insert the valves (13 and 13A—Fig. 247) (all three valves are alike) into bores of body (3). Insert spring (14) (2.89 inches long) in bore farthest from cable wheel end. Insert spring (15) (2.06 inches long) in bore nearest cable wheel end and insert spring (16) (2.47 inches long) in middle bore. Thread the adjusting screws (17) (all screws alike) in flush with valve body. Do not install retainers (18 and 19) until valve has been adjusted as outlined in paragraph 235.

Install retaining plate (23) on valve body (22). On models with PTO, install the two snap rings (38) in grooves of the feathering valve plungers (32 and 33), insert the valves through the return springs (37) and plate (35). Compress return springs and install notched plate (34) over valve plungers. Insert PTO feathering valve (28) (valve with narrow land at hollow outer end) into bore towards end of valve body having four control valve bores and the transmission feathering valve (27) (valve with wide land at hollow outer end) into bore towards end of body having two control valve spool bores. Insert the PTO plunger spring (30) (1.12 inches long) in hollow end of valve (28) and transmission plunger spring (29) (1.140 inches long) in hollow end of valve (27). Install the assembled valve plunger, spring and plate assembly so that plunger (33) with identification groove in outer end is in the PTO valve spool bore (towards end of valve body having four control valve spool bores) and secure with screws.

On models without PTO, install the transmission feathering valve as outlined in preceding paragraph; however, valve spool (26) is installed in place of PTO valve (28), spring (30), plunger (33), return spring (37) and snap ring (38).

On all models except late 4000, drop the six spool return springs (31) in the control valve bores.

Lubricate the control valve spools (26A or 26B) and carefully insert them in their bores with a twisting motion to avoid scraping bores with sharp edges of valve lands. On late model 4000, turn the valve spools so that notches in outer ends of spools are positioned as shown in Fig. 247. On late model 4000,

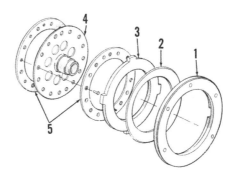

Fig. 252–Exploded view of typical torque limiting clutch assembly. Refer to Fig. 253 for cross-sectional view showing proper installation of the Belleville (spring) washer (2).

1. Clutch housing	4. Clutch disc
2. Spring	5. Facings
3. Pressure plate	

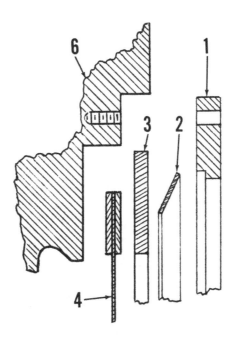

Fig. 253–Cross-sectional view showing placement of Belleville spring washer (2). Flywheel is (6); refer to Fig. 252 for parts identification.

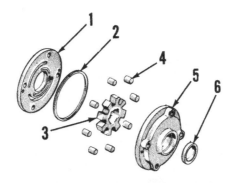

Fig. 254–Exploded view of transmission pump assembly. Unit can be disassembled for inspection as shown; however, only the "O" ring (2) and seal (6) (for models 2000, 3000 & 4000) are available for service.

1. Cover	4. Rollers
2. "O" ring	5. Body
3. Rotor	6. Seal or Plug

place the two valve body halves together so that ends of followers engage notches in valve spools; or, on other models, place valve body halves together so that ends of followers contact ends of valve spools. Install the two cap screws that hold the body halves together and tighten the cap screws to a torque of 6-8 Ft.-Lbs. Refer to paragraph 235 for adjustment of the transmission regulating and relief valves, then install the adjusting screw retainers (18 and 19).

253. TORQUE LIMITING CLUTCH. To remove the torque limiting clutch, first split tractor between engine and transmission as outlined in paragraph 181 or 182. Then, unbolt and remove the clutch assembly from flywheel. Refer to Fig. 252 for exploded view of the clutch assembly.

Renew the clutch disc facings or the clutch disc assembly if facings are glazed, oil soaked or excessively worn. Thickness of new disc with facings is 0.329-0.353 for all models. Note: Late production clutch disc is spring loaded instead of solid disc as shown in Fig. 252.

Renew the Belleville (spring) washer if discolored, cracked, or if it fails to meet the following specifications:

Tractor Model	Color Code	Thickness	Free Height
2000, 3000	Red	0.106-0.109	0.175
4000	Yellow	0.106-0.109	0.189

Note. Free height is measured as dish in the spring washer. Color code is a ⅛-inch wide paint mark on outer edge of the spring washer.

Drive plate thickness on all models should be 0.278-0.282. Renew drive plate if cracked, scored, warped or excessively worn.

Check the splines in the clutch disc hub and on the transmission input shaft for rust or excessive wear. Renew the clutch disc assembly if backlash of disc on input shaft, measured at outer diameter of disc, exceeds ½-inch. Renew the input shaft if backlash of new disc on input shaft exceeds ¼-inch. Before reassembly, be sure splines on input shaft and in clutch disc are clean and free of rust; then, apply a thin film of light silicone grease (Ford part No. M1C-43) to splines of both parts.

When reassembling, be sure that the concave side of the Belleville washer (2) is towards the clutch cover (1) as shown in Fig. 253. The clutch disc is piloted in the flywheel, thus no aligning pilot is required. Tighten the clutch cover retaining cap screws to a torque of 25-30 Ft.-Lbs.

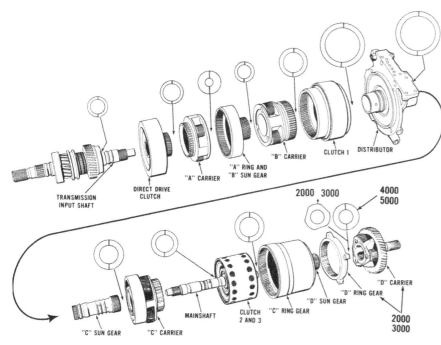

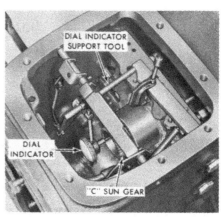

Fig. 256–View showing dial indicator set against "C" sun gear to check transmission front end play on models 2000, 3000 and 4000.

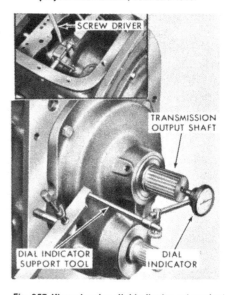

Fig. 255–Exploded view of transmission drive components showing thrust washer placement. Washer between Clutch 1 assembly and Distributor is selective thickness to adjust transmission front end play. Washer between "D" Sun Gear and "D" Carrier (models 2000 & 3000) or between output shaft and rear support (models 4000) is of selective thickness to adjust rear end play.

254. **TRANSMISSION PUMP.** To remove the transmission pump, first split tractor between engine and transmission as outlined in paragraph 181 or 182, then unbolt and remove the pump assembly. Refer to Fig. 254 for exploded view of pump. The transmission input shaft extends through and drives the pump.

The transmission pump, except for rear plate retaining screws, "O" ring (2) and seal (6) is serviced as a complete assembly only. Disassemble, clean and inspect the pump paying particular attention to the cam surface in pump body (5). If cam surface is worn or pitted, renew the pump assembly. If pump is serviceable, reassemble using new "O" ring and seal. Install seal with lip to inside using OTC step plate No. 630-5 or equivalent.

Reinstall pump with new mounting gasket. Be sure pump is correctly positioned and install two retaining cap screws opposite each other. Tighten these screws to torque of 3-5 Ft.-Lbs., then install two remaining screws to torque of 3-5 Ft.-Lbs. Final tighten all four cap screws alternately and evenly to torque of 15-18 Ft.-Lbs.

255. **TRANSMISSION END PLAY.** Transmission end play is controlled by bronze thrust washers placed between each of the rotating members. Refer to Fig. 255. Variation in total length of the transmission components is compensated for during assembly by providing two selective fit thrust washers, one at the rear of each transmis-

sion section, to hold end play within specified limits. Cumulative wear of the thrust washers and of thrust surfaces on other parts will require that the end play be checked at each overhaul, then renewing the selective fit thrust washers with ones of greater thickness to provide correction for wear.

End play should be checked as outlined in paragraphs 256 and 257 before disassembly of the involved transmission components. The overhaul procedures outlined in this manual are based on the supposition that the two selective fit thrust washers are the only ones which will be renewed. However, if any other thrust washers or other transmission parts are renewed, the difference in thickness of the new parts will affect thickness of the selective fit washers to be installed. Thus, it is important to check end play during reassembly to be sure the proper thickness of selective fit thrust washers have been installed.

256. **TRANSMISSION FRONT END PLAY.** To check transmission front end play, all drive components to rear of "C" sun gear and distributor must first be removed and, if the transmission has not been removed from tractor, the steering gear assembly or front compartment cover plate must be removed to provide access to the front transmission components. Loosen Band 1 and Band 2 adjusting screws. Then, mount dial indicator against rear end of "C" sun gear as shown in Fig. 256. Pry "C" sun gear forward and

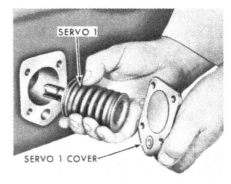

Fig. 257–View showing dial indicator set against transmission output shaft to check transmission rear end play on models 2000, 3000 and 4000. Pry against "C" ring gear as shown in inset to obtain end play reading.

Fig. 258–Removing Servo 1 assembly and cover. Model 5000 is shown for illustrative purposes; however, other models are similar except servo is located at upper side of transmission housing.

set dial indicator to zero while holding slight pressure against the sun gear. Refer to Fig. 255 and insert a screwdriver between the "B" carrier and "A" ring gear and pry "B" carrier rearward. The resulting dial indicator reading is transmission front end play which

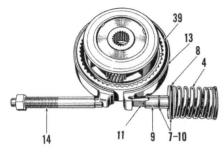

Fig. 259–View showing placement of Servo 1, band 1, adjusting screw and strut to "A" sun gear (direct drive clutch housing).

4. Servo spring	13. Band 1
7. "O" ring	14. Adjusting screw
8. Servo piston	39. "A" sun gear—
9. Piston rod	Direct Drive
10. "O" ring	Clutch housing
11. Strut	

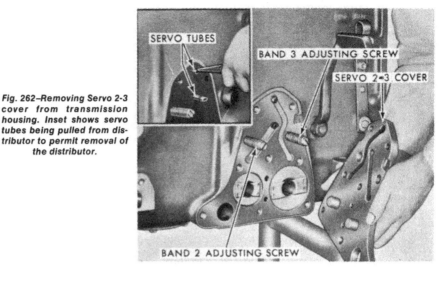

Fig. 262–Removing Servo 2-3 cover from transmission housing. Inset shows servo tubes being pulled from distributor to permit removal of the distributor.

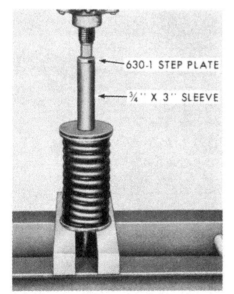

Fig. 260–Exploded view of Servo 1 assembly.

N. Nut	
W. Washer	5. Snap ring
1. Cover	7. "O" ring
2. Gasket	8. Piston
3. Retainer	9. Piston rod
4. Spring	10. "O" ring
	11. Strut

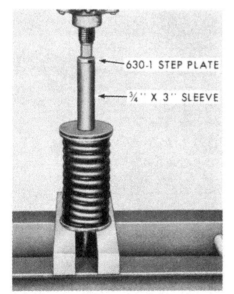

Fig. 261–Compressing servo assembly in hydraulic press to remove nut and washer from piston rod.

should be 0.005 to 0.015. Record the end play measurement for reassembly.

Front end play is adjusted by selecting the proper thickness of thrust washer to be installed between the distributor and clutch 1; refer to Fig. 255. This thrust washer is available in thickness of 0.062 to 0.122 in steps of 0.010.

257. TRANSMISSION REAR END PLAY. Transmission rear end play can be checked after splitting tractor between transmission and the rear axle center housing and removing the top cover. To check rear end play, proceed as follows:

Loosen the Band 3 adjusting screw so that the band is completely released. Mount a dial indicator as shown in Fig. 257. Push in on the transmission output shaft and zero the dial indicator while holding pressure on shaft. Then pry between "C" carrier and "C" ring gear with a screwdriver (see inset in Fig. 257); the resulting dial indicator reading will be transmission rear end play which should be 0.005 to 0.015.

Transmission rear end play is adjusted by selecting the proper thickness of thrust washer to be installed between the shoulder on transmission output shaft and rear support on models 4000, or between the 'D' sun gear and "D" carrier on models 2000 and 3000. This washer is available in thicknesses of 0.102 to 0.142 in steps of 0.010 for models 4000 and in thicknesses of 0.092 to 0.132 in steps of 0.010, and also thickness of 0.140 and 0.150 for models 2000 and 3000.

258. SERVO 1 COVER AND SERVO 1 ASSEMBLY. To remove the Servo 1 Cover and Servo 1 assembly, proceed as follows:

While holding Band 1 adjusting screw from turning, remove the adjusting screw locknut and flat washer.

Then, slowly back adjusting screw out until servo spring tension is relieved. CAUTION: Do not back adjusting screw out farther than necessary as this will allow the Band 1 strut to fall out of place making disassembly of transmission necessary. Remove the four cap screws retaining Servo 1 Cover to transmission and carefully remove cover. Slowly turn adjusting screw in, forcing servo assembly out, until screw is tight. Then, withdraw servo assembly from transmission housing. Refer to Fig. 258.

The servo sealing "O" rings (7 and 10 —Fig. 260) can now be renewed and the servo assembly reinstalled. However, if inspection of unit reveals excessive wear or other defect, proceed as follows to disassemble and overhaul the servo:

NOTE: On early production units, Servo 1 piston rod pilot diameter was 0.730-0.734, and piston outer diameter was 1.993-1.997. If these early parts are encountered, servo operation can be improved by renewing both the rod and piston; service and late production piston rod pilot diameter is 0.734-0.736, and piston outer diameter is 1.981-1.985. The larger rod pilot diameter and the smaller piston diameter reduce the possibility of piston cocking and wedging in bore of transmission housing.

259. To disassemble the servo, place the assembly in a press using sleeve and step plate as shown in Fig. 261. Compress the servo spring and remove the nut (N—Fig. 260) and washer (W) from piston rod (9). Slowly release spring tension, remove unit from press and remove the retainer (3) and spring (4). Remove piston retaining snap ring (6), then push or press piston rod from piston. To reassemble, reverse disassembly procedure. When reinstalling piston, be sure sharp edge of snap ring

is placed away from piston. Install new sealing "O" rings (7 and 10) on piston rod and piston.

260. To reinstall servo, first lubricate the "O" rings with petroleum jelly, then proceed as follows: Insert the assembled servo in bore of housing with notch in end of servo aligned with strut. Push the servo against the strut, then slowly back the band adjusting screw out while pushing servo into housing. When servo contacts inner end of bore and stops moving inward, immediately stop backing the adjusting screw out and reinstall servo cover with new gasket. Tighten the cover retaining capscrews to a torque of 20-25 Ft.-Lbs. Clean the adjusting screw threads, coat threads lightly with a nonhardening plastic lead sealer (Crane Packing Company "Plastic Lead Sealer No. 2" or equivalent); then, turn adjusting screw in to remove free play in linkage. If adjusting screw is equipped with a sealing type nut and flat copper washer, discard the nut and washer and install a plain ⅝-inch flat steel washer (Ford part No. 351502-S) and a plain ⅝-inch X 11 hex nut (Ford part No. 33849-S8). Adjust Band 1 as outlined in paragraph 236.

261. **SERVO 2 AND 3 COVER, TIMING VALVES AND SERVOS 2 AND 3.** The servo cover and servos can be removed with transmission installed in tractor. The servo timing valves are located in the servo cover; refer to exploded view in Fig. 263. Refer to appropriate following paragraphs for removal, overhaul and reinstalling procedure.

262. **R&R SERVO 2 AND 3 COVER.** On all models, drain the transmission and remove the inlet filter screen as outlined in paragraph 231. On all except models 4200, remove the left step plate (foot rest). Hold Band 3 adjusting screw from turning and remove locknut and flat washer from screw. Slowly back the adjusting screw out until Servo 3 spring tension is relieved. CAUTION: Do not back adjusting screw out any farther than necessary to relieve spring tension; if screw is excessively loosened, the band struts may fall out which would require transmission disassembly. Similarly remove Band 2 adjusting screw locknut and flat washer, then back Band 2 adjusting screw out, observing same caution as for Band 3 adjusting screw, until Servo 2 spring tension is relieved. Remove the cap screws retaining Servo 2 and 3 Cover to transmission, insert pry bar in relief behind front or rear edge of cover and pry cover loose; refer to Fig. 262. CAUTION: Do not attempt to remove the

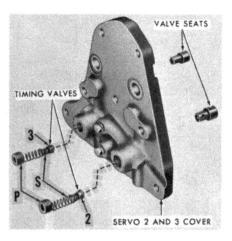

Fig. 263–Exploded view of Servo 2-3 cover assembly. Refer to Fig. 264 for cross-sectional view of timing valves (2 and 3) and valve seats in installed position.

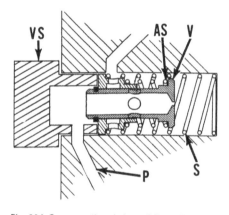

Fig. 264–Cross-sectional view of Servo 2-3 cover showing timing valve (V), valve return spring (S), valve assembly spring (AS), oil passage (P) and timing valve seat (VS). Note that hole in side of valve seat must align with oil passage; refer also to Fig. 265.

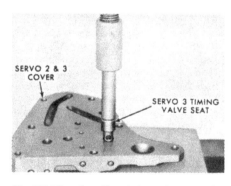

Fig. 265–When installing timing valve seats, be sure that hole in seat is aligned with oil passage in Servo 2-3 cover; refer to cross-sectional view in Fig. 264.

Servos 2 and 3 without first referring to procedure outlined in paragraph 265.

263. Before reinstalling servo cover, clean the threads of the Band 2 and 3 adjusting screws and apply a light coat of non-hardening plastic lead sealer (Crane Packing Company "Plastic

Lead Sealer No. 2" or equivalent) to the screw threads. Install servo cover with new gasket and alternately and evenly tighten the retaining cap screws to a torque of 35-40 Ft.-Lbs. If adjusting screws are equipped with sealing (self-locking) type nuts and copper washers, discard the nuts and washers and install plain steel ⅝-inch flat washers and plain ⅝-inch X 11 hex nuts. Reinstall inlet screen filter with new "O" ring. Tighten the band adjusting screws to remove free travel in the linkage. Refill transmission with proper fluid as outlined in paragraph 231, then adjust the bands as outlined in paragraph 236.

264. **SERVO TIMING VALVES.** The Servo 2 and 3 Cover assembly contains the Servo Timing Valves and Valve Seats; refer to exploded view in Fig. 263. To remove the timing valves, remove the socket head plugs (P) from outer side of cover, then remove the springs (S) and timing valve assemblies (2 and 3). Note that the Servo 3 timing valve is located to front of cover opposite the No. 2 Servo and the Servo 2 timing valve is located towards rear cover opposite No. 3 Servo.

With timing valves removed, remove the valve seats by using a ¼-inch diameter steel rod to press them out towards inside of cover. Refer to cross-sectional view of the cover, timing valve and seat in Fig. 264. When installing new valve seats, press them into servo cover so that hole in side of seat is aligned with passageway (P) in cover. Refer to Fig. 265 for valve seat installation.

The timing valves are serviced as complete assemblies only. Renew the valve assembly if any part is damaged in any way; refer to paragraph 229 for timing valve operation. The color of the valve assembly spring (AS—Fig. 264) correctly identifies the timing valve for tractor model and Servo 2 or Servo 3 application as follows:

Servo 2 Timing Valve Assembly Color Code:

Model 2000	Blue
Model 3000	Blue
Model 4000 (Except 4110 L.C.G.)	Green
Model 4110 L.C.G.	Blue

Servo 3 Timing Valve Assembly Color Code:

Model 2000	Red
Model 3000	Red
Model 4000 (Except 4110 L.C.G.)	Green
Model 4110 L.C.G.	Red

After selecting proper timing valve (spring) color code, install timing

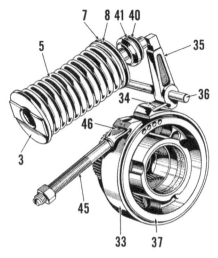

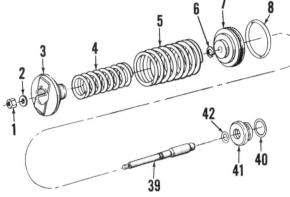

Fig. 269–Exploded view of Servo 3 assembly and piston rod guide.

1. Nut
2. Washer
3. Retainer
4. Inner spring
5. Outer spring
6. Snap ring
7. Piston
8. "O" ring
39. Piston rod
40. "O" ring
41. Guide
42. "O" ring

Fig. 266–View showing Servo 3, band lever, strut, Band 3 and adjusting screw in relationship as they are installed in transmission. Servo 2–Band 2 installation is similar.

3. Retainer	36. Pivot pin
5. Outer spring	37. "C" carrier
7. Piston	40. Guide
8. "O" ring	41. "O" ring
33. Band 3	45. Adjusting screw
34. Strut	46. Strut
35. Lever arm	

valves in servo cover as shown in Fig. 263. Tighten the socket head retaining plugs securely after installing timing valve assemblies (V—Fig. 264), snap ring end first, and springs (S).

265. **R&R AND OVERHAUL SERVOS 2 AND 3.** With Servo 2 and 3 Cover removed as outlined in paragraph 262, proceed as follows: Turn

Band 3 adjusting screw in slowly until tight while pushing against outer end of Servo 3. When Band 3 adjusting screw is tight, withdraw Servo 3 from housing (refer to Fig. 267). With Servo 3 removed, withdraw the piston rod seal guide with hooked wire tool as shown in Fig. 268. Follow same procedure as outlined for Servo 3 and seal guide to remove Servo 2 and seal guide. CAUTION: If either Servo 2 or Servo 3 is removed from housing without first turning the appropriate band adjusting screw in tight, the band struts may fall out of place requiring disassembly of transmission for reinstallation.

266. To disassemble Servo 2 or Servo 3, refer to procedure outlined in paragraph 259 for Servo 1, and also to Fig.

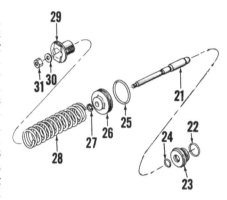

Fig. 270–Exploded view of Servo 2 assembly and piston rod guide.

21. Piston rod	27. Snap ring
22. "O" ring	28. Spring
23. Guide	29. Retainer
24. "O" ring	30. Washer
25. "O" ring	31. Nut
26. Piston	

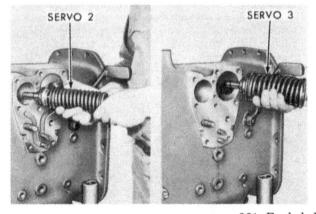

Fig. 267–Removing Servo 2 and Servo 3 from model 5000 transmission. Servos are located at bottom of transmission housing on other models, refer to Fig. 268.

261. Exploded view of the Servo 2 assembly is shown in Fig. 270 and Servo 3 assembly in Fig. 269. Note that Servo 3 has an inner spring (4) and an outer spring (5). Check servo springs against the following specifications:

Models 2000, 3000 and 4110 L.C.G.:
　Servo 2 Spring:
　　Free Length .... 6.54 inches
　　Lbs. @ 4.92 inches 225-275
　Servo 3 Inner Spring:
　　Free Length .... 6.54 inches
　　Lbs. @ 4.92 inches 225-275
　Servo 3 Outer Spring:
　　Free Length .... 9.90 inches
　　Lbs. @ 4.88 inches 467-515

Models 4000 (Except 4110 L.C.G.)
　Servo 2 spring:
　　Free Length .... 7.22 inches
　　Lbs. @ 4.36 inches 375-423
　Servo 3 Inner Spring:
　　Free Length .... 7.22 inches
　　Lbs. @ 4.36 inches 375-423
　Servo 3 Outer Spring:
　　Free Length .... 10.31 inches
　　Lbs. @ 4.88 inches 505-556

To reassemble servos, follow procedure as outlined for Servo 1 in paragraph 259, making sure that sharp edge of snap ring (6—Fig. 269 or 27—Fig. 270) is placed away from piston. Install new sealing "O" rings on pistons and piston rod guides and in guide bores. Lubricate the "O" rings with petroleum jelly and install guides over the piston rods with small diameter of guides away from pistons. Insert the assemblies in their bores. Push against outer end of Servo 3 while slowly backing out the Band 3 adjusting screw; stop turning adjusting screw immediately when servo contacts inner end of bore and stops moving inward. Repeat procedure as outlined to install Servo 2 assembly. CAUTION: If either the Band 2 or Band 3 adjusting screw is

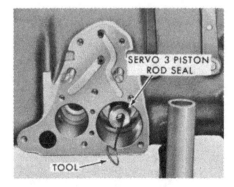

Fig. 268–Removing servo guides with wire hook on models 2000, 3000 & 4000.

Fig. 271–Removing transmission rear support with jack screws.

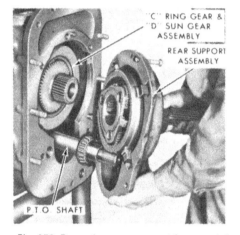

Fig. 272–Removing rear support from model 2000 or 3000 with single speed PTO.

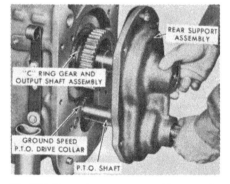

Fig. 273–Removing rear support from model 4000 with two speed PTO

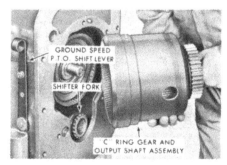

Fig. 274–Removing "C" ring gear and output shaft assembly from model 4000 transmission.

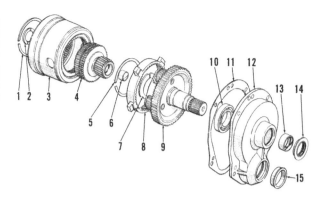

Fig. 275–Exploded view of rear support, "D" planetary and "C" carrier on models 2000 and 3000 with two-speed power take-off. Models with single speed PTO or without PTO are similar except for "D" carrier not having a PTO drive gear.

1. Snap ring
2. Thrust washer
3. "C" ring gear
4. "D" sun gear
5. Snap ring
6. Bushing (not serviced)
7. Selective fit thrust washer
8. "D" ring gear
9. "D" carrier-output shaft
10. Ball bearing

11. Gasket
12. Rear support plate
13. Bushing (not serviced)

13. Oil seal
15. Oil seal or plug

backed out farther than necessary, the band struts may fall out of place. Reinstall Servo 2 and 3 Cover as outlined in paragraph 263.

267. **R&R REAR SUPPORT, OUTPUT SHAFT AND "C" RING GEAR (Includes "D" Planetary On Models 2000 And 3000).** To remove rear support, first split tractor between transmission and rear axle center housing as outlined in paragraph 284 or 299, drain transmission, then proceed as follows:

On models equipped with 2-speed PTO, remove transmission top cover, hold the hex nut retaining PTO interlock cable housing and unscrew the rear support top retaining bolt, then remove the hex nut and lockwasher from inside of housing. On other models, cap screws are used at all five bolting points. Remove the cap screws, then thread two cap screws into the tapped holes in rear support for use as jack screws (See Fig. 271) to remove rear support from transmission housing. Then, remove rear support as shown in Fig. 272 or 273. If not removed with rear support, remove the

output shaft (or "D" sun gear) and "C" ring gear assembly as shown in Fig. 274. Refer to Fig. 275 or 276 for exploded view of the rear support, output shaft ("D" planetary on models 2000 and 3000), and "C" ring gear; be careful not to lose the thrust washer (2) located between output shaft (or "D" planetary) and the Clutch 2 and 3 assembly.

To remove PTO shaft assembly from rear support, refer to paragraph 277 for models with 2-speed PTO. PTO shaft oil seal for all single speed PTO models can be renewed at this time; install new seal with lip to inside (forward) using a suitable driver.

Remove the "D" carrier and "D" ring gear (or output shaft and "C" ring gear) from rear support. On model 4000 with 2-speed PTO, renew needle bearing (10 —Fig. 276) in rear support if bearing is excessively worn or damaged; drive or press on lettered end of bearing cage only when installing new bearing. On models 2000 and 3000, inspect the "D" carrier assembly; planet gear end play should be 0.010-0.028. Renew "D" carrier if any gear is excessively loose or damaged. If "D" carrier is serviceable,

Fig. 276–Exploded view of rear support, output shaft and "C" ring gear assemblies for model 4000 transmission. Top view shows assembly used with two speed PTO; bottom view shows assembly used with single speed PTO or without PTO. Bushings (13) in rear support plate are not serviced separately from plate.

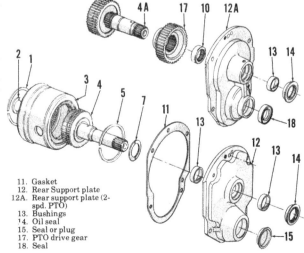

1. Snap ring
2. Thrust washer
3. "C" ring gear
4. Output shaft
4A. Output shaft (2 spd. PTO)
5. Snap ring
7. Thrust washer
10. Needle bearing

11. Gasket
12. Rear Support plate
12A. Rear support plate (2-spd. PTO)
13. Bushings
14. Oil seal
15. Seal or plug
17. PTO drive gear
18. Seal

Fig. 277—View showing selective fit thrust washer installed on output shaft. Model 5000 is used for illustrative purposes.

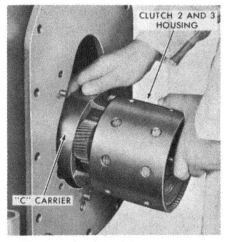

Fig. 278—Removing transmission mainshaft, Clutch 2 and 3 assembly and "C" carrier from transmission.

Fig. 279—Removing "C" sun gear and sealing rings from transmission.

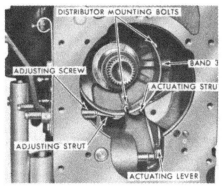

Fig. 280—View showing Band 3 and strut installation.

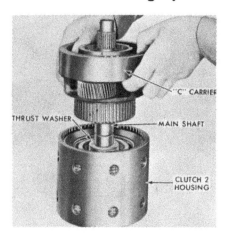

Fig. 281—Removing "C" carrier from mainshaft.

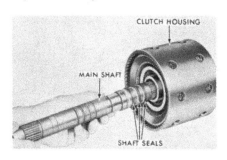

Fig. 282—Removing mainshaft from Clutch 2 and 3 housing.

renew ball bearing (10—Fig. 275) if bearing is worn or rough. Bushings (13—Fig. 275 or 276) are not serviced separately; renew rear support if bushings are excessively worn or scored.

Bushing (6—Fig. 275) in "D" sun gear (models 2000 and 3000) is not serviced separately; renew gear if bushing is excessively worn or scored. Inspect output shaft (model 4000) bearing journals; renew output shaft if journals are excessively worn or rough. To remove "D" sun gear or output shaft, remove rear snap ring (5—Fig. 275 or 276) and push gear or shaft forward out of "C" ring gear. Note: Later production units will have ends of snap ring (5) brazed together; break weld with chisel, then remove and discard the snap ring. To reassemble, install new snap ring (1) in front groove in sun gear or output shaft and insert gear or shaft through "C" ring gear from front side. Install new late type snap ring (5) in rear groove and braze ends of snap ring together using a 1/16-inch bronze rod. Be careful not to overheat ring gear and sun gear or output shaft and remove all brazing flash.

To reassemble transmission, proceed as follows: Install new output shaft oil

seal in rear support using suitable driver so that flange of seal (14—Fig. 275 or 276) contacts rear support evenly. Use a light film of grease to stick the thrust washer (2) over Clutch 2 and 3 retaining snap ring with counterbore of washer over the snap ring. Carefully install the assembled "C" ring gear and "D" sun gear or output shaft over the Clutch 2 and 3 assembly so that "C" ring gear engages gears on "C" carrier. On model 4000, place the selective thickness thrust washer (7—Fig. 276) over end of output shaft as shown in Fig. 277. On models 2000 and 3000, stick the selective thickness thrust washer (7—Fig. 275) in front end of "D" carrier with a light film of grease.

On models 2000 and 3000, install the "D" carrier assembly in rear support taking care not to damage the oil seal, then place "D" ring over the "D" planet gears so that ears on ring gear engage notches in rear support.

Install the rear support using a new gasket (11) taking care not to damage any seals passing over output shaft or PTO shaft. On models with 2-speed PTO, be sure the long cap screw is installed at top and enters PTO interlock cable support bracket. On models with single speed PTO, tighten the lower cap screws at each side of PTO shaft first. On all models, tighten the cap screws to a torque of 35-40 Ft.-Lbs. On 2-speed PTO models, install lock washer and hex nut on top cap screw. Check transmission rear end play as

outlined in paragraph 257 and, if necessary, renew the selective thickness thrust washer with new washer of suitable thickness to bring end play within correct limits. Reinstall top cover as outlined in paragraph 247.

268. **MAINSHAFT, CLUTCH 2 AND 3 ASSEMBLY, "C" CARRIER, "C" SUN GEAR AND BAND 3.** After removing the rear support, "D" planetary (or output shaft) and the "C" ring gear as outlined in paragraph 267, the mainshaft, "C" carrier and Clutch 2 and 3 assembly can be removed as a unit. Back out Band 3 adjusting screw until Band 3 is loose, then remove mainshaft, Clutch 2 and 3 assembly and "C" carrier as shown in Fig. 278. The "C" sun gear and sealing rings can then be removed from the distributor; refer to Fig. 279. Refer to Fig. 280, compress Band 3 and remove the band, adjusting strut and actuating strut. Remove thrust washer (30—Fig. 283) from rear face of distributor.

Remove the four cast iron sealing rings (14—Fig. 284) from front end of mainshaft (13) and slide the "C" carrier (37—Fig. 283) and thrust washer (31) off shaft as shown in Fig. 281. Remove the snap ring (16—Fig. 284) from rear end of mainshaft and remove the mainshaft from Clutch 2 and 3 assembly as shown in Fig. 282. To overhaul the removed Clutch 2 and 3 assembly, refer to exploded view in Fig. 284 and to paragraph 270.

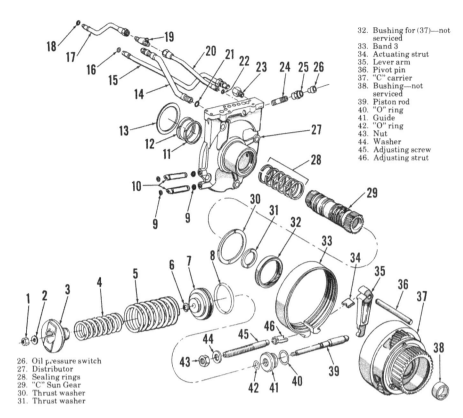

32. Bushing for (37)—not serviced
33. Band 3
34. Actuating strut
35. Lever arm
36. Pivot pin
37. "C" carrier
38. Bushing—not serviced
39. Piston rod
40. "O" ring
41. Guide
42. "O" ring
43. Nut
44. Washer
45. Adjusting screw
46. Adjusting strut

26. Oil pressure switch
27. Distributor
28. Sealing rings
29. "C" Sun Gear
30. Thrust washer
31. Thrust washer

Fig. 283—Exploded view of oil tubes, distributor, "C" Carrier and Sun Gear, Band 3 and Servo 3 assemblies.

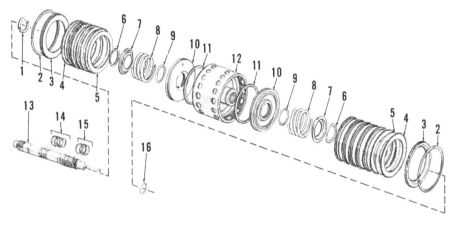

Fig. 284—Exploded view of Clutch 2 and 3 assembly and transmission mainshaft.

1. Thrust washer	5. Internal spline plates	9. "O" rings	13. Mainshaft
2. Snap rings	6. Snap rings	10. Clutch pistons	14. Front sealing rings
3. Pressure plates	7. Retainers	11. "O" rings	15. Rear sealing rings
4. External spline plates	8. Piston return springs	12. Clutch housing	16. Snap ring

The "C" carrier is serviced as a complete assembly only. Check to see that the "C" carrier pinions are in good condition and that pinion shafts are tight in carrier. Pinion end play in carrier should be 0.010-0.028. Also check end thrust surfaces and bushing in carrier; renew carrier as an assembly if defect is noted.

Renew Band 3 if friction material inside band is worn, pitted or eroded or if band has been overheated. Also inspect the strut sockets at band ends for cracks or other damage.

269. To reassemble transmission, proceed as follows: Install Band 3 and the two struts as shown in Fig. 280, making sure that end of flat (actuating) strut having notch is towards the band and the end of band having an identifying notch is towards the adjusting screw strut. Install thrust washer (30—Fig. 283) on rear of distributor. Install new sealing rings (28) on "C" sun gear, lubricate the rings with petroleum jelly, align ring end

gaps along top side of sun gear and carefully insert sun gear through distributor. Rotate sun gear slightly from side to side to align splines on front end of gear with splines in Clutch 1 housing.

Place thrust washer (1—Fig. 284) over front end of mainshaft (13), insert mainshaft through the "C" carrier (37 —Fig. 283), then place thrust washer (31) over front end of mainshaft and against "C" carrier. Carefully install four new sealing rings (14—Fig 284) on front end of mainshaft, lubricate rings with petroleum jelly, align ring end gaps along top side of mainshaft and holding the carrier and shaft assembly with ring end gaps up, carefully insert shaft through "C" sun gear. Work the shaft slowly up and down and from side to side while pushing forward to allow the rings to enter "C" sun gear. Rotate mainshaft slightly from side to side to engage splines on front end of shaft with splines in the "B" carrier. When "C" carrier and mainshaft are in place and fully forward, tighten Band 3 adjusting screw.

Install three new sealing rings (15) on rear end of mainshaft, lubricate the rings with petroleum jelly and align ring end gaps at top side of shaft. Carefully install the Clutch 2 and 3 assembly over mainshaft to avoid breaking sealing rings. Partly support clutch housing and rotate housing slightly from side to side while lightly pushing forward on housing to align Clutch 2 discs with splines on "C" carrier. When housing is in position against the thrust washer and "C" carrier, pull mainshaft rearward while lifting up on clutch housing so that snap ring (16) can be installed in groove on rear end of mainshaft.

270. **OVERHAUL CLUTCH ASSEMBLIES (Except Direct Drive Clutch).** The following service procedure will apply to all transmission clutch assemblies except the Direct Drive Clutch. The illustrations (Figs. 285 through 289) are of the 2-speed PTO clutch; however, the other clutch assemblies are of similar construction. Refer to Fig. 284 for exploded view of the Clutch 2 and 3 assembly; to Fig. 250 for Clutch 1 assembly and to Fig. 305 or 306 for PTO clutch assemblies. To disassemble clutch, proceed as follows:

Remove the pressure plate retaining snap ring as shown in Fig. 285. NOTE: Ends of snap ring in the Clutch 1 and Clutch 2 and 3 assemblies are welded together; break weld with sharp chisel, then remove and discard snap ring. On Clutch 1 assembly, the "B" ring gear also functions as the clutch pressure

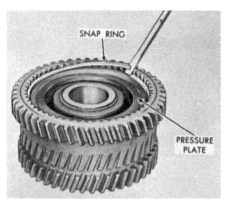

Fig. 285–Removing clutch pressure plate retaining snap ring.

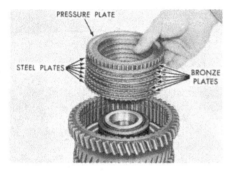

Fig. 286–Bronze plates and steel plates are alternately placed between pressure plate and clutch piston.

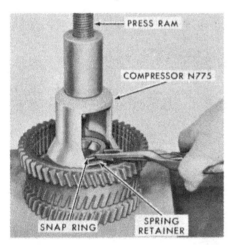

Fig. 287–Clutch piston return spring retaining snap ring being removed. Special tool (Nuday No. N-775) is used in press to compress piston return spring.

plate. After removing snap ring, remove the pressure plate and the bronze and steel discs; refer to Fig. 286.

With the pressure plate and clutch discs removed, place clutch assembly in a press and compress the piston return spring using Nuday tool No. N-775 as shown in Fig. 287. With spring compressed, remove retaining snap ring and slowly release the spring to prevent cocking retaining washer on clutch housing. Remove the return spring, retaining washer and snap

ring, then remove piston using compressed air in port to piston as shown in Fig. 289. Remove the "O" rings from inside and outside diameters of piston.

Clean all parts in solvent, air dry and inspect for wear or other damage. Renew clutch pressure plate if cracked, scored or showing signs of overheating. Fit the piston into clutch housing to check for any binding condition; minor imperfections that cause binding or would cause damage to piston "O" rings can be removed with fine emery cloth. Check the clutch discs for wear, signs of overheating or warping. All clutch discs, except PTO clutch steel discs, should be flat. Renew any bronze discs that will snap over center. Check the PTO clutch steel discs for proper coning as shown in Fig. 290; discs should be coned 0.015 to 0.020. Renew any PTO clutch steel discs coned less than 0.015. Check clutch housing, especially in hub area, and the piston for cracks. Lubricate all parts with transmission fluid and reassemble as follows:

Install new "O" rings in inside diameter and on outside diameter of piston, lubricate with petroleum jelly and install piston, flat side in, in the clutch housing. Be sure piston is seated in housing, then place housing in press. Place piston return spring, retaining washer and snap ring on piston, compress the spring using Nuday tool No. N 775 as shown in Fig. 287 and install the retaining snap ring in groove of clutch housing.

On model 4000 Clutch 3 assembly only, place one of the bronze discs that is lined on one face only next to piston with unlined face down (against piston). Then, alternately place steel and double faced bronze discs. Install the other single faced bronze disc on top of last steel disc with unlined face up (against pressure plate). Install pressure plate with machined surface down (against unlined face of top bronze disc).

On all other clutch assemblies, place a steel disc on top of piston, then alternately install bronze and steel discs. Install pressure plate with machined face down against the top bronze disc. NOTE: On PTO clutch assemblies, install all steel discs with concave (hollow) side up towards pressure plate.

After installing pressure plate, install the retaining snap ring. On all clutch assemblies except the PTO clutch, be sure snap ring end gap straddles two teeth of housing, then weld ends of snap ring together using an AWS 312-16 Electrode. CAUTION: Do

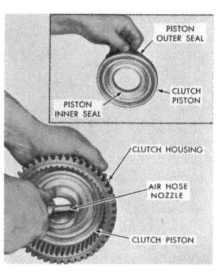

Fig. 289–With piston return spring removed, remove clutch piston with air pressure as shown above. Inset shows piston inner and outer seals ("O" rings).

Fig. 290–Checking steel clutch plates for coning with feeler gage; refer to text.

not attempt to weld snap ring to pressure plate and be careful not to overheat plate or housing. Remove all weld flash.

NOTE: In some early production model 4000 PTO clutch assemblies, a special shim was fitted between the clutch piston and clutch housing; refer to Fig. 291. This shim is not available for service; if present, it must be protected against damage and reinstalled if the PTO clutch housing (driven gear assembly) is to be reused. If new housing is being installed and relief groove (see Fig. 291) is 0.095 or less, the shim may be discarded and the piston be installed without a shim. The shim is necessary only when using a housing with a 0.120 wide relief groove in the model 4000 tractor.

271 **DISTRIBUTOR, CLUTCH 1 AND "B" CARRIER.** The oil distributor, Clutch 1 assembly and the "B" carrier can be removed after removing the control valve assembly as in paragraph 251, the mainshaft "C" sun gear and Band 3 as outlined in paragraph 268 and the Servo 2 and 3 Cover as out-

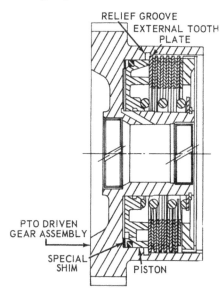

Fig. 291–Cross-sectional view of PTO clutch assembly showing special shim located in some clutch assemblies; refer to text.

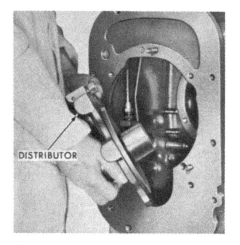

Fig. 292–Removing oil distributor separately from Clutch 1 and "B" carrier assembly; refer to text and also to Fig. 294.

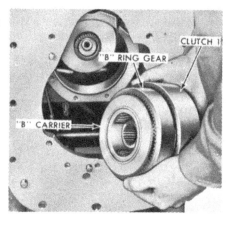

Fig. 293–Removing Clutch 1 and "B" carrier as an assembly after removing oil distributor as shown in Fig. 292. Refer also to Fig. 294.

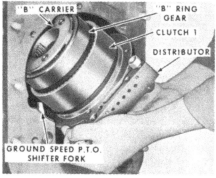

Fig. 294–Removing oil distributor, Clutch 1 and "B" carrier as a unit. Refer also to Figs. 292 and 293.

292. Then, remove Clutch 1 assembly and "B" carrier as a unit as in Fig. 293. On models without PTO or with 2-speed PTO, the distributor, Clutch 1 assembly and "B" carrier can be removed as a unit; refer to Fig. 294.

Remove "B" carrier from Clutch 1 assembly and remove the clutch assembly from distributor if not removed separately. Refer to paragraph 270 for Clutch 1 assembly overhaul instructions. Inspect the pinion gears in "B" carrier for loose or rough bearings or for loose journal pins in carrier. Pinion end play should be 0.010-0.028. Check the thrust and bearing surfaces of distributor for scoring or excessive wear. Renew parts as necessary and reassemble as follows:

Place "B" carrier on work bench, pinion side down, and stick thrust washer (3—Fig. 295) in counterbore of carrier with light film of grease. Place the assembled Clutch 1 and "B" ring gear unit, ring gear side down, over the "B" carrier. While partially supporting the clutch assembly, rotate the clutch assembly back and forth to align notches in clutch discs with splines on "B" carrier and to align teeth of "B" ring gear with pinion gear teeth on "B" carrier. Be sure that thrust washer (1) is in place in hub of "B" sun gear in transmission. Then, holding Clutch 1 and "B" carrier together, install them on "B" sun gear.

lined in paragraph 262. Check transmission front end play as outlined in paragraph 256, then proceed as follows:

Loosen the four distributor retaining cap screws and with a pair of pliers, remove the Servo 2 and 3 pressure tubes from the distributor; do not completely remove tubes from housing. On models with PTO disconnect the fitting on PTO pressure tube at right front corner of distributor; see Figs. 283 and 296. On models with heat exchanger (transmission oil cooler), remove the adapter bolts (see Fig. 296) retaining heat exchanger oil tubes to side of transmission. Remove the cap screws retaining distributor and on models with single speed PTO, insert long screwdriver through center of distributor against Clutch 1 housing. Hold Clutch 1 forward with screwdriver while removing distributor as in Fig.

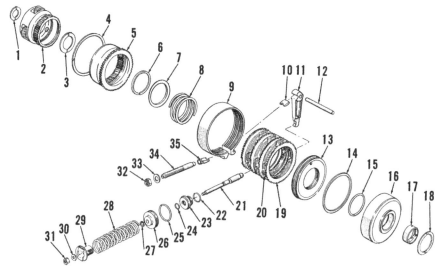

Fig. 295–Exploded view of Clutch 1, "B" carrier and Servo 2 assemblies as installed in model 2000, 3000 and 4000 transmissions.

1. Thrust washer	9. Band 2	18. Thrust washer	27. Snap ring
2. "B" carrier	10. Actuating strut	19. External spline	28. Spring
3. Thrust washer	11. Lever arm	plates	29. Retainer
4. Snap ring	12. Pivot pin	20. Internal spline plates	30. Washer
5. "B" ring gear—	13. Clutch piston	21. Piston rod	31. Nut
Clutch 1 pressure	14. "O" ring	22. "O" ring	32. Nut
plate	15. "O" ring	23. Guide	33. Washer
6. Snap ring	16. Clutch 1 housing	24. "O" ring	34. Adjusting screw
7. Retainer	17. Bushing—not	25. "O" ring	35. Adjusting strut
8. Piston return spring	serviced	26. Piston	

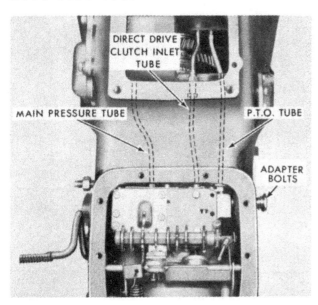

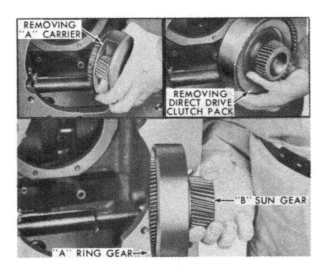

Fig. 297–Removing "B" Sun– "A" Ring Gear, "A" Carrier and Direct Drive Clutch assemblies.

a torque of 20-25 Ft.-Lbs. Recheck end play and if not within limits of 0.005-0.015, remove distributor and renew selective fit thrust washer with one of correct thickness. Tighten the PTO line fitting and heat exchanger adapter bolts on models so equipped. Adapter bolts should be tightened to torque of 20-25 Ft.-Lbs.

Reassemble remainder of transmission as outlined in paragraphs 263, 269 and 251.

272. **"A" RING GEAR—"B" SUN GEAR, "A" CARRIER AND DIRECT DRIVE CLUTCH.** After removing the distributor, Clutch 1 assembly and "B" carrier as outlined in paragraph 271, the "A" Ring Gear—"B" Sun Gear, "A" Carrier and Direct Drive Clutch can be removed as shown in Fig. 297.

NOTE: Two types of Direct Drive Clutch assemblies have been used. Refer to exploded view of Fig. 298. Early production clutch assemblies use only one piston return (Belleville) spring (49). Later production clutch assemblies have two piston return springs and the internally splined bronze discs (45) have radial grooves in faces on disc instead of circumferential grooves as on early production discs. Clutch housings and pistons are not interchangeable between early and late production units and only the new housings and pistons are available for service. If either a piston (51) or a housing (55) of early production assembly requires renewal, then a new type housing, piston and two return springs must be installed. The clutch discs are interchangeable; however, the late type bronze discs with radial grooves should be installed for improved clutch performance.

To disassemble the Direct Drive Clutch, refer to Fig. 299 and remove the snap ring, pressure plate and clutch plates. Then, using the snap ring removed from housing, install the Belleville spring compressor (Nuday tool No. N-488) as shown in Fig. 300. Using the lever (N-488-3), pry the spring compressor down and turn a stepped stud under the snap ring as shown in Fig. 301. Repeat this at the other two studs, then again at all three studs until highest step of studs is under snap ring. Then remove the spiral retainer ring (Fig. 299) from hub of clutch housing and release the compressor by prying it down with lever and turning the studs from under the snap ring.

Remove the compressor tool, refer to Fig. 298 and lift out the small pivot ring (48), piston return spring(s) (49) and large pivot ring (50). Remove the clutch piston (51) with compressed air

If transmission front end play was incorrect when measured before disassembly and no new parts were installed which could change end play, measure old thrust washer (13—Fig. 283) and renew with new washer of proper thickness to bring end play into limits of 0.005-0.015. Place washer on front side of distributor and install new seal rings on distributor with seal ring having the locking ends in the rear groove and ring with square ends in front groove. Note: It is advisable to renew both seal rings with rings having locking ends (Ford part No. 313864) as this will make for easier assembly. Install new "O" rings on main pressure tube and the Servo 1—Direct Drive Clutch tube. Lubricate the sealing rings and "O" rings with petroleum jelly and align ring end gaps at top side of distributor. Insert the front end of distributor through Clutch 1 hub and if square cut end snap ring is used in front groove, compress snap ring with two long screwdrivers and

push the distributor into place. It will be necessary to align the main supply tube, the Servo 1—Direct Drive Clutch tube and, on models so equipped, the PTO pressure tube with the holes and fitting on distributor while the distributor is being pushed into place. It may be necessary to lift distributor up slightly to engage pilot diameter with bore in transmission case. Start the fitting onto PTO connection. Loosely install the four distributor retaining cap screws. Insert the Servo 2 and 3 pressure tubes through left side of transmission case, then install new "O" rings on each end of the tubes. Lubricate the "O" rings, push tubes into bores of distributor and turn the tubes so beveled outer ends are aligned with oil passage slots in Servo 2 and 3 Cover. On models so equipped, install new "O" rings on heat exchanger line adapter bolts, lubricate the "O" rings and install the bolts through lines and housing into the distributor. Tighten the distributor retaining cap screws to

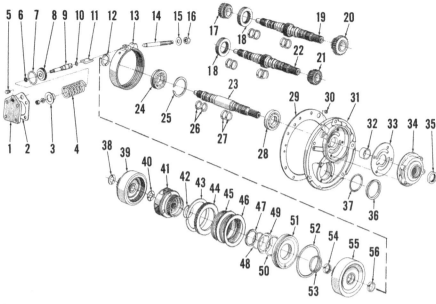

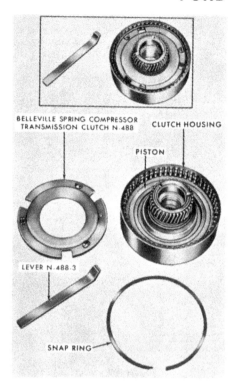

Fig. 298–Exploded view of model 2000, 3000 and 4000 front support plate, input shaft, "A" planetary system and related parts. Input shaft (19) is for models with 2-speed PTO; shaft (22) is for models with single speed PTO; and shaft (23) is for models without PTO.

1. Servo 1 cover	16. Nut	30. Plug
2. Gasket	17. 540 RPM PTO gear	31. Front support plate
3. Retainer	18. Shift collar	32. Bushing
4. Spring	19. Input shaft	33. Gasket
5. Plug	20. 1000 RPM PTO gear	34. Pump assembly
6. Snap ring	21. 540 RPM PTO gear	35. Seal
7. "O" ring	22. Input shaft	36. Plug (w/o PTO)
8. Servo piston	23. Input shaft	37. "O" ring
9. Piston rod	24. Ball bearing	38. Thrust washer
10. "O" ring	25. Snap ring	39. "A" ring—"B" sun
11. Strut	26. Rear sealing rings	gear
12. Thrust washer	27. Front sealing rings	40. Thrust washer
13. Band 1	28. Ball bearing	41. "A" carrier assembly
14. Adjusting screw	29. Gasket	42. Thrust washer
15. Washer		

43. Snap ring	
44. Pressure plate	
45. Internal spline discs	
46. External spline discs	
47. Spiral retainer	
48. Pivot ring	
49. Spring	
50. Pressure ring	
51. Clutch piston	
52. "O" ring	
53. "O" ring	
54. Bushing	
55. Clutch housing	
56. Bushing	

Fig. 300–Special tools used to disassemble the Direct Drive Clutch assembly.

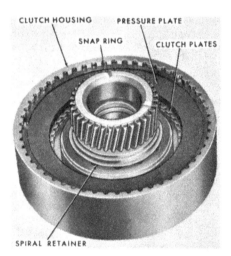

Fig. 299–Direct drive clutch and "A" sun gear assembly.

as shown in Fig. 289.

Carefully inspect all parts and renew as required. The two steel plates (46—Fig. 298), the two bronze plates (45) and the friction surface of the pressure plate (44) should be flat; renew the plates if they are warped, scored or excessively worn. Free height of the piston return (Belleville) spring(s) should be 0.115-0.125; renew if cracked or if free height measures less than 0.115. The bushings in clutch housing

(55) are renewable. Front bushing (56) inside diameter (new) is 1.440-1.441; rear bushing (54) is 1.315-1.316. Bushings are pre-sized and should not require reaming if carefully installed.

To reassemble, install new inner "O" ring (53) and outer "O" ring (52) on piston, lubricate "O" rings with petroleum jelly and install piston in housing. Place large pivot ring (50) on piston, piston return spring(s) on large pivot ring and place small pivot ring on the return (Belleville) spring(s). Compress the piston return spring(s) with compressor tool as during disassembly, then install the spiral retaining ring (47). Remove compressor tool and place steel disc on piston (either side down), install bronze disc (either side down), then the second steel disc and the second bronze disc. Place pressure plate, machined side down, on the top bronze disc and install retaining snap ring (43).

The "A" carrier is renewable as a complete assembly only. Check pinions and thrust surfaces and renew carrier if defect is noted. Pinion end play in carrier should be 0.010-0.028.

Check the two sealing rings (26) and thrust washer (12) on rear end of transmission shaft and renew if worn, scored or broken. Place new thrust washer against shaft rear bearing (24), then

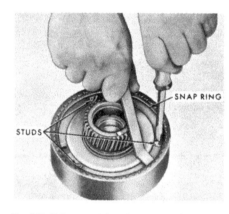

Fig. 301–Using special tools shown in Fig. 300 to compress Direct Drive Clutch spring.

install the two lock-end rings in grooves on shaft. Lubricate the sealing rings with petroleum jelly and align ring ends at top of shaft.

Place the "A" carrier with splines up on work bench. Stick thrust washer (42) in recess of the carrier hub with light film of grease. Set the assembled Direct Drive Clutch on carrier and rotate clutch back and forth while partially supporting it to align splines of clutch discs with splines on carrier. Holding the carrier and clutch together, install them as a unit, taking care to work clutch housing over the sealing rings on transmission shaft. Place thrust washer (40) on shaft against "A" carrier, then install the "A" ring gear—"B" sun gear taking care not to dislodge thrust washer (40).

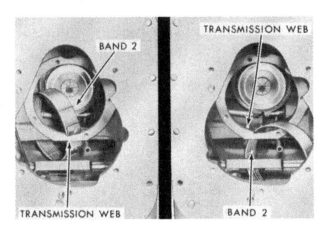

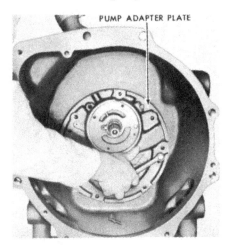

Fig. 302–To remove or install Band 2, it must be rotated around transmission web as shown.

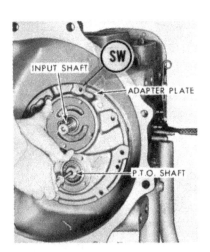

Fig. 303–Using jack screws to remove pump adapter (front support) plate on models 2000, 3000 and 4000.

Install thrust washer (38) in hub of "B" sun gear.

273. **BANDS 1 AND 2, STRUTS AND ACTUATING LINKAGE.** After removing the "A" carrier and Direct Drive Clutch as outlined in paragraph 272, Bands 1 and 2 can be removed as follows: Loosen Band 1 adjusting screw and remove the band and strut. Turn adjusting screw in to remove from housing. Compress Band 2 and remove band struts. Remove Band 2 from housing by rotating it around cast web in top of transmission as shown in Fig. 302. With Band 2 removed, the actuating lever pivot pin for both Bands 2 and 3 can now be removed and the Bands 2 and 3 actuating levers removed from housing. However, it is usually not necessary to remove the pivot pin and actuating levers. Note: Servo 3 actuating lever is the longest lever.

274. **PUMP ADAPTER PLATE (TRANSMISSION FRONT COVER).** To remove the pump adapter plate, the tractor must be split between engine and transmission as outlined in paragraph 181 or 182, then proceed as follows:

On models so equipped, remove the PTO pressure tube from front cover and PTO bearing retainer, then unbolt and remove PTO front bearing retainer. Take care not to lose or damage any of the shims located between PTO front bearing retainer and pump adapter plate. Unbolt transmission pump, bump pump body with soft faced hammer to break gasket seal and remove the pump assembly.

Remove the six cap screws and nut securing adapter plate to transmission housing and thread two jack screws into the tapped holes in pump adapter plate as shown in Fig. 303. Tighten the jack screws evenly to pull adapter plate loose from gasket. CAUTION: Do not allow input shaft to move forward while removing pump adapter plate, or thrust washer may drop from between

"A" carrier and "A" ring gear requiring disassembly of transmission to reinstall thrust washer.

On all models, inspect inside of sleeve (32—Fig. 298) located in pump adapter plate and renew sleeve if worn or scored. Press damaged sleeve out, align oil holes in new sleeve and adapter plate and press sleeve in so that end with chamfered inside diameter is flush with recessed surface at rear side of plate.

NOTE: During complete transmission overhaul, the main supply pressure tube, Servo 1 and on models so equipped, the PTO clutch pressure tube should be removed and new sealing "O" rings installed on the tube ends. The oil distributor must be removed as well as the front plate to allow removal of all tubes.

Remove the oil passage tubes as state of transmission disassembly will permit, renew the sealing "O" rings on tube ends and reinstall in housing. Note: Lubricate "O" rings with petroleum jelly before reinstalling tubes.

Install new sealing rings on front end of input shaft, lock ring ends together, lubricate with petroleum jelly and align ring ends at top side of shaft. Carefully install pump adapter plate with new gasket over input shaft to avoid damage to the sealing rings; lift PTO shaft up on models so equipped using a screwdriver as shown in Fig. 304. Install new seal on the top pump adapter plate retaining cap screw (see SW—Fig. 304) and tighten the cap screws and nut to a torque of 25-30 Ft.-Lbs. On models so equipped, install PTO shaft front bearing cup and a new sealing ring on front end of PTO shaft. Lock ends of ring together, lubricate ring with petroleum jelly and turn ring so that ends are at top of shaft. Carefully install PTO front bearing retainer with shims and new "O" ring and tighten the retaining cap screws to a torque of 14-17 Ft.-Lbs. Check PTO shaft preload as outlined in paragraph 278 and adjust preload by varying shim

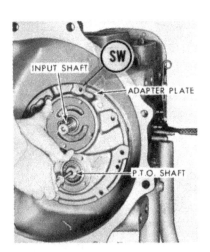

Fig. 304–Aligning PTO shaft with screwdriver inserted in end of shaft when installing adapter (front support) plate on models 2000, 3000 and 4000. Sealing washer (SW) is installed on one plate retaining bolt.

thickness under PTO bearing retainer as necessary. Install PTO pressure tube with new "O" rings. Install transmission pump as outlined in paragraph 254.

275. **INPUT SHAFT, BEARINGS AND PTO DRIVE GEARS.** To remove the transmission input shaft, it is necessary to remove the transmission from tractor as outlined in paragraph 249. Then, remove all transmission components as outlined in paragraphs 247 through 272 and the front cover (pump adapter) plate as outlined in paragraph 274. The input shaft can then be removed from transmission.

Remove the four cast iron sealing rings (26 and 27—Fig. 298) from input shaft (19, 22 or 23) and if renewal of shaft, bearings, PTO drive gear(s) and/or sliding coupling is indicated, remove the ball bearings (24 and 28) from shaft and slide off the PTO gear (21) and coupling (18) (single speed PTO) or PTO gears (17 and 20) and

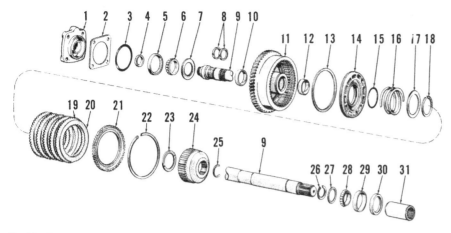

Fig. 305–Exploded view of single speed PTO clutch and shaft assembly. Refer to Fig. 306 for 2-speed PTO clutch.

1. Bearing retainer	9. PTO countershaft	17. Retainer	24. Clutch hub
2. Shims	10. Bushing	18. Snap ring	25. Snap ring
3. "O" ring	11. Clutch housing	19. External spline	26. Snap ring
4. Sealing ring	12. Bushing	plates	27. Washer
5. Bearing cup	13. "O" ring	20. Internal spline plates	28. Bearing cone & roller
6. Bearing cone & roller	14. Piston	21. Pressure plate	29. Bearing cup
7. Washer	15. "O" ring	22. Snap ring	30. Oil seal
8. Sealing rings	16. Piston return spring	23. Thrust washer	31. Coupling

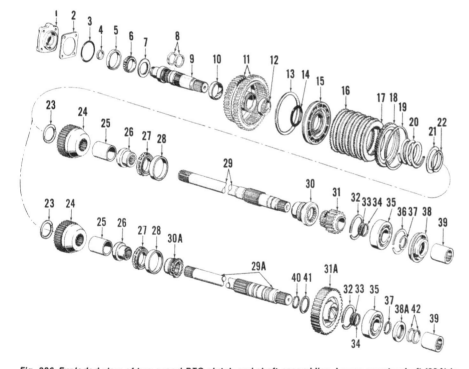

Fig. 306–Exploded view of two-speed PTO clutch and shaft assemblies. Lower countershaft (29A) is for model 4000; other models use shaft (29). Refer to Fig. 305 for single speed PTO.

1. Bearing retainer	14. "O" ring	25. Spacer	32. Snap ring
2. Shims	15. Piston	26. Bearing adapter	33. Snap ring
3. "O" ring	16. External spline	27. Bearing cone & roller	34. Washer
4. Sealing ring	plates	28. Bearing cup	35. Ball bearing
5. Bearing cup	17. Internal spline plates	29. Countershaft, rear	36. Snap ring
6. Bearing cone & roller	18. Pressure plate	(models 2000 & 3000)	37. Snap ring
7. Washer	19. Snap ring	29A. Countershaft, rear	38. Oil seal
8. Sealing rings	20. Piston return spring	(model 4000)	38A. Oil seal
9. Countershaft	21. Retainer	30. Shift collar	39. Coupling
10. Bushing	22. Snap ring	30A. Shift collar	40. Snap ring
11. Clutch housing	23. Thrust washer	31. Driven gear	41. Thrust washer
12. Bushing	24. Clutch hub	31A. Driven gear	42. Sealing rings
13. "O" ring			

sliding coupling (2-speed PTO models). To reassemble and reinstall input shaft, proceed as follows:

On 2-speed PTO models, install large PTO gear (20) on front end of shaft (19) with clutch teeth to rear. Press front bearing (28) onto shaft with ball filling notches in race forward. Slide the coupling (18) over splines on shaft, then install small PTO gear (17) on rear end of shaft with clutch teeth forward. Press rear bearing (24) onto shaft with

snap ring (25) groove in outer race forward. Be sure that both bearings are pressed firmly against shoulders on shaft; press on center race of bearings only.

On single speed PTO models, install PTO gear (21) on front end of shaft with clutch teeth to rear. Press front bearing (28) onto shaft with ball filling notch in race forward. Slide the coupling (18) onto splined section of shaft, then press rear bearing (24) onto shaft with snap ring (25) groove in outer race forward.

On models without PTO, install bearing on shaft (23) as outlined for 2-speed or single speed PTO models.

Reinstall input shaft assembly in transmission making sure that on PTO equipped models, the PTO shifter fork engages groove in sliding coupling (18) and that snap ring (25) on bearing (24) is seated against rear wall of front compartment in transmission. Then, install pump adapter plate as outlined in paragraph 274 and install other components to reassemble transmission as outlined in appropriate paragraphs.

276. **PTO CLUTCH.** To remove the PTO clutch assembly, the transmission input shaft must first be removed as outlined in paragraph 275. Then, proceed as follows:

On 2-speed PTO transmissions, the clutch and front PTO shaft can be removed as a unit. Disassemble the unit by removing rear bearing adapter (26 —Fig. 306) and pulling shaft (9) forward out of assembly. Pull front bearing cone and roller and thrust washer from shaft, and pull bearing cone and roller from rear bearing adapter if renewal is indicated.

To remove the single speed PTO clutch, first remove the bearing cone and roller (28—Fig. 305), washer (27) and snap ring (26) from rear end of shaft (9). Using a small punch or screwdriver, drive the snap ring (25) out of groove on shaft behind the splined clutch hub (24). Then pull shaft forward out of clutch assembly and lift the clutch assembly from transmission.

Bushings (10 and 12—Fig. 305 or Fig. 306) in clutch housing are renewable. Inside diameter of new front bushing is 1.874-1.875; inside diameter of new rear bushing is 1.624-1.625. Bushings are pre-sized and reaming should not be necessary if carefully installed. For servicing remainder of clutch assembly, refer to paragraph 270.

After clutch is reassembled, stick the thrust washer (23—Fig. 305 or Fig. 306) in recess in front side of splined hub (24) of PTO clutch using light film

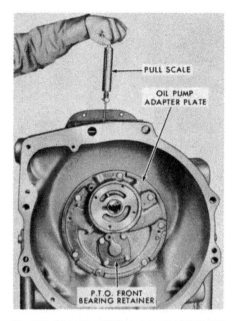

Fig. 307–Checking PTO countershaft bearing preload with pull scale and cord wrapped around shaft.

Fig. 308–Exploded view of feathering pedal and linkage and two-speed PTO interlock linkage used on all models except 4200. Refer to Fig. 239 for model 4200 feathering pedal and linkage.

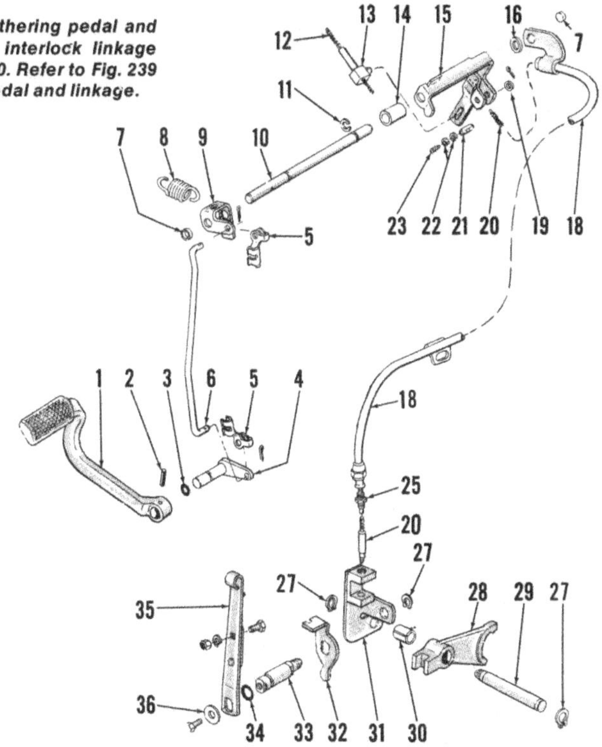

1. Foot pedal
2. Roll pin
3. "O" ring
4. Shaft and arm
5. Retainers
6. Link
7. Plugs
8. Return spring
9. Rocker
10. Pivot shaft
11. Snap rings
12. PTO control cable
13. PTO slider
14. Spacer
15. Rocker assembly
16. Spacer
18. Conduit assembly
19. Washer
20. Interlock cable
21. Retaining pin
22. Jam nuts
23. Set screw
25. Connector
27. Snap ring
28. Shift fork
29. Shift rail
30. Spacer
31. Interlock
32. Lever
33. Shaft
34. "O" ring
35. Shift lever
36. Washer

of grease. Insert the hub into clutch discs until thrust washer contacts thrust surface of clutch housing.

On 2-speed PTO, assemble the clutch and shaft on bench as follows: Place thrust washer (7—Fig. 306) on front end of shaft and press front bearing cone and roller (6) tightly against washer. Install the two sealing rings (8) from rear end of shaft, lubricate the rings with petroleum jelly and align ring end gaps. While holding the bevel gear securely in clutch, insert shaft, with snap ring ends up, through the clutch housing and bevel gear. Then, install spacer sleeve (25) and rear adapter (26) with rear bearing cone and roller (27) installed. Reinstall unit in transmission.

On single speed PTO, install thrust washer (7—Fig. 305) on front end of PTO shaft and press front bearing cone and roller (6) tightly against washer. Install two sealing rings (8) from rear end of shaft, lubricate the rings with petroleum jelly and align ring end gaps. Holding the splined hub in clutch assembly, place clutch in transmission and insert shaft through clutch with sealing ring end gaps at top of shaft. When thrust washer (7) is against clutch housing, drive the snap ring (25) around shaft into groove at rear of splined hub. Install the snap ring (26), washer (27) and bearing cone and roller (28) on rear end of shaft.

277. 2-SPEED PTO GROUND DRIVE SPEED GEARS. All 2-speed PTO transmissions also have a ground drive speed PTO ratio, in which the PTO shaft is driven from a gear on the transmission output shaft (model 4000)

or on the "D" carrier (models 2000 and 3000). The driven gear (31 or 31A—Fig. 306) is located on the rear transmission PTO shaft (29 or 29A).

To remove the ground drive gears, remove the transmission rear support as outlined in paragraph 267. On models 2000 and 3000, the drive gear is serviced as an assembly with the "D" carrier only; on model 4000, drive gear is serviced separately. To remove the driven gear, proceed as follows: Remove the PTO shaft oil seal (38 or 38A) from rear support and then remove snap ring (37) from shaft (29 or 29A). The PTO shaft can then be bumped forward out of the ball bearing (35). Remove the thrust washer (34) and snap ring (33); then, slide PTO driven gear (31 or 31A) from rear end of shaft. If necessary to renew bearing (35), remove snap ring (32) and drive bearing forward out of rear support. Reassemble by reversing disassembly procedure. Be sure seal surface of shaft (29 or 29A) is not damaged before installing new seal (38 or 38A).

278. PTO SHAFT END PLAY (BEARING PRELOAD). End play (bearing preload) of the power take-off clutch and shaft unit is controlled by varying the thickness of the shim stack (2—Fig. 305 or 306) between the transmission front cover (pump adapter plate) and the PTO front bearing retainer (1). Shims are available in thicknesses of 0.002, 0.003, 0.010 and 0.030. Check bearing preload as follows:

On 2-speed PTO, bearing preload is

checked with the rear support and rear PTO shaft removed from the transmission. On single speed PTO, check bearing preload with PTO shaft oil seal removed from transmission rear support.

Tighten the PTO front bearing retainer (1) cap screws to a torque of 14-17 Ft.-Lbs. Wrap a cord around the PTO shaft and attach a pull scale as shown in Fig. 307. Bearing preload is correct when a steady pull of 18 to 25 pounds is required to keep the shaft rolling. Add or remove shims as required to obtain proper bearing preload.

279. TRANSMISSION CONTROL LINKAGE. Usually, the control linkage components do not need to be disassembled when overhauling transmission. However, if necessary to renew linkage or sealing "O" rings at PTO shift levers or inching pedal shaft, proceed as follows:

280. INCHING PEDAL AND FEATHERING VALVE LINKAGE. Refer to Fig. 308 for exploded view of inching pedal and feathering valve linkage. (Refer to paragraph 281 for information on PTO interlock linkage shown in Fig. 308). Inching pedal (1) can be renewed by driving out pin (2) and removing transmission top cover (paragraph 247) and holding pry bar against lever shaft (4) while installing new pedal. If necessary to renew the "O" ring (3), shaft (4) or link (6), it will be necessary to remove the Clutch 2 and 3 assembly, mainshaft and "C" carrier as outlined in paragraph 268.

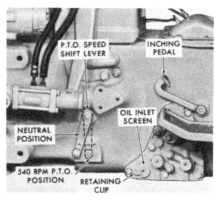

Fig. 309–View showing neutral and engaged positions for single speed PTO shift lever. Refer to Fig. 310 for 2-speed PTO.

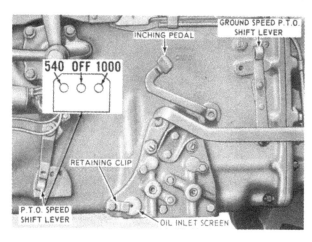

Fig. 310–View showing neutral, 540 RPM and 1000 RPM positions for 2-speed PTO shift lever. Refer to Fig. 309 for single speed PTO.

able to synchronize movement of sliding collar (30 or 30A—Fig. 306) with detents in transmission case for shift lever tang.

282. PTO SPEED SHIFT LINKAGE. On single speed PTO models, the power take off drive is in neutral position with lever to front as shown in Fig. 309, or is in engine drive (540 RPM) position when lever is engaged in center hole; rear hole is not used on single speed PTO models.

On 2-speed models, the PTO speed shift lever has three positions; 540 RPM, neutral and 1000 RPM as shown in Fig. 310.

On both the single speed and 2-speed PTO models, speed shift lever can be

moved only when engine is not running. Refer to Fig. 311 for exploded view of the linkage. After removing front cover as outlined in paragraph 274, the linkage can be removed by removing pin (6), sliding the shaft (3) out of housing, then removing arm (5) and fork (7). Bushing (4) in transmission housing can be renewed if necessary. Install new sealing "O" ring (2),

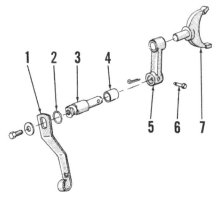

Fig. 311–Exploded view of PTO shift linkage for both single speed and 2-speed PTO models. Refer also to Figs. 309 and 310.

1. Shift lever
2. "O" ring
3. Shaft
4. Bushing
5. Shift arm
6. Pin
7. Shift fork

lubricate ring with petroleum jelly and reassemble by reversing disassembly procedure.

283. **SPEED SELECTOR ASSEMBLY.** To remove the speed selector assembly, refer to paragraph 245. To disassemble, refer to exploded view in Fig. 312 and proceed as follows:

First, remove the two side covers (1 and 17) and the three Phillips head screws securing the left hand selector housing (2). Remove the hex nut (3) from the shaft (10) and remove the snap ring (7) and retainer plate (8) to expose the wheel and cable assembly. Note: Impact torque is usually required to loosen nut (3) without twisting the shaft (10). Further disassembly procedure is obvious from inspection of unit and reference to exploded view in Fig. 312. To reassemble, reverse disassembly procedure. Adjust the speed selector assembly as outlined in paragraph 237 after reinstalling assembly as in paragraph 245.

To remove the feathering valve lever shaft (10) or components located on shaft, remove top transmission cover, disconnect inching pedal return spring (8), remove snap rings (11) from shaft and drive against either sealing plug (7) with thin punch. Shaft need not be driven all the way out. After reinstalling shaft and components, insert sealing plugs and center the feathering valve levers on the valves by driving against either plug as required. Note: On 2000, 3000 and 4000 transmissions with single speed PTO, item (18) and items (20 through 36) are not used; on models without a PTO, items (12 through 36) are not used.

281. PTO INTERLOCK LINKAGE. On models 2000, 3000 and 4000 with 2-speed PTO, interlock linkage is used to prevent both the ground drive PTO speed and the engine drive PTO from being engaged at the same time. Refer to items (18) and (20 through 36) in Fig. 308. To service the ground speed shift and interlock mechanism, the "C" carrier, mainshaft and Clutch 2 and 3 assembly must be removed as outlined in paragraph 268. Refer to exploded view in Fig. 308 for disassembly and reassembly guide. Refer to paragraph 240 for adjustment procedure for interlock mechanism. Lever (35) is adjust-

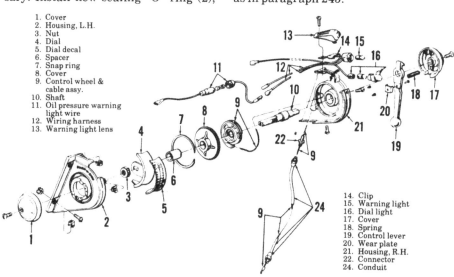

1. Cover
2. Housing, L.H.
3. Nut
4. Dial
5. Dial decal
6. Spacer
7. Snap ring
8. Cover
9. Control wheel & cable assy.
10. Shaft
11. Oil pressure warning light wire
12. Wiring harness
13. Warning light lens
14. Clip
15. Warning light
16. Dial light
17. Cover
18. Spring
19. Control lever
20. Wear plate
21. Housing, R.H.
22. Connector
24. Conduit

Fig. 312–Exploded view of transmission speed (gear ratio) selector assembly.

DIFFERENTIAL, BEVEL GEARS AND REAR AXLE

(Models 2000, 3000 and Model 4110 L.C.G.)

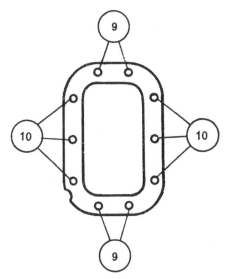

Fig. 313–When reconnecting rear axle center housing to transmission on models 2000 and 3000, tighten bolts at location (9) to a torque of 65-86 Ft.-Lbs. and bolts at (10) to a torque of 40-55 Ft.-Lbs.

SPLIT TRACTOR BETWEEN TRANSMISSION AND REAR AXLE CENTER HOUSING

284. To remove final drive bevel pinion and perform other jobs, it will be necessary to split tractor between transmission and rear axle center housing. Proceed as follows:

Drain lubricant from rear axle center housing. Note: On models with hydraulic lift, move selector lever to draft control, control lever to bottom of quadrant and force lift arms to lowest position. Disconnect wiring to rear light and brake pedal return springs under right step plate. Unbolt and remove both step plates from fenders, transmission and rear axle center housing. On models with transmission type PTO, move PTO shift lever to neutral position. Disconnect clutch release rod at front end on models with engine clutch. On models with Select-O-Speed transmission, disengage tractor coupling.

If tractor has been split between engine and transmission, proceed as follows: Remove hydraulic pump to center housing tube and manifold assembly. Attach hoist to transmission, support rear axle center housing, then unbolt and remove transmission.

If performing transmission to rear axle center housing split on assembled tractor, proceed as follows: Disconnect horizontal exhaust pipe from muffler, if

so equipped, at left side of tractor. Unbolt hydraulic tube manifold from rear axle center housing. Drive wood wedges between front axle and front support and place floor jack or work stand under rear of transmission housing. Support rear axle center housing with suitable hoist or rolling floor jack, then unbolt transmission from rear axle center housing and move the rear axle unit away from transmission.

To reconnect tractor between transmission and rear axle center housing, reverse disassembly procedure using a new gasket between transmission and center housing. Refer to Fig. 313 for bolt tightening torques. Refill rear axle center housing (hydraulic sump) with proper lubricant as outlined in paragraph 339. Bleed the piston type hydraulic pump as in paragraph 361.

DIFFERENTIAL AND BEVEL GEARS

285. **R&R DIFFERENTIAL ASSEMBLY.** To remove the differential assembly, first drain rear axle center housing and hydraulic system, then proceed as follows: Block up under center housing to raise rear wheels, then remove left rear wheel and disconnect brake linkage and light wire. Unbolt the left rear axle housing from center housing and remove the axle, brake, housing and fender assembly from tractor. The differential assembly can then be removed from the rear axle center housing.

NOTE: On all models, differential carrier bearing preload is adjusted by varying the thickness of gasket used between the left rear axle housing and the center housing. Therefore, when removing left axle housing, determine thickness of gasket installed before destroying or discarding the old gasket. However, if the rear axle center housing, either axle housing, the differential housing or the differential carrier bearings are renewed, differential carrier bearing preload should be readjusted as outlined in paragraph 286. If a different thickness of axle housing to center housing gasket is installed, axle shaft bearing adjustment should be checked as outlined in paragraph 294. If an early production unit having paper (composition) axle housing to center housing gaskets is encountered, both axle housings should be removed and the later steel gaskets installed; refer to paragraph 286.

Proceed as follows to reinstall differential assembly: If removed, reinstall bevel pinion assembly as outlined in paragraph 288 and reinstall right rear axle housing using one 0.006-0.008 gasket; tighten retaining cap screws to a torque of 125-155 Ft.-Lbs. Install the differential assembly in center housing, then reinstall left rear axle housing with gasket thickness as determined as outlined in paragraph 286, or if no new parts affecting carrier bearing preload were installed, with same gasket thickness as removed. Tighten axle housing retaining cap screws to a torque of 125-155 Ft.-Lbs. If axle assemblies were removed or if a different gasket thickness was installed between left axle housing and rear axle center housing than was removed on disassembly, refer to paragraph 294 for axle installation.

After final installation of left rear axle housing on all models, check to be sure that the differential assembly can be turned by hand and that some backlash exists between differential ring gear and bevel pinion gear.

286. **ADJUST DIFFERENTIAL CARRIER BEARING PRELOAD.** Differential carrier bearing preload is adjusted by varying the thickness of gasket installed between the left rear axle housing and rear axle center housing. To determine the proper thickness of gasket to be installed, proceed as follows:

Remove the rear axle center housing from transmission as outlined in paragraph 284. If removed, reinstall bevel pinion assembly as outlined in paragraph 288 and reinstall right rear axle housing with a 0.006-0.008 thick steel gasket; tighten housing retaining cap screws to a torque of 125-155 Ft.-Lbs. Turn the axle housing and center housing assembly so that axle housing is pointing down and place the differential assembly in the center housing. Lower the left rear axle housing, without a gasket and without axle assembly installed, down over the differential assembly and rotate the differential to be sure bearings are seated. Install four axle housing retaining cap screws equally spaced and finger tight. Using a feeler gage, check gap between axle housing and center housing and equalize the gap, if necessary, by loosening and tightening alternate cap screws. Be sure not to tighten any of the cap screws more than finger tight.

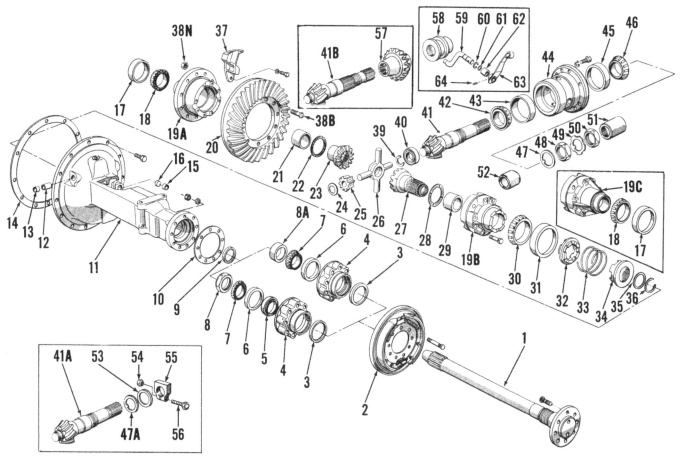

Fig. 314–Exploded view of final drive and rear axle assemblies for models 2000, 3000 and model 4100 L.C.G. Model 2000 tractors with rear axle housing (11) having round cross-section instead of square type housing shown will have rear axle bearing (7) retained by lock collar (8A); other models will have axle bearing retained by nut (8). Select-O-Speed models will have disconnect collar (58). Exploded view shows models with different lock (32 through 36) which locks differential case half (19B) to axle gear (27); other models have plain axle gear (23) and differential case half (19C).

1. Axle shaft	13. Plug	20. Bevel ring gear	30. Bearing cone & roller
2. Brake assembly	14. Gasket	21. Bushing	31. Bearing cup
3. Grease seal	15. Brake shaft bushing	23. Axle gear, L.H. (also	32. Adapter
4. Axle bearing retainer	16. Plug	R.H. without	33. Spring
5. Spacer	17. Bearing cup	differential lock)	34. Coupling
6. Bearing cup	18. Bearing cone & roller	24. Thrust washers (4)	35. Washer
7. Bearing cone & roller	19A. Differential case half	25. Differential pinions	36. Snap ring
8. Self-locking nut	(L.H.)	(4)	37. Ring gear thrust
8A. Lock collar	19B. Differential case half	26. Spider	block
9. Oil seal	(R.H. with	27. Axle gear, R.H. with	38B. Ring gear bolts
10. Shims	differential lock)	lock	38N. Ring gear nuts
11. Axle housing	19C. Differential case half	28. Thrust washer	39. Snap ring
12. Bushing, differential	(R.H. without	29. Bushing	40. Pilot bearing
lock shaft	differential lock)		

41. Bevel pinion	50. Hex nut
41B. Bevel pinion (with	51. Drive shaft coupling
handbrake)	52. Spacer (used on 41B
42. Bearing cone & roller	less handbrake)
43. Bearing cup	57. Handbrake gear
44. Pinion bearing	58. Disconnect coupling
retainer	59. Disconnect shaft arm
45. Bearing cup	60. Snap ring
46. Bearing cone & roller	61. "O" ring
47. Thrust washer (not	62. Bushing
used with 52 or 57)	63. Lever
48. Hex nut	64. Roll pin
49. Locking washer	

Carefully measure the resulting gap with feeler gage to determine thickness of steel gasket to be installed between left axle housing and center housing. After determining the proper gasket thickness, remove the left rear axle housing, then reinstall housing with gasket as outlined in paragraph 285. The steel gaskets are available in thicknesses of 0.006-0.008, 0.011-0.013 and 0.015-0.017. Install only one gasket of thickness nearest to measured gap. **Do not** install more than one gasket.

287. OVERHAUL DIFFERENTIAL. With the differential assembly removed as outlined in paragraph 285, proceed as follows:

If equipped with differential lock, remove the snap ring (36—Fig. 314), washer (35), coupling (34), coupling

spring (33) and adapter (32). Place correlation marks on the two halves of the differential case (19A and 19B) so that they may be reassembled in same relative position. Cut the cap screw locking wire and remove the cap screws evenly while lifting up right half of differential case (19B).

On models without differential lock, cut the cap screw locking wire, remove the cap screws and lift right differential case half (19C) from the differential assembly.

On all models, remove the thrust washer (28), right hand side gear (27), spider (26), pinion (25) and washer (24) assembly, left hand side gear (23) and left thrust washer (22). Check the differential carrier bearings (30 and/or 18) and renew if worn or damaged. To renew the carrier bearing cups (31

and/or 17), the right hand axle housing must be removed from center housing and the axle assemblies removed from both axle housings to provide clearance for removing the cups.

The differential case left half (19A) and the bevel ring gear (20) are riveted together during factory assembly. If renewing either the case or ring gear separately, drill through rivet heads with a ½-inch drill, then remove rivets with punch. Assemble new ring gear and/or differential case with special bolts (38B) and self locking nuts (38N) that are available for service. Be sure the ring gear is not cocked on differential case and tighten the nuts to a torque of 40-45 Ft.-Lbs. Note: Bevel ring gear and bevel pinion are available as a matched set only; when installing new ring gear, refer to para-

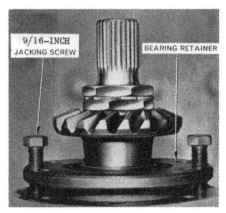

Fig. 315–Removing bevel pinion bearing retainer with jack screws.

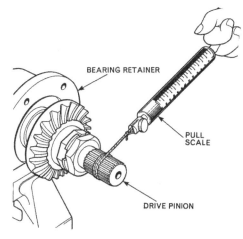

Fig. 316–Checking bevel pinion bearing preload with pull scale and cord wrapped around pinion shaft.

graph 288 and install the matched bevel pinion.

Renew the thrust washers (22, 24 and 28) if worn or scored and other parts if excessively worn or damaged. The differential case bushings (21 and 29) may be renewed if the case halves are otherwise serviceable. Bushings are pre-sized and should not require reaming if carefully installed.

Reassemble the differential by reversing disassembly procedure and tighten the differential case retaining cap screws to a torque of 65-75 Ft.-Lbs. and secure with locking wire.

288. **BEVEL PINION.** To remove the main drive bevel pinion, remove the differential assembly as outlined in paragraph 285 and split rear axle center housing from transmission as outlined in paragraph 284. Remove the hydraulic power lift cover, if so equipped, as outlined in paragraph 347 and proceed as follows:

Remove the cap screws and lock washers retaining bevel pinion bearing carrier (44—Fig. 314) to rear axle center housing, then remove the carrier and pinion assembly with jack

screws as shown in Fig. 315. Straighten tabs on locking washer (49 —Fig. 314) and remove nuts (50) and (48); then press shaft (41) and bearing (42) out of retainer (44) and associated parts.

The bevel pinion is available only in a matched set with the bevel ring gear; if necessary to renew the pinion, refer to paragraph 287 and install the matched ring gear. Renew bearings if rough or excessively worn and adjust bearing preload as follows: Tighten the nut (48) until pull required to rotate the pinion in the bearings is 16 to 21 pounds when checked with pull scale and cord as shown in Fig. 316. When bearing preload adjustment is correct, install new tab washer (49—Fig. 314) tighten the nut (50) and bend tabs against nuts. Then recheck bearing preload as shown in Fig. 316 and readjust if necessary.

When reinstalling the pinion, bearing and carrier assembly, tighten retaining cap screws to a torque of 80-100 Ft.-Lbs.

289. **BEVEL RING GEAR.** Procedure for renewing the bevel ring gear is outlined in paragraph 287, OVERHAUL DIFFERENTIAL.

290. **DIFFERENTIAL LOCK.** The differential lock consists of a dog type coupling (32 and 34—Fig. 314) which can be engaged to lock the right hand axle gear (27) to the differential case (19B). This results in both rear wheels being turned at the same speed, regardless of any difference in traction. In operation, when one rear wheel starts to spin, the foot pedal (78—Fig. 317) is depressed which applies spring pressure to the sliding coupling. When the dogs on the coupling are aligned with notches in the coupling adapter, the spring pressure applied through the operating fork will snap the coupling into engaged position. The foot pedal can then be released and the differential lock will remain engaged until the traction on the rear wheels becomes equalized. As there will then be no side pressure on the coupling dogs, the coil spring (33—Fig. 314) between the sliding coupling (34) and the coupling adapter (32) will push the sliding coupling to disengaged position. If necessary to make a turn before the differential lock is automatically disengaged, the lock can be manually disengaged by momentarily depressing the transmission clutch pedal or applying the brakes on the wheel with least traction.

291. ADJUST DIFFERENTIAL LOCK. Proper adjustment of the differential lock mechanism requires minimum clearance between the operating

fork and face of the sliding coupling when the lock is in disengaged position and that the lock be fully engaged when the foot pedal is depressed until it contacts the right hand foot rest. Before attempting to adjust differential lock, be sure foot rest is not bent out of position and there is nothing on the foot rest to prevent full travel of foot pedal. Then, disconnect the spring loaded operating rod (75—Fig. 317) from operating lever (74). Back-off the locknut (68) on adjusting screw (69) in axle housing. Position operating lever so that operating fork (67) just contacts face of sliding coupling (34—Fig. 314). Turn adjusting screw in until screw contacts operating fork (operating lever will start to move), then back screw out ¼-turn from this point.

Block up right rear wheel and turn the wheel while pushing down on operating lever to fully engage the differential lock. Hold operating lever in this position and with foot pedal against foot rest, adjust length of spring loaded operating rod by loosening locknut (76 —Fig. 317) and turning rod (75) so that pin can just be inserted through rod and the operating lever. Shorten the rod one turn from this position and reinstall the clevis pin and retaining cotter pin.

292. OVERHAUL DIFFERENTIAL LOCK. The differential lock foot pedal pivots on an extended lower hydraulic link shaft. Pedal has renewable bushing (79—Fig. 317), be sure to align grease hole in bushing with hole in pedal. The spring loaded operating rod is renewable only as a complete assembly. The operating lever and fork shaft pivots in renewable bushings in right rear axle housing. To renew the operating shaft (71), bushings (12 and 70), fork (67) and/or pedal pivot shaft (66), remove the right axle housing as follows: Drain oil from differential and hydraulic lift compartments. Disconnect differential lock operating rod from operating lever and remove foot pedal and hydraulic lower link from pivot shaft. Block up under center housing and unbolt and remove right rear wheel and fender, then remove axle assembly from axle housing. Be careful not to lose or damage shims (see Fig. 319) located between bearing retainer axle housing. Then, unbolt and remove axle housing from center housing. Renew shaft bushings, fork and/or pedal pivot shaft as necessary. Procedure is evident from inspection of unit and reference to Fig. 317. The differential lock sliding coupling, spring and coupling adapter can also be inspected and, if necessary, renewed at this time. Note: If necessary to renew right differential carrier bearing (30—

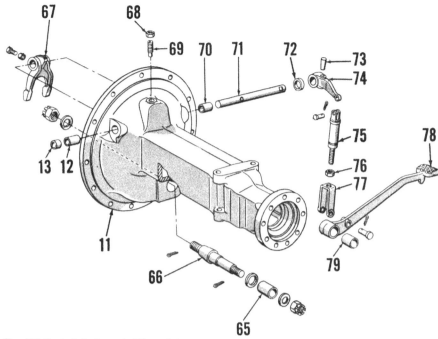

Fig. 319–Axle shaft bearings are adjusted by varying shims located between axle housing and bearing retainer.

Fig. 317–Exploded view of differential lock linkage for models 2000, 3000 and 4110 L.C.G. so equipped. Fork (67) contacts coupling (34–Fig. 314) to engage coupling with adapter (32) when foot pedal is depressed.

11. Axle housing	66. Pedal & lift link shaft
12. Bushing	67. Fork
13. Plug	68. Jam nut
65. Sleeve (w/o 3-point hitch)	69. Adjusting screw
	70. Bushing

71. Shaft	76. Jam nut
72. Oil seal	77. Clevis
73. Pin	78. Pedal
74. Lever	79. Bushing
75. Link assembly	

Fig. 314) and cup (31), carrier bearing preload should be adjusted as outlined in paragraph 286. If service of the differential unit is indicated, remove left axle housing and differential as outlined in paragraph 285.

When reassembling tractor, install one 0.006-0.008 thick steel gasket between axle housing and center housing and tighten the retaining cap screws to a torque of 125-155 Ft.-Lbs. Install axle assembly in axle housing with same number of shims as were removed and tighten retaining nuts to a torque of 90-100 Ft.-Lbs. Note: If axle housing, axle or axle shaft bearings were renewed, axle shaft end play should be readjusted as outlined in paragraph 294.

REAR AXLE SHAFTS, BEARINGS AND SEALS

293. **RENEW WHEEL DISC BOLTS.** To renew a broken or damaged wheel bolt, remove wheel and brake drum from axle and refer to Fig. 318; a knockout plug is provided in brake backing plate so that wheel bolts can be driven out of axle shaft flange without disassembly of axle unit. After driving in new bolt(s), seal the opening in brake backing plate with rubber plug (Ford part No. C5NN-2N214-A) which is available for service. Then, reinstall brake drum and rear wheel.

294. **BEARING ADJUSTMENT.** The rear axle shafts are carried on one tapered roller bearing at the outer end of each shaft. The bearings are retained in their cups by contact of the inner ends of the two axle shafts in the differential assembly. The recommended axle shaft end play of 0.004-0.012 is obtained by varying the shims installed between the axle housing and axle bearing retainer as shown in Fig. 319.

Fig. 320–To check axle shaft bearing preload (endplay), hold one axle shaft outwards with chisel or other wedge, then check endplay of other shaft as shown in Fig. 321.

To measure axle end play, block up under rear end of tractor and remove both rear wheels and brake drums. Use a chisel or other suitable tool to wedge one axle shaft outward as shown in Fig. 320; then, measure end play of other axle with dial indicator as shown in Fig. 321. If the measured end play is not within limits of 0.004-0.012, correct the adjustment by adding or removing shims between the bearing retainers and axle housings. Adjustment may be made on either axle shaft bearing, but

Fig. 318–View showing knockout plug in brake backing plate that can be removed to facilitate wheel bolt renewal. Replace with rubber plug.

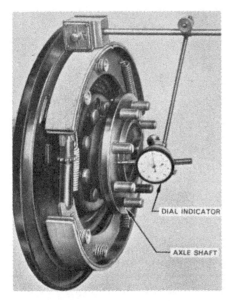

Fig. 321–Checking axle shaft endplay with dial indicator. Other shaft must be held outwards as shown in Fig. 320 while checking endplay.

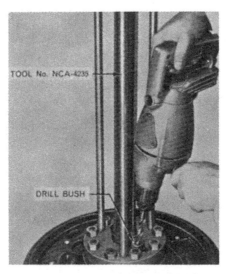

Fig. 323–Drilling axle bearing collar with drill guide bushing inserted in special axle puller (Nuday tool No. NCA-4235). If special tool is not available, refer to Fig. 326 and text.

Fig. 325–Pressing axle shaft from bearing retainer after drilling collar (Fig. 323), then cracking collar with chisel (Fig. 324).

Fig. 322–On model 2000 tractors with round axle housing, axle shaft bearing is retained by lock collar as shown. On other models, a self-locking nut is used; refer to Fig. 327.

Fig. 324–Cracking bearing collar with chisel after drilling hole as shown in Fig. 323.

an effort should be made to keep the total shim pack equally divided between both axle shaft bearing retainers. To add or remove shims, it will be necessary to remove the axle shaft(s) as outlined in following paragraph 295. Shims are available in thicknesses of 0.004-0.006, 0.008-0.010, 0.015-0.017, 0.020-0.022, 0.030-0.032 and 0.049-0.051.

295. **R&R AXLE ASSEMBLY.** To remove either rear axle assembly, support rear of tractor, disconnect brake linkage and wiring to fender mounted light and remove the wheel and tire assembly and the brake drum. Then, unbolt axle bearing retainer from axle housing and remove the axle shaft, retainer and bearing assembly, brake

assembly and brake camshaft from axle housing as a unit. Take care not to lose or damage the shims located between bearing retainer and axle housing. Disassemble, if necessary, as outlined in paragraph 296 or 297.

The axle shaft inner oil seal (9—Fig. 314) can now be renewed. Pry out old seal and drive new seal into housing with lip of seal to inside. Lubricate seal before reinstalling axle assembly.

When reinstalling axle assembly, reverse the removal procedure, adjusting axle end play as outlined in paragraph 294. Tighten axle bearing retainer nuts to a torque of 90-100 Ft.-Lbs.

296. **OVERHAUL AXLE ASSEMBLY (model 2000 with Round Axle Housing).** On model 2000 with round axle housing, the rear axle shaft is retained with a steel collar (see Fig. 322) which is a heat-shrink fit on rear axle shaft. To remove the lock collar, refer to Figs. 323, 324 and 325 for pro-

cedure using special rear axle puller (Nuday tool No. N-4235) and cracking chisel (Nuday tool No. N-4235-A1). If the special puller is not available, proceed as follows:

Center punch upper face of lock collar near the outer edge, then drill a ⅜-inch pilot hole at approximately a 10 degree angle from axle shaft as shown in Fig. 326. The drill will speed up when collar is drilled through and the drill contacts the hard bearing inner race. Follow up with a ½-inch drill bit to enlarge the pilot hole, then crack the lock collar with a chisel. Press the axle shaft from the collar, bearing and retainer with a heavy duty press.

Axle shaft outer seal (3—Fig. 314) can be renewed without removing bearing cup (6). Install new seal with lip inward, and lubricate lip of seal. Install new bearing cup in retainer if bearing is worn or damaged. Assemble brake backing plate to bearing retainer and place the assembly over axle shaft. Pack bearing with about 3 ounces of wheel bearing grease and use a clean 2⅛-inch I.D. pipe, approximately 30 inches long, to drive bearing cone and roller onto axle shaft. Be sure cone is firmly seated against shoulder on shaft. Heat a new locking collar to 800° F. and install collar on shaft next to bearing. A temperature indicating crayon such as "Therm-O-Melt" or "Tempilstick" should be used; the crayon will smear when rubbed against the collar when collar is heated to proper temperature. Hold the collar in place while cooling with the pipe used to drive bearing onto shaft.

297. **OVERHAUL AXLE ASSEMBLY (All Models Except 2000 With Round Axle Housing).** The rear

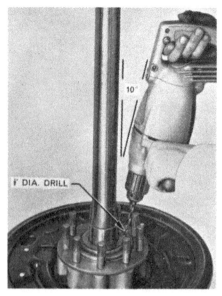

Fig. 326–If special axle puller (Figs. 323, 324 & 325) is not available, drill through collar with 3⁄8-inch drill, then enlarge hole with 1⁄2-inch drill as shown. Drill will speed up when it contacts bearing cone.

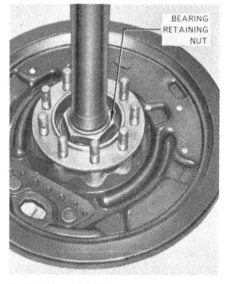

Fig. 327–On models 2000 with square cross-section rear axle housing and on models 3000 and 4110 L.C.G., rear axle bearing is retained with self-locking nut.

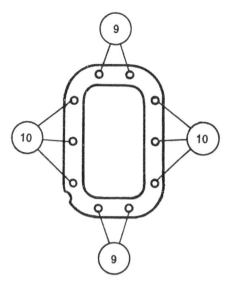

Fig. 328–When reconnecting rear axle center housing to transmission on model 4000 tractors, tighten bolts at (9) to a torque of 65-86 Ft.-Lbs. and bolts at (10) to a torque of 40-55 Ft.-Lbs.

axle shaft is retained with a self-locking nut (see Fig. 327) on all models except model 2000 with round axle housing.

To disassemble, unscrew the self-locking nut and press axle shaft from bearing and retainer. If available, special axle puller (Nuday tool No. N-4235) can be used to remove the axle shaft. The axle shaft outer seal may be renewed without removing bearing cup from retainer. If renewing bearing cup, be sure cup is driven into retainer until firmly seated. Install new seal with lip of seal inward.

Assemble brake back plate and bearing retainer onto axle shaft, pack bearing cone and roller assembly with about 3 ounces of wheel bearing grease, then drive bearing onto shaft with a clean 2⅛-inch I.D. pipe approximately 30 inches long. Install a new self-locking nut and tighten nut to a torque of 230-250 Ft.-Lbs.

NOTE: Set axle in wheel disc to hold axle while removing or installing nut. Use of Nuday Tool No. N-4235-D is recommended. This special tool multiplies torque wrench reading by 2; therefore, 115-125 Ft.-Lbs. torque reading is required.

298. **AXLE HOUSINGS.** Either rear axle housing may be removed after draining the rear axle center housing and hydraulic system and removing the rear axle assembly as outlined in paragraph 295. The differential carrier bearings should be adjusted as outlined in paragraph 286 when installing left axle housing.

remove the step plates, or on rowcrop models the operator's platform. On model 4000 with transmission type PTO, move PTO shift lever to neutral position. On models with Select-O-Speed transmission disconnect traction coupling.

If tractor has been split between engine and transmission, proceed as follows: On model 4000 with engine driven hydraulic pump, unbolt and remove hydraulic tube and manifold assembly. Attach a hoist to transmission, support rear axle center housing, then unbolt and remove transmission assembly. Note: On model 4000 with independent type PTO, be careful not to lose or damage any shim(s) which may be located on transmission PTO shaft between pump drive gear retaining snap ring and PTO clutch shaft.

If performing split on assembled tractor, proceed as follows: On model 4000 with engine driven hydraulic pump, disconnect hydraulic tube manifold from right side of center housing.

DIFFERENTIAL, MAIN DRIVE BEVEL GEARS, FINAL DRIVE AND REAR AXLE

(Model 4000 Except 4110 L.C.G.)

SPLIT TRACTOR BETWEEN TRANSMISSION AND REAR AXLE CENTER HOUSING

299. To remove the main drive bevel pinion and perform other jobs, it is necessary to split tractor between transmission housing and rear axle center housing as follows:

Drain lubricant from rear axle center housing. Note: On models with hydraulic lift, move selector lever to draft control, push selector valve in on models so equipped, move control lever to bottom of quadrant and force lift arms to lowest position. Disconnect wiring to rear light, disconnect brake pedal return springs and brake rods and, on models with engine clutch, disconnect clutch release rod. Unbolt, and

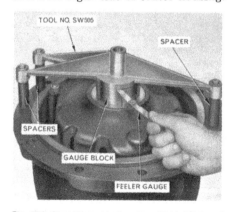

Fig. 329–Checking axle housing, final drive and brake assembly for proper differential carrier bearing spacer shim thickness. Refer to text.

On models with horizontal exhaust, disconnect pipe from muffler. On row-crop models, adequately support tricycle front end to keep it from tipping. On models with wide front axle, drive wood wedges between front axle and front support and place a jack or work stand under rear end of transmission. Support rear axle with suitable moving hoist or floor jack, then unbolt rear axle center housing from transmission and move rear unit away. Note: On model 4000 with independent type PTO, be careful not to lose or damage any shims which may be located on transmission PTO shaft between pump drive gear snap ring and PTO clutch shaft.

NOTE: When reassembling model 4000 with independent type PTO, be sure same shim thickness (See Fig. 363) as removed is reinstalled on transmission PTO shaft against pump drive gear retaining snap ring. However, if any part that would affect PTO clutch shaft end play has been renewed, recheck end play and install required shim thickness as outlined in paragraph 331.

To reconnect transmission and rear axle center housing, reverse disassembly procedure. For transmission to center housing bolt tightening torque, refer to Fig. 328. Refill center housing with proper lubricant as outlined in paragraph 340. On model 4000 with engine driven hydraulic pump, bleed pump as outlined in paragraph 361.

DIFFERENTIAL AND BEVEL GEARS

300. **R&R DIFFERENTIAL ASSEMBLY.** To remove the differential assembly, the left axle and housing assembly must be removed on model 4000 tractors. To remove and reinstall axle assembly, refer to paragraph 306.

With axle assembly removed, remove the differential assembly from rear axle center housing. Refer to paragraph 302 for differential overhaul procedure.

301. **DIFFERENTIAL CARRIER BEARING ADJUSTMENT.** The differential carrier bearing preload is non-adjustable, however bearing preload can be considered correct when differential carrier bearing cup position in axle housing is properly established by installation of correct shim thickness behind the bearing cup in bore of inner brake housing. The right carrier bearing cup on model 4000 with differential lock is retained in a bore in the rear axle center housing and does not require shimming. Once proper shim thickness is established for a particular assembly of axle housing, final drive ring gear, brake outer housing

and brake inner housing, the shim, if removed, may be retained for reassembly. However, if the shim is lost or when renewing an axle housing, final drive ring gear, brake outer housing and/or brake inner housing, the assembly must be gaged and proper shim thickness be installed in the new assembly. To gage the axle housing, ring gear and brake housing assembly for proper shim thickness, proceed as follows:

Assemble the axle housing, final drive ring gear, brake outer housing and brake inner housing without the brake assembly and final drive sun gear and with carrier bearing cup and shim removed from the inner brake housing. Tighten the brake housing retaining stud nuts to a torque of 72-90 Ft.-Lbs. and install gage frame and block as shown in Fig. 329.

Measure the resulting gap between gage frame and gage block with feeler gage taking care to obtain accurate measurement. For final assembly of the axle housing, axle, final drive and brake assembly, select a shim of thickness nearest to the measured gap between gage frame and block. Eight different shim thicknesses are available from 0.038-0.040 to 0.080-0.082 in steps of 0.006.

302. **OVERHAUL DIFFERENTIAL.** Remove differential as outlined in paragraph 300. Service procedure for disassembly and reassembly of the model 4000 differential is the same as that for models 2000 and 3000; therefore, refer to paragraph 287 for overhaul of the removed unit. Exploded view of the model 4000 differential is shown in Fig. 330.

303. **BEVEL PINION.** To remove the main drive bevel pinion, first remove the differential assembly as outlined in paragraph 300, the hydraulic lift cover as outlined in paragraph 347, the gear type hydraulic pump as outlined in paragraph 366 and split tractor between transmission and rear axle center housing as outlined in paragraph 299.

Removal, overhaul and reinstalling procedure for bevel pinion on model 4000 is same as outlined in paragraph 288 for models 2000 and 3000.

304. **ADJUST DIFFERENTIAL LOCK.** To adjust linkage on model 4000, proceed as follows:

Disconnect adjustable (spring loaded) link (77—Fig. 331) from operating lever (76) and loosen locknut (78). Jack up right rear wheel, push down on operating lever (76) and turn rear wheel until dog clutch is aligned and differential lock is fully engaged. With foot pedal (80) resting on plat-

form (foot rest), adjust length of link (77) so that pin can just be inserted through link and operating lever while holding lever in engaged position. Shorten the link one turn from this length, reinstall pin through link and lever and tighten locknut.

305. **OVERHAUL DIFFERENTIAL LOCK.** The differential lock clutch can be removed from the differential assembly after removing the right axle and housing assembly. Refer to differential overhaul as outlined for model 2000 and 3000 tractors with differential lock in paragraph 287. To remove the axle and housing assembly, disconnect differential lock linkage, then refer to paragraph 306.

On model 4000 tractors, the differential lock operating lever shaft, shaft seal and fork can be removed after removing the right axle and housing assembly; refer to Fig. 332. Install new oil seal in bore of axle housing with lip of seal inward.

REAR AXLE AND FINAL DRIVE

306. **R&R REAR AXLE ASSEMBLY.** First, drain the differential center housing and hydraulic system, then proceed as follows: Remove wheel weights if so equipped. Disconnect hydraulic lift lower link, brake linkage, wiring to rear fender light and differential lock linkage on right axle assembly. Remove the fender, support tractor under center housing and attach hoist to rear wheel and axle unit. Remove the cap screws and the stud nuts retaining axle housing to center housing, then remove the wheel and axle assembly from tractor. Lay the wheel flat on floor with axle housing up to provide solid base for overhauling brakes, final drive and rear axle unit. When removing left axle assembly the differential assembly should also be removed from center housing to prevent the assembly from accidentally falling out and becoming damaged.

When reinstalling, place new sealing "O" ring on axle housing. Tighten the axle housing retaining cap screws to a torque of 110-135 Ft.-Lbs. and the retaining stud nuts to a torque of 130-160 Ft.-Lbs.

307. **OVERHAUL FINAL DRIVE AND REAR AXLE ASSEMBLY.** With the rear axle and housing assembly removed as outlined in paragraph 306, refer to exploded view in Fig. 330 and proceed as follows:

Unbolt and remove ring gear thrust block (61) from left axle housing. On model 4000 with differential lock, refer to Fig. 331 and remove the locknut (71) and set screw (70) retaining the actuating fork (72) and remove shaft (73)

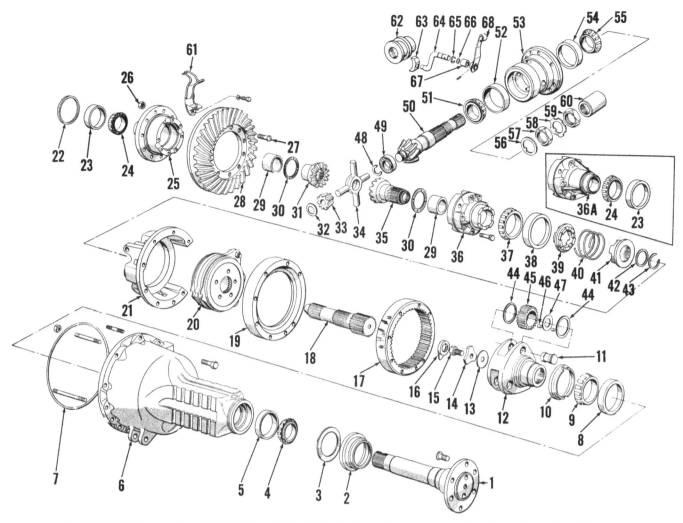

Fig. 330–Exploded view of model 4000 differential, final drive, brake and rear axle assemblies. On models without differential lock, right hand differential case half (36A) is used instead of (36).

1. Axle shaft	15. Cap screw	27. Ring gear bolts	36A. R.H. differential case
2. Oil seal	16. Cap screw lock	28. Bevel ring gear	half (w/o lock)
3. Gasket	17. Planetary ring gear	29. Bushings (2)	37. Bearing cone & roller
4. Bearing cone & roller	18. Planetary sun gear	30. Thrust washers (2)	38. Bearing cup
5. Bearing cup	19. Outer brake housing	31. Differential side gear	39. Adapter
6. Axle housing	20. Brake assembly	32. Thrust washers (4)	40. Spring
7. "O" ring	21. Inner brake housing	33. Differential pinions	41. Coupling
8. Bearing cup	22. Spacer shim	(4)	42. Washer
9. Bearing cone & roller	23. Bearing cup	34. Differential spider	43. Snap ring
10. Retainer	24. Bearing cone & roller	35. Differential side gear	44. Thrust washers
11. Pinion shafts	25. L.H. differential case	(with lock)	45. Planetary pinions
12. Planetary carrier	half	36. R.H. differential case	46. Needle rollers
13. Spacer shim	26. Ring gear nuts	half (with lock)	47. Spacer washer
14. Retaining washer			

48. Snap ring	60. Drive shaft coupling
49. Pilot bearing	61. Ring gear thrust
50. Bevel pinion gear	block
51. Bearing cone & roller	62. Select-O-Speed or
52. Bearing cup	torque converter
53. Pinion bearing	transmission
retainer	disconnect coupling
54. Bearing cup	63. Thrust block
55. Bearing cone & roller	64. Disconnect shaft &
56. Thrust washer	arm
57. Hex nut	65. Snap ring
58. Locking washer	66. "O" ring
59. Hex nut	67. Bushing
	68. Disconnect lever

and fork from housing. Remove the stud nuts retaining brake inner housing (21—Fig. 330) and remove the inner housing, brake assembly (20) and brake outer housing (19). Note: Refer to Fig. 347 for exploded view of model 4000 brake assembly. Lift out the final drive sun gear (18—Fig. 330) and the locking plate (16). On model 4000, it may be necessary to turn axle housing to align planetary (see Fig. 341) so that lock plate can be removed. Remove the cap screw (15—Fig. 330) retaining final drive planetary (12) to axle shaft and remove the retainer (14), spacer shim (13) and planetary assembly. Drive axle shaft seal (2) from outer end of axle housing and lift housing from

shaft.

To remove the bearing cone (9) and retainer (10) from planetary carrier (12) and to disassemble the early type planetary unit, refer to Fig. 334 and press out all three pinion shafts at one time which will also press retainer and bearing cone from carrier. Each planetary pinion (45—Fig. 330) is fitted with two rows of needle rollers (46) which are separated by a washer (47). On model 4000, there are 16 rollers in each row. The bearings are serviced in kit of 32 rollers. Renew planetary pinions, bearing rollers, thrust washers and/or bearing cone (9) as necessary. The carrier (12) is serviced only as a complete assembly including pinions and bear-

ings. To reassemble, proceed as follows: Coat bore of pinion gear with heavy grease, stick row of bearings in one side of pinion, insert center thrust washer (47) and stick second row of bearing rollers in pinion. Place a thrust washer (44) on each side of pinion, place the pinion, bearings and thrust washers in carrier and insert pinion shaft (11) with flat on end towards center of carrier. Press the pinion shaft into carrier as shown in Fig. 338. Repeat the procedure for remaining two pinions to complete assembly of the carrier, then press retainer (10) and bearing cone (9) onto carrier housing.

In September, 1969, a new type planetary unit was introduced. Refer to Fig.

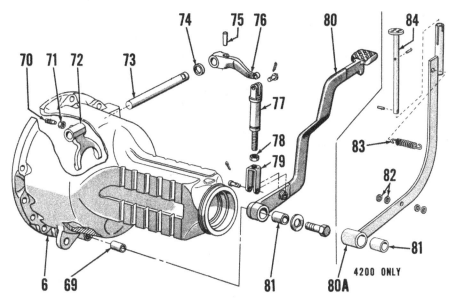

Fig. 331–Exploded view of model 4000 differential lock linkage. Model 4200 has different pedal (80A) than other models.

Fig. 334–Pressing the three planetary pinion shafts, retainer and bearing from early type planetary carrier. A flat plate and three pins of equal length are used to press shafts and retainer evenly. On late models, refer to Figs. 335 & 336.

6. Axle housing	73. Shaft	77. Spring loaded link	80A. Pedal (4200)
69. Pivot sleeve	74. Oil seal	78. Jam nut	81. Bushing
70. Set screw	75. Pin	79. Clevis	82. Spacers
71. Jam nut	76. Arm	80. Pedal (except 4200)	83. Coil spring
72. Fork			84. Foot pedal (4200)

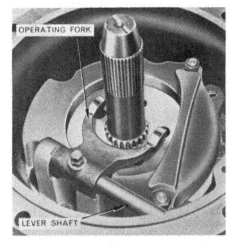

Fig. 332–View of model 4000 axle assembly showing differential lock operating fork and lever shaft.

Fig. 333–Removing brake assembly from axle housing.

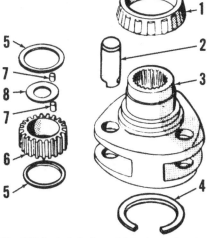

Fig. 335–Exploded view of late type planetary unit. Refer to Fig. 336 for disassembly techniques.

1. Bearing	5. Thrust washer
2. Pinion shaft	6. Planet pinion
3. Planet carrier	7. Needle roller
4. "C" ring	8. Spacer washer

335. Planet pinion shafts (2) are a slip fit in carrier (3) and are retained by "C" ring (4) which rotates into a slot in carrier and shafts as shown in Fig. 336. To disassemble the unit, straighten end of "C" ring and rotate ring until shafts are free. Withdraw shaft and slide out pinion and bearing, being careful not to lose any of the loose needle rollers (7—Fig. 335). Assemble by reversing the disassembly procedure. Bend the ends of "C" ring into depression in carrier as shown in Inset, Fig. 336. Planet pinions and bearings are interchangeable in old and new units.

If necessary to renew final drive ring gear, position removal tool (Nuday tool No. SW6-56) at outer side of gear as shown in Fig. 339, insert suitable bar through outer end of axle housing and press ring gear out of housing. To install new gear, attach the removal tool to gear and position the gear and tool on retaining studs as shown in Fig. 340, then press gear into housing. Remove the tool and, using a thin feeler gage, check to be sure ring gear is fully seated against shoulder of housing.

Using suitable bearing pullers, remove the bearing cone (4—Fig. 330) from axle shaft (1), then remove seal assembly (2) and gasket (3). Inspect sealing surface of shaft and remove all dirt, rust, paint and/or burrs which may damage new seal. Coat the shaft

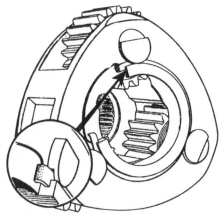

Fig. 336–Planet pinion shafts are retained by "C" ring as shown. Bend ends of ring into depression in carrier (inset) to secure the ring.

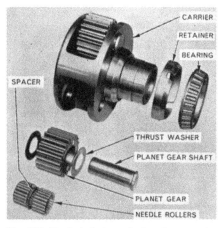

Fig. 337–Exploded view of planetary carrier showing two pinions already reinstalled; refer to Fig. 338.

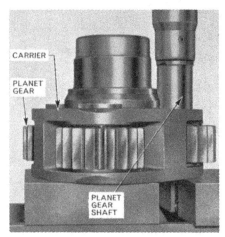

Fig. 338–Installing planetary gear unit in planetary carrier.

Fig. 339–View showing special planetary ring gear tool (Nuday tool No. SW6-56) placed behind ring gear in position to press gear from axle housing. Refer also to Fig. 340.

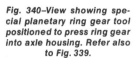

Fig. 340–View showing special planetary ring gear tool positioned to press ring gear into axle housing. Refer also to Fig. 339.

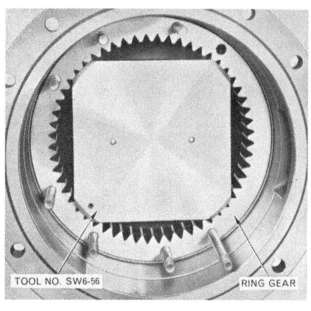

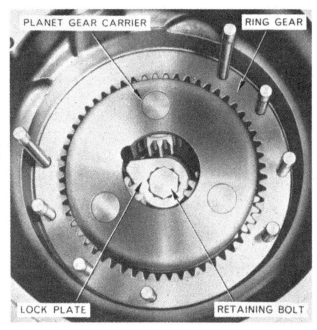

Fig. 341–View of planetary carrier installed in model 4000 rear axle housing.

and lip of new seal with grease, then install new seal and gasket over shaft. Inspect the bearing cups (5 and 8) in housing and renew if worn, eroded or chipped. Lubricate the bearing cone (4) and press cone tightly against shoulder on axle shaft. Lubricate bearing cups and lower the axle housing over shaft, seal and bearing. Lubricate bearing (9) and install the planetary carrier assembly into axle housing and over the axle shaft.

Axle shaft end play/bearing preload must be adjusted at this time; proceed as follows: Obtain thickest spacer shim (13) available and install this shim with retainer (14) on inner end of axle shaft. Install axle retaining cap screw and tighten to torque of 350-440 Ft.-Lbs. Mount a dial indicator with extended plunger against head of axle retaining cap screw as shown in Fig. 342 and zero the indicator dial. Lift the axle housing upward and note the dial indicator reading. Remove the cap screw, retainer and spacer shim, measure spacer shim thickness using a micrometer, then subtract dial indi-

cator reading from measured shim thickness. For proper axle bearing adjustment (axle end play of 0.001 to bearing preload of 0.003), select a shim of thickness not more than 0.001 larger or 0.003 smaller than the figure resulting from subtracting dial indicator reading from thick shim measured thickness. Note: Shim thickness may vary 0.001, so measure selected shim to be sure it is of appropriate thickness. Shims are available in eleven different thicknesses of 0.049-0.050 to 0.089-0.090 in steps of 0.004.

Install the selected shim, retainer and cap screw, tighten cap screw to a torque of 350-440 Ft.-Lbs. and install locking plate (16—Fig. 330). It may be necessary to tighten or loosen cap screw slightly in order to install locking plate.

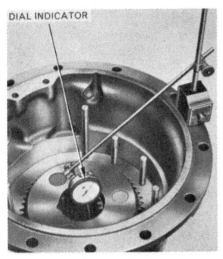

Fig. 342–View showing method of checking rear axle shaft bearing adjustment (end play) with dial indicator. Refer to text.

Be sure that axle seal (2—Fig. 330) is fully over end of axle housing, then stake rim of seal down into groove around axle housing in at least four equally spaced points. Reinstall final drive sun gear, outer brake housing, brake assembly and inner brake housing. If renewing differential carrier bearing cup (23) in brake inner housing, be careful not to lose or damage the shim (22) located between bearing cup and the brake housing. If shim is lost, or if the brake inner housing, brake outer housing, final drive ring gear and/or axle housing have been renewed, a new shim (22) should be selected as outlined in paragraph 301. Reinstall ring gear thrust block (61), where removed, and the differential assembly, if removed. Then, reinstall axle and wheel assembly as outlined in paragraph 306.

BRAKES

ADJUSTMENT
Models 2000, 3000 and
4110 L.C.G.

308. Refer to Fig. 343 and jack up tractor so that both rear wheels are free. Remove adjusting slot cover from brake backing plate and, using screwdriver, turn adjuster wheel towards rear of tractor until brake drags when wheel is turned. Back off adjuster wheel until brake drags very slightly when rear wheel is turned and reinstall adjusting slot cover. Repeat adjustment procedure on opposite brake.

Remove pin from brake rod clevis on right brake tie-rod and turn clevis so that pin can just be reinstalled when

Fig. 343–Brake shoe adjustment on models 2000, 3000 and 4110 L.C.G.

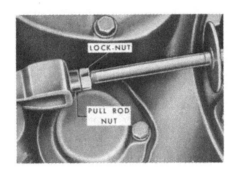

Fig. 344–To adjust brake linkage on model 4000, loosen lock-nut and turn pull rod nut as required; refer to text.

brake camshaft lever is pulled forward to take up clearance. Then, adjust left brake tie-rod so that pedals are in line when both brakes are engaged.

Model 4000 Except 4110 L.C.G.

309. Refer to Fig. 344 and disconnect the return spring on the right brake pedal allowing pedal to drop. Push pedal down to take up any slack in the linkage. Loosen lock nut on pull rod and turn pull rod nut until right brake

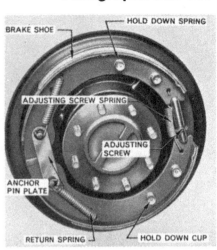

Fig. 345–View of brake assembly for models 2000 and 3000 with wheel and brake drum removed. Refer to Fig. 346 for exploded view.

pedal is 1½ to 1¾ inches below the left pedal; then, tighten lock nut. Disconnect return spring on left brake pedal, allowing pedal to drop, and push pedal down to take up any slack in linkage. Loosen lock nut on left pull rod and turn pull rod nut until left pedal is level with the right pedal. Check the adjustment by engaging the brake pedal lock; lock should engage easily if pedals are aligned. Tighten the lock nut and reinstall both pedal return springs.

R&R BRAKE SHOES
Models 2000, 3000 and
4110 L.C.G.

310. Jack up rear of tractor and remove rear wheels. Remove brake drums from axle shafts. It may be necessary to loosen brake adjustment slightly in order to remove drums.

Remove the brake shoe hold down spring cups, springs and pins. Remove

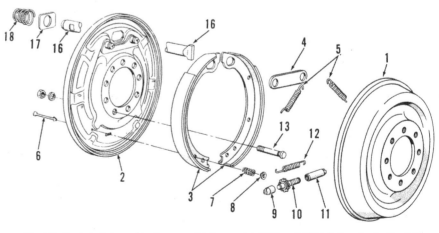

Fig. 346–Exploded view of brake assembly for models 2000 and 3000. Refer also to Fig. 345.

1. Brake drum	5. Return springs	9. Adjusting screw cup	13. Retainer bolts
2. Back plate	6. Hold down pins	10. Adjusting screw	16. Camshaft
3. Brake shoes	7. Hold down springs	11. Adjusting screw nut	17. Plate
4. Anchor pin plate	8. Hold down clips	12. Adjusting spring	18. Retainer spring

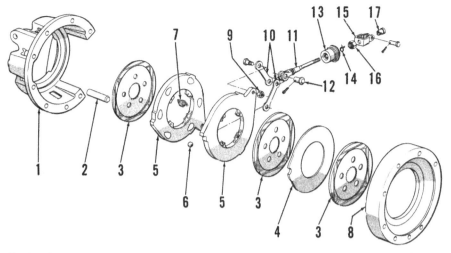

Fig. 347–Exploded view of model 4000 brake assembly. Brake inner housing (1) is same as (21–Fig. 330) and outer housing (8) is same as (19–Fig. 330).

1. Brake inner housing
2. Torque pin
3. Brake discs
4. Stationary disc
5. Actuating discs
6. Steel balls
7. Return springs
8. Brake outer housing
9. Jam nuts
10. Actuating links
11. Pull rod
12. Clevis pin
13. Oil seal
14. Clip
15. Clevis
16. Lock nut
17. Pull rod nut

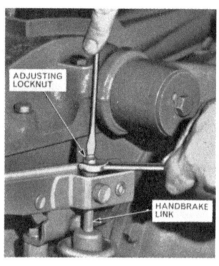

Fig. 348–Adjusting handbrake unit on models so equipped.

R&R BRAKE DISCS AND ACTUATING ASSEMBLY
Model 4000

311. Brakes are of the multiple disc wet type and are located in the rear axle housings. To gain access to the brakes, proceed as follows:

Remove rear fenders and disconnect electrical wiring and brake linkage. Drain oil from center housing. Remove rear wheel weights if installed. Support tractor under rear axle center housing and adequately brace tractor so that it will not tip with rear axles removed. Attach hoist to rear wheel and axle housing unit, unbolt and separate axle housing from center housing. Lay the wheel and tire unit down on floor with axle housing upward to provide a solid base for further disassembly of unit. Remove opposite wheel and axle housing unit in similar manner. Note: When left axle housing is removed it is suggested that the differential assembly be removed from the rear axle center housing to prevent the assembly from accidentally falling out.

On 4000 models with differential lock, it will be necessary to remove differential lock fork from right axle housing. Loosen differential lock fork

lock nut and set screw, then remove lever shaft and fork from axle housing.

Refer to Fig. 347 and proceed as follows: Unbolt and remove inner disc brake housing assembly from inner end of axle housing. Remove brake control rod fastener and, if damaged, remove brake rod seal. To remove seal, place a sharp tool between seal flange and housing and pry seal out. Remove the brake disc assemblies, intermediate discs, the actuating disc assembly and the outer brake housing.

NOTE: Where differential carrier bearing cup is installed in inner brake housing, differential carrier bearing adjustment is made by adding or removing shims from between bearing cup and brake housing. Therefore, if either the inner or outer brake housing are renewed, differential carrier bearing adjustment may be affected. Refer to paragraph 301.

The brake actuating assembly can be disassembled, if necessary, as follows: Remove clevis pin to disconnect control rod from the actuating links. Remove the four actuating disc return springs, separate the actuating discs and remove the six steel balls. Inspect all parts and renew as necessary. To reassemble, lay one disc on bench with inner side up and lay the six balls in the ramped seats. Lay the other disc on top of the balls so that the lugs for the torque pin are about one inch apart and install the four return springs. If self-locking nuts have been removed from links, install new nuts. Reconnect control rod to links.

To reinstall brake components in axle housing, proceed as follows: Install outer brake housing on the eight studs in axle housing and insert torque pin in brake housing. Then, refer to

Fig. 347 for reassembly of brakes. After inner brake housing is installed, tighten the retaining nuts to a torque of 72-90 Ft.-Lbs.

Reinstall axle housing to center housing using new "O" ring on axle housing. Tighten axle housing retaining cap screws to a torque of 110-135 Ft.-Lbs. and stud nuts to a torque of 130-160 Ft.-Lbs. Refill center housing to proper level with correct lubricant; refer to paragraph 340.

BRAKE CROSS SHAFT AND SEALS
All Models

312. On all models, the brake pedals are supported by a cross shaft running through the rear axle center housing. The right brake pedal pivots on outer right end of shaft and is retained by a snap ring and washer. The left brake pedal is clamped and keyed to the shaft. The shaft pivots in renewable bushings in the bores in center housing and an oil seal at outer side of each bushing prevents lubricant from leaking past the shaft. On all models with engine clutch the clutch release pedal pivots on left end of brake cross shaft.

To renew the cross shaft, shaft bushings, pedal bushings and/or cross shaft seals, proceed as follows: Drain rear axle center housing. Disconnect brake pedal return springs, brake rods and the clutch release rod. Remove snap ring and washer from right end of cross shaft and remove right brake pedal. Loosen clamp bolt in left brake pedal, remove pedal from shaft, extract the key from the shaft keyway and remove thrust washer. Remove all burrs from right end of shaft, then withdraw the shaft from center housing and, if so fitted, the clutch pedal. Remove clutch pedal.

Pry the oil seals from each side of center housing. Inspect the bushings in clutch pedal, right brake pedal and in bores of center housing. Renew any bushing that is excessively worn. To remove bushings from center housing, drive the bushings inward and remove through PTO shift lever or inspection plate opening. Bushings are presized and should not require reaming if carefully installed. Install new shaft oil seals with lip inward. Before inserting shaft, remove all burrs and lubricate shaft and seals. Position the clutch pedal, if so equipped, at left side of center housing. If not equipped with clutch pedal, place spacer on cross shaft, then insert the cross shaft through center housing. Be particularly careful when inserting shaft through right seal. Complete reassembly of tractor by reversing disassembly procedure. Refill center housing with proper lubricant.

TRANSMISSION HAND BRAKE
All Models So Equipped

NOTE: The transmission hand brake is not a regular production item, but will be found on some 3000 or other models imported from Ford production facilities outside the United States.

313. **BRAKE ADJUSTMENT.** Refer to Fig. 348 and proceed as follows: Push in release button and lower the hand brake lever to fully released position. Using a screwdriver to prevent link from turning, turn the adjusting locknut down (clockwise) until brake becomes solid, then back locknut off two turns to provide running clearance.

314. **R&R BRAKE ASSEMBLY.** Refer to the exploded view of brake assembly in Fig. 349 and proceed as follows: Drain center housing and hydraulic system and remove left step plate. Remove the adjusting nut (2) and spacer (3); then, unbolt and remove handle assembly (1) from brake housing. Remove cover plate (11), outer brake disc (12) as shown in Fig. 350, then slide actuating assembly from housing and remove the inner brake disc. The cap screws retaining brake housing (17—Fig. 349) and the bearing retainer (23) and shaft (29) assembly to center housing can then be removed. Be careful not to lose or damage shims (25) located between bearing retainer and center housing.

Except when parts which would affect brake drive pinion backlash have been installed, reinstall hand brake assembly with same number of shims between brake bearing retainer and

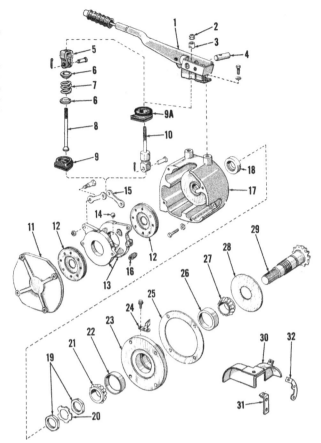

Fig. 349–Exploded view of transmission handbrake assembly. Brake shaft (29) is driven from bevel gear mounted on bevel pinion shaft. Rod (10) and seal (9A) are early production; late units use items (5) through (9).

1. Brake lever
2. Adjusting nut
3. Seat
4. Pin
5. Clevis
6. Collars
7. Spring
8. Rod (late)
9. Boot seal
10. Rod (early)
11. Cover
12. Discs
13. Actuating discs
14. Steel balls
15. Actuating links
16. Return springs
17. Housing
18. Oil seal
19. Hex nuts
20. Locking washer
21. Bearing cone & roller
22. Bearing cup
23. Bearing retainer
24. Oil trough
25. Shim
26. Bearing cup
27. Bearing cone & roller
28. Oil slinger
29. Brake shaft
30. Shroud
31. Mounting bracket
32. Mounting bracket

rear axle center housing. If parts have been renewed which would affect backlash, check for proper shim pack thickness as follows: Install the assembled bearing retainer and pinion shaft assembly in center housing without any shims and hold the pinion gears in mesh position. Measure the resulting gap between the bearing retainer and center housing with a feeler gage, then remove the bearing retainer and shaft assembly and reinstall it with a shim pack 0.013 thicker than the measured gap on models 2000 or 3000 and 0.009 thicker than the measured gap on model 4000.

Reinstall brake discs and actuating assembly, brake cover and brake lever, then adjust brake as outlined in paragraph 313.

315. **OVERHAUL BRAKE ASSEMBLY.** Remove the brake assembly as outlined in paragraph 314. The pinion shaft seal (18—Fig. 349) can be renewed at this time; install new seal with lip facing inward.

To disassemble actuating unit, remove clevis pin retaining clevis (5) or link (10) to operating links (15) and disconnect return springs (16). Be careful not to lose the steel balls (14) as the plates (13) are separated. Renew parts as necessary and reassemble by reversing disassembly procedure. If removed, stake the nuts on bolts retaining links (15) to plates (13).

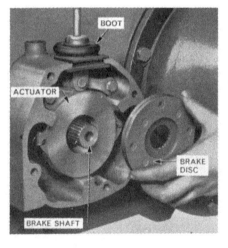

Fig. 350–View showing handbrake unit partly disassembled.

Clamp the flange of bearing retainer in soft jawed vise and straighten tabs on locking washer (20), then remove the nuts (19) and locking washer and push or bump shaft (29) out of retainer. Remove bearing cups (22 and 26) from retainer and the inner bearing cone and roller assembly (27) from shaft if renewal of bearings is indicated. Remove oil slinger (28) from shaft if necessary. Renew parts as required and reassemble by reversing disassembly procedure. Tighten inner nut (19) until a steady pull of 20-26 pounds is required to rotate shaft from cord wrapped around splined portion of shaft, then install locking washer and outer nut and recheck adjustment.

When adjustment is correct with outer nut tightened, bend tabs of locking washer against flats on nuts.

POWER TAKE-OFF

(Transmission Type)

All Models With 4-Speed Transmission

316. On models equipped with 4-speed transmission, power take-off is driven from rear end of transmission countershaft and disengaging the transmission clutch stops power flow to both the transmission and power take-off. Refer to exploded view of the power take-off unit in Fig. 351.

317. **R&R AND OVERHAUL OUTPUT SHAFT.** To remove the output shaft (9—Fig. 351), first drain lubricant from rear axle center housing. Then, unbolt retainer (4) and withdraw the output shaft, bearing and retainer assembly.

To disassemble, remove snap ring (8) from front of retainer (4) and remove shaft and bearing from the retainer. Oil seal (2) can now be removed from the retainer; install new oil seal with lip forward. Pull or press the bearing (6) and sleeve (3) from rear end of shaft and remove snap ring (7).

To reassemble, install snap ring in groove on output shaft, then press ball bearing tightly against snap ring.

Press new sleeve (3) onto shaft with chamfer on outside diameter to rear and press sleeve tightly against bearing. Lubricate oil seal and install shaft and bearing assembly in retainer, then install snap ring in retainer at front side of the ball bearing. Install the assembly into rear axle center housing using a new gasket and tighten the retaining cap screws to a torque of 35 to 47 Ft.-Lbs. Refill rear axle center housing and hydraulic system with proper lubricant.

318. **R&R AND OVERHAUL SHIFT COVER.** To remove the shift lever (20—Fig. 351) and cover (21), first drain rear axle center housing to below level of cover, remove the left step plate (foot rest) and then unbolt and remove shift lever and cover assembly from side of housing.

To disassemble, remove rivet from lever (20), remove lever from shift arm (23) and pull shift arm from cover. When reassembling, insert shift arm through cover, place new "O" ring on shaft, insert cap screw through top bolt hole in cover and then reinstall shift lever on shift arm shaft. With lever and

arm in relative position shown in Fig. 351, install lever to shaft retaining rivet.

Install lever and cover assembly in housing with new gasket (22) and with shift arm engaged in slot of fork (11). Tighten cover retaining cap screws to a torque of 35-47 Ft.-Lbs. Refill rear axle center housing and hydraulic system with proper lubricant.

319. **R&R AND OVERHAUL SHIFTER MECHANISM.** Power take-off shifter housing (17—Fig. 351) is also the transmission countershaft bearing retainer; refer to paragraph 200 for information on removal, replacement and countershaft bearing preload adjustment. To disassemble the removed unit, proceed as follows:

Remove nut and washer from front end of shift rail (13), pull rail from fork (16) and housing (17); catch detent ball (14) and spring as rail is withdrawn. Remove fork (16) and sliding coupling (10) from housing. Remove countershaft rear bearing cup (19) and seal (18) from front end of housing.

To reassemble, install new seal and bearing cup in front end of housing using proper size step plates and press. Insert sliding coupling into rear end of housing with wide land forward and place fork in groove on sleeve. Insert spring and detent ball in housing, hold ball down with punch and insert rail through housing and fork. Install fork retaining nut and washer. Install shift fork (11) on rear end of rail if removed. Reinstall unit as outlined in paragraph 200.

All Models with 6-Speed and 8-Speed Transmissions

320. Power take-off is driven from the PTO countershaft which rotates inside hollow transmission countershaft. On all models with single clutch, power flow to both transmission and power take-off is stopped when transmission clutch is disengaged. On 3000 models with dual clutch, power flow to transmission is stopped when clutch pedal is depressed about half-way and power flow to both transmission and power take-off is stopped when clutch pedal is fully depressed. Refer to Fig. 352 for exploded view of shifter mechanism, to Fig. 354 for exploded view of model 4000 output shaft assembly and to Fig. 353 for exploded view of output shaft assembly for other models.

321. **R&R AND OVERHAUL OUTPUT SHAFT ASSEMBLY.** For model 4000 only, refer to following paragraph 322; for all other models, proceed as follows:

First, drain lubricating oil from rear axle center housing and be sure that PTO shift lever is in engaged position.

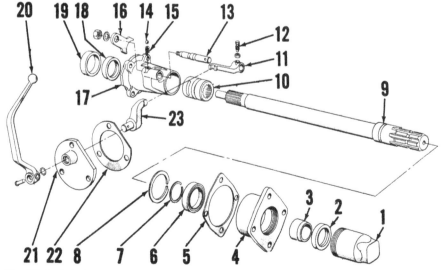

Fig. 351–Exploded view of power take-off unit used with four-speed transmission. Shifter support (17) is also retainer for transmission countershaft oil seal (18) and bearing (19).

1. PTO shaft cover	7. Snap ring	13. Shift rail	19. Bearing cup
2. Oil seal	8. Snap ring	14. Detent ball	20. Shift lever
3. Sleeve	9. Output shaft	15. Spring	21. Shift cover
4. Bearing retainer	10. Shift collar	16. Shift fork	22. Gasket
5. Gasket	11. Connector	17. Support	23. Shift arm
6. Ball bearing	12. Set screw	18. Oil seal	

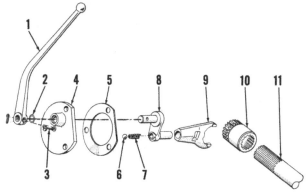

1. Shift lever
2. "O" ring
3. Pin
4. Shift cover
5. Gasket
6. Detent ball
7. Detent spring
8. Shift arm
9. Shift fork
10. Shift collar
11. Output shaft

Fig. 353–Exploded view of PTO output shaft assembly for model 2000, 3000 and 4110 L.C.G. tractors.

11. Output shaft
12. Snap ring
13. Snap ring
14. Ball bearing
15. Gasket
16. Bearing retainer
17. Sleeve
18. Oil seal
19. PTO shaft cover

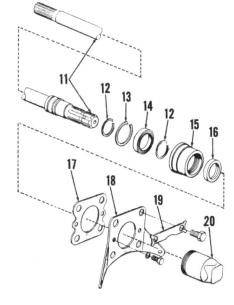

Fig. 354–Exploded view of PTO output shaft assembly for model 4000 (except 4110 L.C.G.) tractors.

11. Output shaft
12. Snap rings
13. Snap ring
14. Ball bearing
15. Bearing retainer
16. Oil seal
17. Gasket
18. Retainer plate
19. Lock plate
20. PTO cover

Then, unbolt retainer (16—Fig. 353) from rear axle center housing and withdraw the shaft, bearing and retainer assembly. Overhaul unit as outlined in paragraph 317 for 4-speed transmission equipped models.

NOTE: If the output shaft has been removed with the PTO shift mechanism in neutral position, or if the PTO shift lever has been moved while PTO output shaft is removed, the shift collar (10—Fig. 352) can fall out of shift fork (9). In either event, remove the shift cover and lever assembly as outlined in paragraph 323 and place collar on PTO countershaft in transmission before reinstalling output shaft.

Reinstall the PTO output shaft, bearing and retainer assembly with new gasket (15—Fig. 353) and tighten retaining cap screws to a torque of 35-47 Ft.-Lbs. Refill rear axle center housing with proper lubricant.

322. To remove PTO output shaft assembly on model 4000 tractor, proceed as follows: Drain lubricant from rear axle center housing and disconnect hydraulic lift lower link sway chains from retainer plate (18—Fig. 354). Be sure that the PTO shift lever is in engaged position, then unbolt and remove retainer plate from rear axle center housing and withdraw the output shaft, bearing and bearing support assembly.

To disassemble, remove snap ring (13) from front of support (15) and remove shaft and bearing from the support. Remove rear snap ring (12) from shaft, then pull or press the bearing

(14) from rear end of shaft. Remove front snap ring (12) if necessary. Remove seal (16) from bearing support.

To reassemble, install front snap ring on shaft and drive or press bearing tightly against snap ring and install the rear snap ring. Install new seal in support with sealing lip forward, lubricate seal and install shaft and bearing assembly in support. Install bearing retaining snap ring in front end of support, then install the assembly in rear axle center housing. Note: If output shaft was removed with PTO shift lever in disengaged position or if shift lever was moved while PTO output shaft was removed, remove shift lever and cover assembly as outlined in paragraph 323 and place shift collar (10—Fig. 352) on PTO countershaft in transmission before installing output shaft.

With output shaft assembly installed, reinstall retainer plate with new gasket (17—Fig. 354), tighten the large cap screws to a torque of 170-200 Ft.-Lbs. and the small cap screws to a torque of 35-47 Ft.-Lbs. Refill rear axle center housing and hydraulic system with proper lubricant.

323. R&R AND OVERHAUL SHIFT MECHANISM. First, drain rear axle center housing lubricant to below level of shift cover (4—Fig. 352) and remove the left step plate (foot rest). Then, unbolt cover from rear axle center housing and remove the cover, shift lever and shift fork assembly.

To disassemble, remove fork (9) from shift arm (8). Remove cotter pin from drilled pin (3), remove shift lever and pull shift arm from cover. Be careful not to lose detent ball (6) and spring (7) from shift arm. To reassemble, stick spring and detent ball in bore of arm with heavy grease, insert arm through housing and install new "O" ring (2). While holding arm in cover against detent spring pressure, install shift lever, retaining pin and cotter pin.

Place shift fork on arm and reinstall unit with new gasket (5). Tighten the retaining cap screws to a torque of 35-47 Ft.-Lbs. and refill rear axle center housing and hydraulic system with proper lubricant.

POWER TAKE-OFF

(Select-O-Speed)

All Select-O-Speed Equipped Models

324. On all Select-O-Speed equipped models, the PTO clutch is an integral part of the transmission assembly; refer to paragraph 276 for clutch service information. For service information on the output shaft assembly, refer to following paragraph 325.

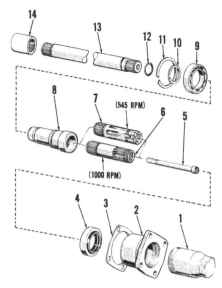

Fig. 355–Exploded view of PTO output shaft assembly used with Select-O-Speed transmission with 2-speed PTO clutch on models 2000 and 3000.

1. PTO cover	8. Connector
2. Bearing retainer	9. Ball bearing
3. Gasket	10. Snap ring
4. Oil seal	11. Snap ring
5. Socket head bolt	12. "O" ring
6. 1000 RPM output	13. Intermediate shaft
shaft	14. Coupling
7. 545 RPM output	
shaft	

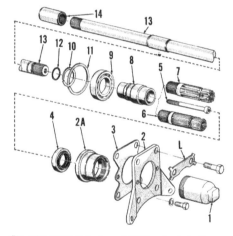

Fig. 356–Exploded view of PTO output shaft assembly for model 4000 tractor with Select-O-Speed transmission with 2-speed PTO option.

L. Lock plate	7. 545 RPM output
1. PTO cover	shaft
2. Retainer plate	8. Connector
2A. Bearing retainer	9. Ball bearing
3. Gasket	10. Snap ring
4. Oil seal	11. Snap ring
5. Socket head bolt	12. "O" ring
6. 1000 RPM output	13. Intermediate shaft
shaft	14. Coupling

325. R&R AND OVERHAUL OUTPUT SHAFT ASSEMBLY. To remove the output shaft assembly, first drain lubricant from rear axle center housing, then proceed as follows: On model 4000 (except 4110 L.C.G.), disconnect hydraulic lift lower link sway chains from retainer plate (2—Fig. 356), then unbolt and remove the retainer plate and withdraw output shaft, bearing and retainer assembly

from rear axle center housing. On all other models, remove cap screws retaining the output shaft assembly in rear axle center housing and withdraw the shaft, bearing and retainer assembly.

On models 2000, 3000 and 4110 L.C.G. with single speed PTO, the removed unit is similar to that used with other transmissions; refer to Fig. 351 or 353 for exploded view of unit and to paragraph 317 for service information. On model 4100 (except 4110 L.C.G.), the single speed PTO output shaft assembly is similar to that used with 8-speed transmission; refer to Fig. 354 for exploded view of unit and to paragraph 322 for service information.

To service 2-speed PTO output shaft assemblies, proceed as follows: Refer to Fig. 355 for models 2000, 3000 and 4110 L.C.G. and to Fig. 356 for other 4000 models. Remove the Allen head cap screw (5) and remove either the 540 RPM or 1000 RPM stub shaft (6 or 7). The output shaft (13) can then be removed from the sleeve (8) and the sealing "O" ring (12) can be renewed at this time. Remove snap ring (11) and press or drive sleeve and bearing from front end of retainer (2 or 2A). Remove

snap ring (10) and press sleeve from bearing assembly (9). Remove oil seal (4) from retainer. Reassemble by reversing disassembly procedure and using new "O" ring (12) and seal (4); install seal with crimped edge of steel shell forward.

When removing the shaft and bearing assembly from rear axle center housing, the connecting sleeve (14) will usually remain on the transmission PTO shaft. After reinstalling shaft assembly and before tightening retaining cap screws, check to see that the output shaft and transmission shaft are connected; if the output shaft will turn, remove output shaft and cover or handbrake assembly from left side of rear axle center housing, extract connector from bottom of housing and place it on transmission PTO shaft. Then reinstall the output shaft assembly. On model 4000 (except 4110 L.C.G.), tighten large cap screws to torque of 170-200 Ft.-Lbs. and small cap screws to a torque of 35-47 Ft.-Lbs. On all other models, tighten all cap screws to a torque of 35-47 Ft.-Lbs. Refill the rear axle center housing and hydraulic system with proper lubricant.

POWER TAKE-OFF

(Model 4000 With Independent PTO)

OPERATING PRINCIPLES

326. Model 4000 tractors equipped with an eight speed transmission are available with an independent type PTO. On models so equipped, the transmission clutch cover is fitted with a splined hub which drives the PTO input shaft. Reduction gearing in front end of transmission housing provides a constant running PTO clutch input shaft speed of 540 RPM.

When PTO control lever is in engaged position, the PTO control valve directs high pressure oil to the clutch piston (2—Fig. 357) which, due to valve design for gradual build-up in hydraulic pressure, gradually applies pressure to the clutch discs to slowly engage (feather) the clutch. At the same time, oil pressure is directed to the brake piston (25) which compresses spring (28) and releases pressure on brake arm (22).

When PTO control lever is in disengaged position, oil pressure to clutch piston and brake piston is relieved and the clutch piston return spring (5) disengages clutch discs and the brake

spring (28) moves brake piston to apply pressure to brake arm (22) stopping the clutch housing and PTO output shaft.

TROUBLE SHOOTING

327. **OPERATING CHECKS.** When trouble shooting problems are encountered with the model 4000 independent PTO, refer to the following:

PTO CLUTCH WILL NOT ENGAGE OR SLIPS UNDER LOAD. Could be caused by:
  A. Low rear axle oil level.
  B. Failure of hydraulic pump.
  C. Failure of connecting pipe.
  D. Control valve stuck open.
  E. Control valve spring broken.
  F. Cast iron sealing rings on clutch broken.
  G. Clutch piston sealing rings leaking.
  H. Brake piston sealing rings leaking.

PTO CLUTCH WILL NOT DISENGAGE. Trouble could be caused by:
  A. Control valve stuck.
  B. Clutch piston return spring broken.

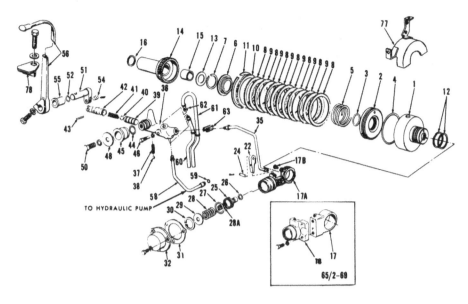

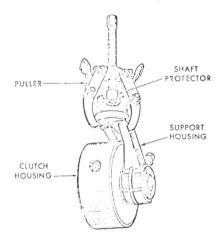

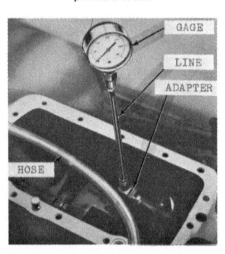

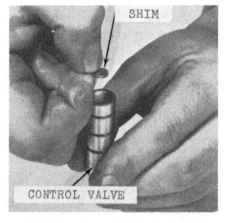

Fig. 357–Exploded view of independent PTO clutch, brake and control valve assemblies for model 4000 tractor.

1. Clutch housing	14. PTO clutch shaft	29. Washer	45. Bushing
2. Clutch piston	15. Bushing	30. Snap ring	46. Spigot bolt
3. "O" ring	16. Shims	31. Gasket	48. Washer
4. "O" ring	17. Brake support	32. Cover	50. Mounting bolt
5. Spring	17A. Support	35. Pressure tube	51. Shaft & Arm
6. Spring retainer	17B. Connector	36. Control valve body	52. "O" ring
7. Snap ring	18. Brake housing	37. Detent spring	55. Bushing
8. External spline	22. Brake lever	38. Detent ball	56. Control lever
plates	24. Pivot pin	39. Control valve	58. Pump to valve tube
9. Internal spline plates	25. Piston	40. Shims	59. "O" ring
10. Pressure plate	26. "O" ring	41. Spring	60. Discharge tube
11. Snap ring	27. "O" ring	42. Plunger	61. Discharge tube
12. Sealing rings	28. Outer spring	43. Pin	62. Tube seals
13. Thrust washer	28A. Inner spring	44. "O" ring	63. Tube clamp

Fig. 357A–Use puller attached as shown to remove and install brake piston in one-piece support. Refer to text.

Fig. 358–View showing pressure gage connected to PTO control valve for checking PTO system relief pressure. Refer to text.

Fig. 359–Placing adjusting shim inside control valve.

PTO CLUTCH DISENGAGES, BUT PTO STILL TURNS. Could be caused by:

A. Cold oil.
B. Brake piston stuck.
C. Brake piston return spring weak or broken.
D. Clutch piston return spring weak or broken.
E. Worn brake pad.
F. Clutch discs distorted or seized.

328. **PTO HYDRAULIC PRESSURE CHECK.** To check PTO clutch hydraulic pressure, first operate tractor until rear axle and hydraulic oil is at normal operating temperature. Then, stop tractor and remove the hydraulic lift cover as outlined in paragraph 347. Disconnect the control valve to clutch pressure tube (35—Fig. 357) at the control valve (36) and install a 0-400 psi hydraulic gage in valve assembly as shown in Fig. 358. Connect a tube from hydraulic lift cover feed tube as shown or mount the accessory cover from top of lift cover over tube opening in edge of center housing so that hydraulic pump flow is directed back into hydraulic sump in center housing. Start engine and operate at slightly above low idle speed. Engage and disengage the PTO clutch several times. Pressure gage reading should be nearly equal each time and within the limits of 175-200 psi. If gage reading is not within limits, remove control valve assembly as outlined in paragraph 329 and add shims (40—Fig. 357) (see Fig. 359 also) to increase pressure. If pressure cannot be increased by adding shims, remove and overhaul the hydraulic lift pump assembly; refer to paragraph 367. If pressure is within limits, but PTO clutch slips under load, remove and overhaul clutch as outlined in paragraphs 330 and 332.

OVERHAUL

329. **R&R AND OVERHAUL CONTROL VALVE.** To remove the PTO clutch control valve, first remove the hydraulic lift cover assembly as outlined in paragraph 347, then proceed as follows:

Disconnect the pump to control valve tube (58—Fig. 357) and the control valve to clutch support tube (35) at the control valve ends. Disconnect valve lever (51) from valve plunger (42), then unbolt and remove the valve and exhaust tubes (60 and 61) from center housing as an assembly. Remove the clamp (63) from exhaust tubes and remove the tubes from valve assembly.

To disassemble valve, use suitable pin punch to drive the plunger retaining roll pin (43) from valve housing, then carefully remove plunger to avoid losing the spring loaded detent ball (38). Remove the detent ball and spring (37) and the control valve spring

(41). Be careful not to lose any adjustment shims from the hollow control valve spool. If valve spool will not readily slide from housing, close the exhaust ports and one of the pressure ports with fingers and apply compressed air to other pressure port as shown in Fig. 360.

Clean all parts in suitable solvent, air dry and lightly oil. Inspect valve, plunger and valve bore in housing for excessive wear or scoring. If housing is not suitable for further use, install a new valve assembly. Parts other than the housing are available separately for service. Check the valve spring for cracks or distortion and against the following specifications: Valve spring free length should be 1.70-1.72 inches and a 43 pound load should compress spring to a length of 1.50 to 1.52 inches. Check detent spring for cracks or distortion; detent spring free length should be 1.19 to 1.21 inches.

To reassemble, insert shims removed in the control valve as shown in Fig. 359, or if a change in operating pressure is necessary (paragraph 328), insert additional shims to increase pressure or less shims to decrease pressure. Shims are available in one thickness (0.030) only. Insert control valve in housing with hollow end out, then insert valve spring inside valve. Place detent spring and ball in housing (see Fig. 361), hold the detent ball down with a pin punch and insert valve plunger in housing. Attach exhaust tubes to valve housing and install the valve and tube assembly with new "O" ring on valve boss.

Connect a hydraulic pressure gage to control valve as shown in Fig. 358 and reconnect pump to control valve pressure line. Check system pressure as outlined in paragraph 328; if pressure is not within limits, remove valve and change number of shims as required. When system pressure is correct, remove pressure gage, reconnect valve to

clutch pressure tube and reinstall hydraulic lift cover assembly.

330. R&R PTO CLUTCH AND BRAKE ASSEMBLY. To remove the PTO clutch and brake assembly, it is first necessary to remove the hydraulic lift cover as outlined in paragraph 347, the hydraulic pump as in paragraph 366, the PTO output shaft assembly as in paragraph 334 and to split tractor between transmission and rear axle center housing as outlined in paragraph 299. Note: Be careful not to lose any shim(s) that may be located on rear end of transmission PTO shaft or stuck to the PTO clutch shaft (see Fig. 363) when separating tractor. Then, proceed as follows:

NOTE: It is possible to remove the brake piston (25-Fig. 357) to renew "O" rings (26 & 27) without removing the assembly from transmission. Disconnect the left brake rod, remove cover (32) and using a 3-bolt wheel puller, compress spring (28) as shown in Fig. 362 and remove the snap ring (30—Fig. 357). Then, remove pullers, washer (29), spring (28) and piston (25). Reinstall by reversing removal procedure.

Disconnect left wheel brake link from cross shaft through center housing and right wheel brake link from brake pedal, remove the brake pedals from cross shaft and withdraw the shaft and clutch pedal from center housing. Unbolt and remove the clutch shroud (77—Fig. 357) and disconnect the pressure lines (35 and 58) from control valve and clutch support. Remove the PTO brake cover (32) from outside of center housing, then lift the clutch and brake assembly from center housing.

To reinstall the PTO clutch and brake assembly, reverse removal procedure. Renew the brake cross shaft oil seals if damaged in any way and remove all burrs, rust or rough spots on shaft and lubricate shaft before inserting it through bores in center

housing. Be sure to reinstall the same shim thickness on transmission PTO shaft (between pump drive gear retaining snap ring and clutch shaft) as was removed unless components have been renewed that would affect clutch shaft end play. If necessary to check clutch shaft end play, refer to following paragraph 331. Check PTO clutch pressure before reinstalling hydraulic lift cover; refer to paragraph 328.

331. ADJUST PTO CLUTCH SHAFT END PLAY. Shims are placed between the hydraulic pump drive gear retaining snap ring and the PTO clutch shaft (see Fig. 363) to limit clutch shaft end play to 0.016 to 0.025. If new components have been installed that would affect clutch shaft end play, the end play must be measured and a new shim stack thickness installed to bring end play within specified limits. To measure end play, proceed as follows:

If the clutch has been serviced as outlined in paragraph 332, reconnect transmission to rear axle center housing without installing the lift cover or hydraulic pump and without any shims between hydraulic pump drive gear retaining snap ring and the clutch shaft, but with gasket placed between transmission and rear axle center housing. Using a feeler gage, measure the gap between snap ring and clutch shaft (see Fig. 363), then separate tractor as before. When rejoining, place shim thickness equal to measured gap less 0.016 to 0.025 on the transmission PTO shaft against the snap ring.

If the clutch and brake assembly have not been removed, there is a possibility of some clearance between the clutch housing and support which would affect clutch shaft end play. Remove the plate from left side of rear axle center housing and using a feeler

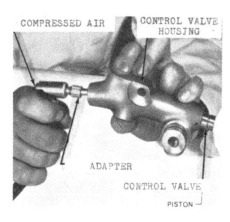

Fig. 360—Removing control valve from housing with air pressure.

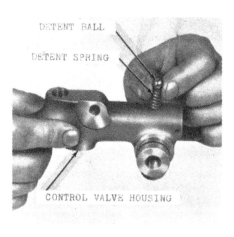

Fig. 361—Installing detent ball and detent spring in control valve housing.

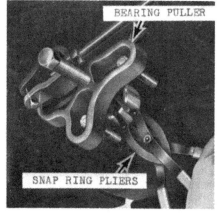

Fig. 362—Compressing PTO clutch brake spring to remove spring retaining snap ring.

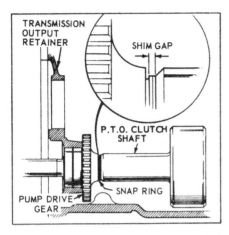

Fig. 363–On model 4000 with independent PTO, shims are placed between hydraulic pump drive gear retaining snap ring and PTO clutch shaft. Refer to text.

gage, measure any gap existing between PTO clutch housing and front face of support. If not already removed, remove the hydraulic pump assembly and rejoin transmission to rear axle center housing without any shims between hydraulic pump drive gear retaining snap ring and clutch shaft, but with a gasket between transmission and rear axle center housing. Working through pump opening, use feeler gage to measure gap between clutch shaft and snap ring as in Fig. 363. Add the measured gap to that measured between clutch housing and support. Separate the tractor, then rejoin placing shim thickness equal to total measured gap less 0.016 to 0.025 on the transmission PTO shaft.

332. OVERHAUL PTO CLUTCH AND BRAKE ASSEMBLY. With the clutch and brake assembly removed as outlined in paragraph 330, refer to exploded view of unit in Fig. 357 and proceed as follows:

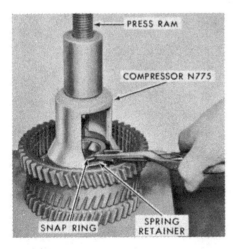

Fig. 364–Compressing clutch piston return spring to permit retaining snap ring removal. Select-O-Speed clutch assembly is shown; however, procedure is same.

NOTE: Beginning with production date of March 1969, a one-piece support and brake housing assembly (17A-Fig. 357) replaced the two-piece assembly (17 and 18) used on earlier models. A different service procedure should be used with the one-piece support assembly as outlined in the following paragraphs.

On models prior to production date of March 1969, unbolt the brake housing (18) from support (17) and remove the support from clutch housing (1) rear hub. Place brake housing in press and push washer (29) inward far enough to remove the snap ring (30), then release spring pressure and remove washer and springs (28 and 28A). Remove brake piston (25) from housing and remove "O" rings (26 and 27) from piston. Remove pin (24) and brake arm (22) from housing.

On models with production date of March 1969 and later, remove brake assembly from support before removing support from clutch housing hub as follows: Attach a puller as shown in Fig. 357A, push washer (29) inward far enough to remove snap ring (30), then release spring pressure and remove washer and springs (28 and 28A). Pull support (17A) from clutch housing hub, then remove brake piston (25) from support and remove "O" rings (26 and 27) from piston. Remove pin (24) and brake arm (22) from housing.

Remove the cast iron rings (12) from rear hub of clutch housing. Remove snap ring (11), clutch pressure plate (10) and clutch discs (8 and 9). Place clutch housing in press and using Select-O-Speed transmission clutch spring compressor (Nuday tool No. N775) as shown in Fig. 364, compress clutch spring (5—Fig. 357) and remove snap ring (7). Gradually release spring, then remove the compressor tool, retaining washer (6) and spring. Apply air pressure to port in rear hub of housing to remove clutch piston (2), then remove sealing rings (3 and 4) from piston.

Clutch spring free length should be 2.34 inches and spring should exert a force of 200-225 pounds when compressed to 1.88 inches. Clean all parts in suitable solvent, air dry and carefully inspect for cracks, scoring, excessive wear or other defect such as warped or overheated discs.

Brake outer spring (28) has a free length of 1.54 inches and should exert 195-205 pounds force when compressed to a length of 1.11 inches. Brake inner spring (28A) has a free length of 1.52 inches and should exert 95-105 pounds force when compressed to a length of 1.11 inches.

To reassemble, proceed as follows:

Install new sealing rings on clutch piston, lubricate the sealing rings and inside diameter of clutch housing, then install piston in housing. Place spring and spring retainer on piston, compress spring as during disassembly and install snap ring on clutch hub. Remove housing from press and install clutch disc with external splines next to piston, then alternately install internally splined and externally splined discs. Note: Units prior to March 1969 use seven externally splined discs whereas later units use six. Install clutch pressure plate and retaining snap rings. Install new cast iron rings on rear hub of housing, and lubricate the rings. On models prior to production date of March 1969, install support over the rings with chamfered inside diameter toward housing. Install new sealing rings on brake piston, lubricate the rings and insert piston in housing. Place brake housing in a press, install brake springs and install snap ring. Install brake arm in housing, be sure that machined face of support contacts rear face of clutch housing, then bolt the brake housing to support and tighten retaining cap screws to a torque of 14-17 Ft.-Lbs.

On models production date March 1969 and later, install brake arm (22) and pin (24) in support (17A), then install support over rings on clutch housing hub with oil port on top side. Install new sealing rings on brake piston, lubricate the rings and insert piston into support. Install brake springs and washer, attach puller (Fig. 357A), compress springs and install snap ring.

Reinstall clutch and brake assembly as outlined in paragraph 330.

333. R&R PTO OUTPUT SHAFT ASSEMBLY. To remove the PTO output shaft assembly, first drain the rear axle center housing and then proceed as follows:

NOTE: The PTO output shaft can be removed as outlined in the following paragraph. However, it is likely that the thrust washer (13—Fig. 357) located between the hub of the clutch housing (1) and the clutch shaft (14) will drop down preventing the PTO shaft from entering the bushing (15) in clutch shaft when reinstalling shaft. For that reason, the procedure outlined in paragraph 335 includes splitting the tractor between transmission and rear axle center housing so that the washer can be lifted up and held in position from the front while inserting output shaft from rear.

334. Remove PTO shaft shield if so equipped and disconnect hydraulic lift lower link check chains from the PTO output shaft bearing retainer plate (3

—Fig. 365). Unbolt and remove the retainer plate, then withdraw the output shaft and bearing assembly from rear end of rear axle center housing.

335. To reinstall PTO output shaft assembly, proceed as follows: Split tractor between transmission and rear axle center housing as outlined in paragraph 299. Insert a long screwdriver or similar tool through front end of clutch shaft (14—Fig. 357) and lift thrust washer (13) up even with bore in bushing (15) while inserting PTO shaft from rear. If assistance is not available, use heavy grease to hold thrust washer in position, then install the PTO shaft. Install support plate with new gasket, tighten large cap screws to a torque of 175-200 Ft.-Lbs., then tighten small cap screws to a torque of 35-47 Ft.-Lbs. Reconnect check chains to support plate and rejoin transmission to rear axle center housing as outlined in paragraph 299.

336. **OVERHAUL PTO OUTPUT SHAFT.** With the shaft assembly removed as outlined in paragraph 334, refer to exploded view in Fig. 365 and proceed as follows:

Remove snap ring (9) and press shaft and bearing assembly from retainer (6). Remove rear snap ring (7), then pull or press bearing (8) from shaft. Drive or press oil seal (5) from retainer.

If necessary, remove remaining snap ring (7) from shaft.

Renew parts as required, then reassemble as follows: Install front snap ring (7) if removed, then press or drive bearing firmly against the snap ring. Install rear snap ring (7) at rear side of bearing. Install new oil seal in retainer with lip of seal forward, lubricate seal and install retainer over shaft and bearing. Install snap ring (9) in retainer at rear side of bearing.

The front end of PTO output shaft is supported in a bushing that is pressed into bore in rear axle center housing. If necessary to renew the bushing, first remove the PTO clutch and brake assembly as outlined in paragraph 330 and the differential assembly as outlined in paragraph 300, then drive old bushing from bore and install new bushing with piloted arbor. Bushing is pre-sized and should not require reaming if carefully installed.

BELT PULLEY

OVERHAUL BELT PULLEY ASSEMBLY
Models 2000 and 3000

337. Refer to exploded view of unit in Fig. 366 and proceed as follows: Drain lubricant and remove cover (12). Straighten tabs on washer (10) and remove nuts (17), then remove shaft (9), outer bearing (7) and seal (8) from housing. Withdraw driven gear (4) and bearing (6) from housing. The drive gear (3) can now be removed. To renew seal (1), remove bearing (2) cup and drive seal to inside of housing. Install new seal with spring loaded lip to inside. To renew oil seal (8), pull bearing (7) cone and roller from shaft (9) and remove seal. Fit new seal over shaft

with lip to inside and reinstall bearing cone and roller. Reinstall shaft, seal and bearing as unit is being reassembled.

Adjust pulley shaft end play to 0.002 by means of the adjusting nuts (17) and bend tabs of washer (10) against nuts after they are tightened. Adjust drive gear shaft to 0.002 end play by installing shim gaskets (13) as required between cover and housing. Gaskets are available in thicknesses of 0.008-0.015, 0.013-0.020 and 0.018-0.023.

Model 4000

338. Refer to exploded view of unit in Fig. 367 and proceed as follows: Drain lubricant and unbolt pulley from hub (3). Unbolt pulley shaft bearing retainer (8) from housing (12) and re-

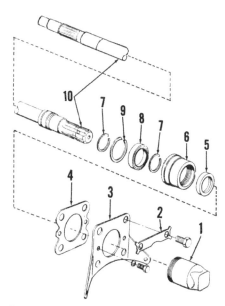

Fig. 365–Exploded view of model 4000 independent type PTO output shaft assembly. Front end of output shaft (10) is carried in bushing (15—Fig. 357) in PTO clutch shaft (14).

1. PTO cover	6. Bearing retainer
2. Lock plate	7. Snap rings
3. Retainer plate	8. Ball bearing
4. Gasket	9. Snap ring
5. Oil seal	10. PTO output shaft

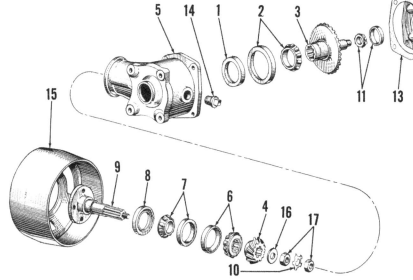

Fig. 366–Exploded view of belt pulley assembly available for models 2000 and 3000.

1. Oil seal	5. Housing	8. Oil seal	12. Cover plate
2. Tapered roller	6. Tapered roller	9. Pulley shaft	13. Shim gasket
bearing	bearing	10. Locking washer	14. Oil level plug
3. Drive gear	7. Tapered roller	11. Tapered roller	15. Belt pulley
4. Driven gear	bearing	bearing	

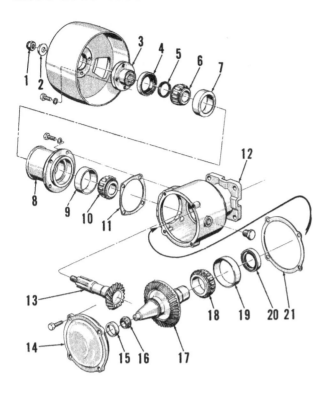

Fig. 367–Exploded view of belt pulley assembly available for model 4000 tractor.

1. Hex nut
2. Washer
3. Pulley hub
4. Oil seal
5. Seal ring
6. Bearing cone & roller
7. Bearing cup
8. Bearing retainer
9. Bearing cup
10. Bearing cone & roller
11. Gasket shim
12. Housing
13. Driven gear
14. Cover
15. Bearing cup
16. Bearing cone & roller
17. Drive gear
18. Bearing cone & roller
19. Bearing cup
20. Oil seal
21. Gasket shim

move the assembly. Unscrew nut (1) from pulley shaft and remove hub from shaft and shaft and bearing (10) from retainer. Remove seals (4 and 5) and bearing (6) from outer end of retainer. Remove bearing cups (7 and 9) if worn or scored.

Remove cover (14) and drive gear (17) and bearing assembly. Remove bearing cup (19) and oil seal (20) from housing. Remove bearing cup (15) from cover if cup is worn or scored.

Reassemble by reversing disassembly procedure. Install new seals (4 and 20) with lip to inside. Select a gasket shim (21) that will give 0.002 end play of drive gear (17). Tighten self-locking nut (1) to provide 0.002 end play of pulley shaft in bearings, then install bearing retainer (8) to housing with proper thickness of shim gasket (11) so that there is some backlash between pulley shaft pinion and drive gear. Shim gaskets (11 and 21) are available in thicknesses of 0.002, 0.005 and 0.010. Refill housing to oil level plug with SAE 80 EP gear lubricant.

HYDRAULIC LIFT SYSTEM

The hydraulic lift system for models 2000, 3000 and 4000 incorporates automatic draft control, automatic position control and, except on model 2000, pump flow (rate of lift) control. An auxiliary service port is provided for operation of remote single acting cylinders in conjunction with the 3-point hitch lift cylinder. Provision is made for installation of optional selector valve (see paragraph 368) for operating single acting remote cylinders either in conjunction with or independently of the 3-point hitch lift cylinder or for installation of either a single or dual spool remote control valve (see paragraphs 370, 377 and 385) for operation of either single or double acting remote cylinders. Fluid for the system is common with the differential and final drive lubricant and, on model 4000, the wet brake lubricant; fluid is separated from the transmission lubricant by oil seals. Hydraulic power is supplied by an engine driven piston type hydraulic pump except on model 4000 equipped with independent PTO. Models with independent PTO are equipped with a gear type hydraulic pump driven by a shaft splined into the engine clutch cover. The hydraulic system is protected by a wire mesh filter screen on pump intake and renewable paper element filter on sump return tube.

Fig. 368–View showing location of hydraulic controls on models 2000, 3000 and 4000.

HYDRAULIC FLUID

Models 2000, 3000 and 4110 L.C.G.

339. Recommended hydraulic fluid and rear axle final drive lubricant is Ford Hydraulic Transmission Fluid. Fluid capacity of model 4110 L.C.G. and all model 2000 is 23.5 quarts; fluid capacity of model 3000 is 20.5 quarts. Rear axle center housing should be drained and the system refilled with new lubricant after every 1200 hours of service or yearly. Drain system with 3-point hitch in lowered position and any remote cylinders retracted; check oil

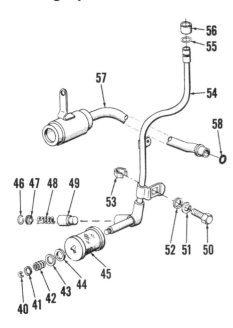

Fig. 369—Exploded view of suction and return line hydraulic fluid filters for models 2000, 3000 and 4000. System back-pressure valve is (49).

40. Snap ring	50. Mounting screw
41. Washer	51. Lock washer
42. Spring	52. Seal
43. Washer	53. Captive nut
44. Seals (2)	54. Return line
45. Filter element	55. "O" ring
46. Snap ring	56. Sleeve
47. Plate	57. Suction line & screen
48. Spring	58. "O" ring
49. Back pressure valve	

level with 3-point hitch in raised position and with any remote cylinders extended. Oil level plug is located in right side of rear axle center housing at front side of rear axle housing.

Model 4000 (Except Model 4110 L.C.G.)

340. The fluid used in rear axle center housing and hydraulic lift system must be compatible with the wet type disc brakes used on model 4000 tractor. The recommended fluid is Ford Hydraulic Transmission Fluid. Capacity of models with independent PTO is 28 quarts; capacity of other model 4000 tractors is 26¼ quarts. Rear axle center housing should be drained and refilled with new lubricant after each 1200 hours of service or yearly. Drain system with 3-point hitch in lowered position and with any remote cylinders retracted; check oil level with 3-point hitch in raised position and with any remote cylinders extended. Oil level plug is located in right side of rear axle center housing just in front of rear axle housing.

HYDRAULIC FILTERS

All Models

341. The hydraulic system filters are accessible after removing the hydraulic lift cover as outlined in paragraph 347.

To remove the paper filter element, refer to Fig. 369 and remove the snap ring (40), then remove washer (41), spring (42), washer (43), seal (44), the element (45) and second seal. When installing new filter element, a new seal (44) should be placed at each end of filter.

The wire mesh filter on pump intake line (57) should be cleaned and the paper filter element remewed after each 2400 hours of service or each two years to coincide with oil drain and refill period as outlined in paragraph 339 or 340.

TROUBLE SHOOTING

All Models

342. When trouble shooting problems encountered with the hydraulic lift system, refer to the following malfunctions and possible causes:

A. FAILURE TO LIFT UNDER ALL CONDITIONS. Could be caused by:
1. Low oil level in rear axle center housing.
2. Flow control valve binding.
3. Hydraulic piston pump not primed.
4. Hydraulic pump pressure low.
5. Check valve damaged or worn.
6. Draft control or position control linkage damaged.
7. Unload valve or back pressure valve faulty.
8. Lift piston seals damaged.
9. Unload valve plug worn.

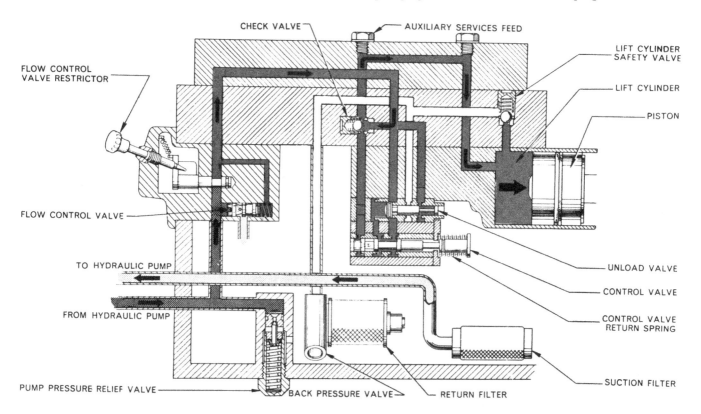

Fig. 370—Schematic diagram of model 3000 and 4000 hydraulic system. Model 2000 is similar, but does not include flow control valve and valve restrictor.

Fig. 371–Connect pressure gage (see Fig. 373) at plug (P) opening in piston type hydraulic pump.

Fig. 372–Connect pressure gage at plug (3) opening on model 4000 with gear type hydraulic pump. On all models, reverse accessory cover (2) to block oil pressure from lift cover when checking pump relief pressure.

1. Control lever	3. Pressure port
2. Accessory cover	4. Cap screws

10. Lift cylinder, lift cover or pressure transfer tube cracked.

B. FAILURE TO LIFT UNDER LOAD. Could be caused by:
1. Hydraulic pump pressure low.
2. Damaged "O" rings between lift cylinder and lift cover or between accessory cover and lift cover.
3. Damaged "O" rings on hydraulic pump pipes.
4. Damaged lift cylinder safety valve.
5. Faulty lift piston seals.
6. Cracked or porous lift cylinder or lift cover casting.

C. EXCESSIVE CORRECTIONS (BOBBING OR "HICCUPS") IN RAISED OR TRANSPORT POSITION. Could be caused by:

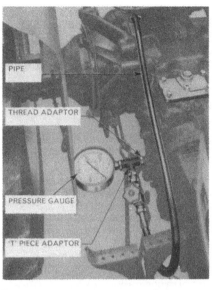

Fig. 373–View showing pump relief pressure check set up. Connect gage to plug (P–Fig. 371) on models with piston type pump or to plug (3– Fig. 372) on model 4000 with gear type pump.

1. Worn or damaged check valve ball or seat.
2. Selector valve worn or damaged.
3. Unload valve plug worn.
4. Lift cylinder safety valve damaged.
5. Faulty lift piston seals.
6. Control valve worn.
7. Damaged "O" rings between lift cylinder and lift cover or between lift cover and the accessory cover.
8. Cracked or porous lift cylinder or lift cover castings.

D. OCCASIONALLY FAILS TO LIFT NOT DUE TO LOAD. Could be caused by:
1. Worn or loose selector valve.
2. Unload valve sticking.
3. Faulty back pressure valve.
4. Control valve incorrectly adjusted.

HYDRAULIC PRESSURE CHECK
All Models

343. On models with piston type (engine driven) hydraulic pump, refer to Fig. 371 and remove plug (P) from rear cover of hydraulic pump.

On model 4000 with gear type hydraulic pump mounted in right side of rear axle center housing, refer to Fig. 372 and remove plug (3) from lower side of flow control valve housing.

Connect a 0-4000 psi hydraulic pressure gage, a shut-off valve and a ½-inch hose long enough to reach to hydraulic oil filler plug opening. Refer to Fig. 373 for typical gage, valve and hose installation.

On all models, refer to Fig. 372 and remove the accessory plate (2). Note: Place selector lever in draft control and

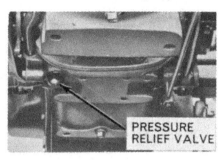

Fig. 374–View showing location of pressure relief valve assembly on models 2000 and 3000.

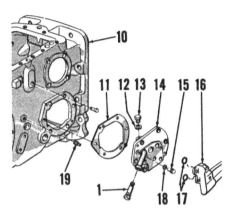

Fig. 375–On model 4000 (except 4110 L.C.G.) with piston type hydraulic pump, pressure relief valve assembly (1) is located in adapter plate (14).

1. Pressure relief valve	14. Adapter
10. Rear axle center housing	15. Plug
11. Gasket	16. Manifold
12. Seal	17. "O" rings
13. Plug	18. "O" ring
	19. Oil level plug

lower the 3-point hitch before removing plate. Fabricate a gasket that will cover the two oil passages in outer edge of lift cover between the cap screw (4) holes and with holes in gasket for the cap screws. Turn accessory cover end for end and reinstall with gaskets and the two long cap screws (4); it will be necessary to place flat washers under the cap screw heads to keep them from bottoming.

Open the shut-off valve, start engine and set engine speed at 1650 RPM. Gradually close the shut-off valve while observing pressure gage; pressure should increase to 2500 psi, then remain steady. If pressure gage reading with valve closed is not approximately 2500 psi, refer to Fig. 374 for models 2000 and 3000, or to Fig. 375 for model 4000 with piston type pump and remove the pressure relief valve assembly. On model 4000 with gear type hydraulic pump, remove pressure relief valve assembly from bottom flange of the pump assembly. Note: It will be necessary to drain the hydraulic system or to plug the opening while the pressure relief valve assembly is removed.

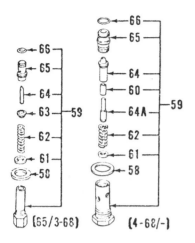

Fig. 376–Exploded view of early (65-3/68) and late (4/68 & up) hydraulic system relief valve for all models.

58. Seal	63. Spring seat
59. Relief valve assy.	64. Relief valve
60. Seal	64A. Guide pin
61. Shim	65. Valve seat
62. Spring	66. "O" ring

Refer to exploded view of pressure relief valve in Fig. 376 and unscrew the relief valve seat (65), then remove the valve (64), spring seat (63); or seal (60), guide pin (64A), and spring (62). Add one shim to shim stack (61) to increase relief valve setting or remove a shim to decrease setting; a difference in shim stack thickness of 0.010 should change relief valve setting approximately 100 psi. CAUTION: Measure total shim stack thickness and do not install a total thickness of more than 0.080. Shims are available in thicknesses of 0.010, 0.015 and 0.025. Reassemble the relief valve, reinstall with new "O" ring (66) and recheck pressure.

If adding a shim did not materially change the pressure gage reading, remove and overhaul the hydraulic pump assembly. Refer to paragraph 360 for piston type pump or to paragraph 366 for the model 4000 gear type pump assembly.

ADJUSTMENTS

344. **MAIN DRAFT CONTROL SPRING.** On models 3000 and 4000, the main draft control spring assembly (Fig. 377) is designed for draft control of mounted implements with the top link under either tension or compression. Note: If draft control from top link tension is not desired on model 3000 or 4000, refer to Fig. 379 and install washer (W). On model 2000, draft control is obtained with top link under compression only; refer to Fig. 378.

NOTE: Beginning with hydraulic lift unit production code date of September 16, 1969, the following changes have been incorporated into the hydraulic lift units. A

Fig. 378–Adjusting main draft control spring on model 2000 hydraulic system.

solid type quadrant has been installed and the quadrant retainer is no longer used. The quadrant control lever now hooks over the quadrant and locates on flats machined on a new control lever shaft instead of being keyed to the shaft. See Fig. 380.

A complete range of new control valve and sleeve assemblies have been intro-

Fig. 379–A washer (W) can be placed between main draft control front spring seat (6) and housing (2) to prevent draft control reaction from tension on top link. Shims (5) do not need to be removed.

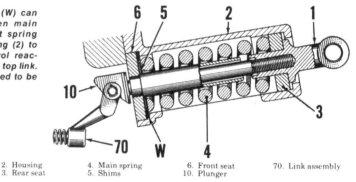

W. Washer	2. Housing	4. Main spring	6. Front seat	70. Link assembly
1. Yoke	3. Rear seat	5. Shims	10. Plunger	

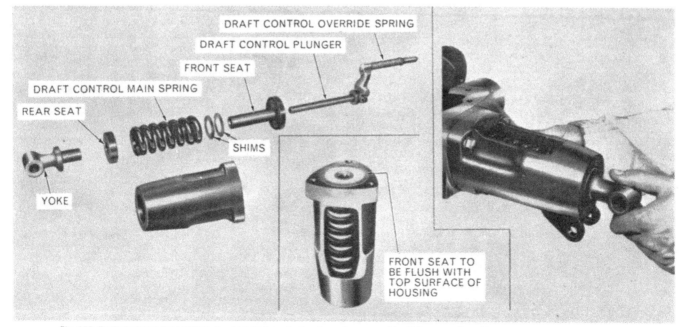

Fig. 377–Exploded and assembled views of draft control main spring assembly. Shims are placed between main spring and front seat to bring front seat flush with housing (early units) as shown in center view. View at right shows turning yoke to adjust main spring.

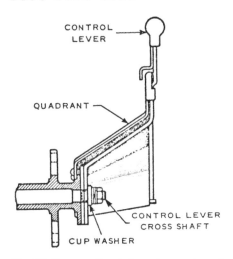

Fig. 380–Cross-sectional view of current quadrant assembly. Control lever shaft has flats instead of Woodruff key.

duced. The new control valves have a stepped extension (pilot) at front end that provides a restriction to the flow of exhaust oil and smooths the lowering of heavy implements. The counterbore inside the sleeves has been shortened to accommodate the new control valves. See Fig. 395.

The previous parts will continue to be available for service and although previous parts are not interchangeable with current parts, the current parts can be installed in earlier units providing they are installed as a matching set.

When the new type control valves and sleeves are installed, adjusting dimensions for control valve differ. Since the new control valve is longer, the dimensions from end of valve to edge of bushing is 0.030 instead of the previous 0.200. A new valve setting gage SW-508A is available which has a 0.030 gage at one end and a 0.200 gage at the other end and can be used for both style valves.

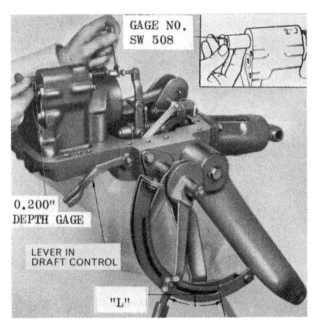

Fig. 381–Adjusting draft control linkage on early units. Refer to text for procedure and specifications.

The quadrant lever positions used during adjustment on previous units were specified as measured along the arc of the quadrant. While the current quadrant lever positions remain basically the same, they are now measured in a straight line (Fig. 382) for ease of measuring.

Prior to adjusting the main draft control spring on models 3000 and 4000, unscrew the yoke (Fig. 377) and unbolt and remove the spring housing, rear seat, draft control main spring and front seat. Place the rear seat, spring, shims and front seat in spring housing and check the front seat which should be flush with the top surface of spring housing for previous units using the shorter control valves (Fig. 377); or 0.010 above the top surface of spring housing for current units using the longer control valves. If necessary, remove the front seat and add or remove shims as necessary. Shims are available in thicknesses of 0.015, 0.020 and 0.025. Reinstall assembly, tighten retaining cap screws securely and adjust main draft control spring as follows:

On all models, with pin removed from rocker and main draft control spring adjusting yoke, refer to Fig. 377 or 378 and turn yoke in until all free play is removed. Then, unscrew yoke to nearest position that pin hole is horizontal and reinstall pin through rocker and yoke.

345. **ADJUST DRAFT CONTROL LINKAGE.** With lift cover and cylinder assembly removed as outlined in paragraph 347, turn main draft control spring yoke in until all spring free play is removed, then remove the baffle from front end of cylinder and proceed as follows:

Move the selector lever to draft con-

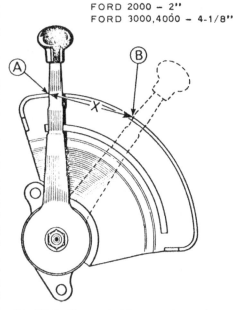

A–B DIMENSION
FORD 2000 – 2"
FORD 3000,4000 – 4-1/8"

Fig. 382–Position control lever as shown when adjusting draft control linkage on current units. Refer to text for procedure and specifications.

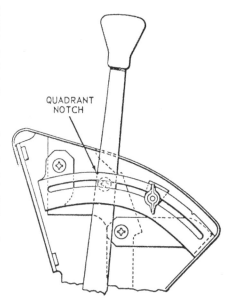

Fig. 383–Position control lever at top notch in quadrant when adjusting draft control linkage on model 4200. Refer to text for procedure and specifications.

trol and the lift arms to fully raised position. Move the quadrant control lever to position on quadrant so that the measurement (L—Fig. 381) or (X—Fig. 382) between top quadrant stop and control lever is 2 inches on model 2000, 4¼ inches on early 3000 and 4000 models or 4⅛ inches on current 3000 and 4000 models. (Note: Use top notch in quadrant for model 4200. See Fig. 383). The front end of the control valve spool should then be 0.200 (previous models) or 0.030 (current models) below flush with the front end of con-

Fig. 384–View showing special gage (Nuday tool No. SW-508) being used to adjust draft control linkage. Use gage No. SW-508A when adjusting current units. Refer to text.

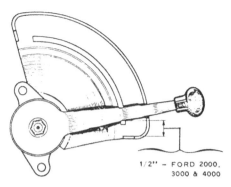

1/2″ – FORD 2000, 3000 & 4000

Fig. 386–Position control lever as shown when adjusting position control linkage on current units. Refer to text for procedure and specifications.

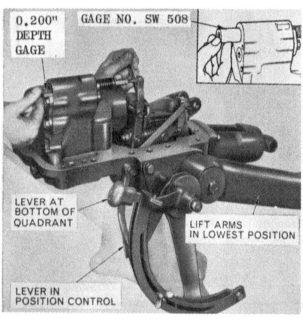

0.200″ DEPTH GAGE

GAGE NO. SW 508

LEVER AT BOTTOM OF QUADRANT

LIFT ARMS IN LOWEST POSITION

LEVER IN POSITION CONTROL

Fig. 385–Adjusting position control linkage. Refer to text for procedure and specifications. Also, refer to exploded view of linkage in Fig. 388.

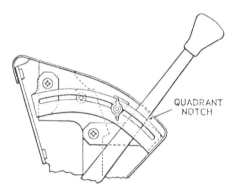

QUADRANT NOTCH

Fig. 387–Position control lever at bottom notch in quadrant when adjusting position control linkage on model 4200. Refer to text for procedure and specifications.

trol valve bushing. Measure control valve position with an accurate depth gage or with special adjustment gage (Nuday tool No. SW-508A). NOTE: The Nuday SW-508A adjustment tool has a 0.200 gage on one end and a 0.030 gage on the other and can be used to adjust previous and current units.

If necessary to adjust, loosen lock nut on control valve turnbuckle and turn the turnbuckle in or out as required until correct control valve spool position is established, then tighten lock nut and recheck adjustment. See Fig. 384.

With draft control linkage correctly adjusted, adjust the position control linkage as outlined in the following paragraph 346.

346. ADJUST POSITION CONTROL LINKAGE. With draft control linkage adjusted as outlined in paragraph 345, refer to Fig. 385 and proceed as follows:

Move the selector lever to position control and the lift arms to fully lowered position. Move the quadrant control lever to bottom of quadrant for previous units (Fig. 385), ½-inch from bottom of quadrant for current units (Fig. 386) or bottom notch in quadrant for model 4200 (Fig. 387). The front end of the control valve spool should then be 0.200 (previous units) or 0.030 (current units) below flush with the front end of control valve bushing. The control valve position should be measured with an accurate depth gage or special adjusting gage as outlined in paragraph 345.

If necessary to adjust position control linkage, refer to Fig. 388, hold plate (65) from turning with a wrench while loosening nut (64); failure to hold plate from turning may cause pin (66) to be sheared or bent. With nut loosened, turn position control rod (57) in or out as necessary to bring control valve spool to proper position. When properly adjusted, tighten nut (64) while holding plate (65) and rod (57) from turning, then recheck adjustment and readjust if necessary.

LIFT COVER AND CYLINDER ASSEMBLY

347. R&R LIFT COVER AND CYLINDER ASSEMBLY. On model 4200 rowcrop, remove the operator's seat, platform and control console assembly; on other models, remove the operator's seat. Thoroughly clean the lift cover and adjacent parts, then on all models, proceed as follows:

Move the selector lever to draft control and the control lever to bottom of quadrant to fully lower the 3-point hitch lift arms. Remove the pins that connect main draft control spring yoke to top link rocker and the lift shaft arms to lift links. If so equipped, remove any remote control valves and related remote control hose. Unbolt the lift cover assembly from rear axle center housing and remove from tractor. If lift cover or cylinder are to be serviced, thread two long bolts into the operator's seat mounting holes in top of lift cover and clamp the bolts in a

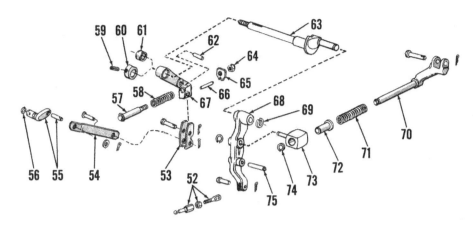

Fig. 388–Exploded view of hydraulic system control linkage. Spacer (61) is used on model 2000; cam (60) is used on models 3000 and 4000.

52. Turnbuckle	58. Spring	64. Jam nut	70. Link
53. Cam	59. Roll pin	65. Cam plate	71. Override spring
54. Link	60. Cam	66. Pin	72. Bushing
55. Selector arm	61. Spacer	67. Position control arm	73. Swivel
56. "O" ring	62. Pin	68. Actuating arm	74. Snap ring
57. Position control rod	63. Control lever shaft	69. Snap ring	75. Pin

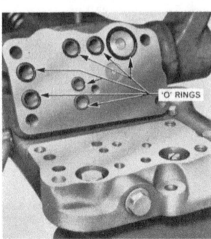

Fig. 389–Flow control override adjuster and cam follower rides in bore in edge of rear axle center housing. Cam follower contacts cam (60–Fig. 388) on control lever shaft (63–Fig. 388). Cam action returns flow control valve to full flow position when lift control lever is in raise position. Note gap between adjuster and cam follower; place steel rule in gap while installing lift cover.

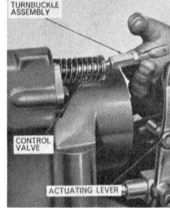

Fig. 391–Disconnecting turnbuckle from actuating lever. Turnbuckle can now be pulled from rear end of control lever.

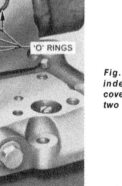

Fig. 390–View showing accessory cover removed from hydraulic lift cover assembly. New "O" rings should be installed before reinstalling cover.

Fig. 392–Removing lift cylinder assembly from lift cover. Note hollow dowels at two of the lift cylinder retaining bolt holes.

heavy duty vise with bottom of lift cover upward and so that all components are accessible. If lift cover is to be stored for other service, be sure that no weight is placed on control linkage or levers. Refer to paragraphs 348 to 355 for lift cover and/or cylinder overhaul.

When reinstalling the lift cover, position a new gasket on the rear axle center housing and install a new "O" ring on pressure supply tube. On models 3000 and 4000, turn the flow control knob fully in and insert a steel rule in the flow control adjuster and plunger (see Fig. 389) so that the plunger cannot be forced downward when installing cover. Then, carefully install the lift cover and cylinder assembly on the rear axle center housing and, on models 3000 and 4000, withdraw the steel rule supporting the flow control plunger. Install the retaining cap screws and alternately and evenly tighten to a torque ot 40-45 Ft.-Lbs.

348. **R&R LIFT CYLINDER ASSEMBLY.** With the lift cover and cylinder assembly removed as outlined in paragraph 347, proceed as follows:

Unbolt and remove the accessory cover (see Fig. 390) or selector valve, if so equipped, from top of lift cover and remove the control valve turnbuckle assembly as shown in Fig. 391. Remove the four cap screws from top of lift cover that retain the cylinder assembly, then remove the cylinder assembly from cover and lift piston connecting rod. See Fig. 392.

When reinstalling lift cylinder, renew all of the "O" rings in counterbores in top of cylinder, install the cylinder over piston rod and to lift cover so that the dowels enter holes in cover. Tighten the retaining cap screws to a torque of 50-55 Ft.-Lbs. Place new "O" rings in counterbores of accessory cover, install cover assembly on top of

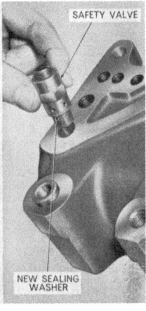

Fig. 393–View at left shows safety valve assembly for model 4000 tractor; view at right is safety valve being installed on model 2000 and 3000 lift cylinder.

a small rod inserted into valve opening, push piston from rear end of cylinder.

Inspect the lift piston and cylinder bore for scoring or excessive wear and inspect cylinder casting for cracks. If cylinder bore is deeply scored, the cylinder must be renewed. Renew piston if worn or scored.

To install piston seals, soak the leather back-up ring in water for about two minutes, then install the ring in groove on piston with rough side of leather forward (toward closed end of piston). Lubricate the "O" ring in hydraulic fluid and install it in groove at front side of leather back-up ring. After allowing the leather back-up ring to shrink to original size, lubricate the piston and cylinder bore and install piston in cylinder.

lift cover and tighten retaining cap screws to a torque of 40-45 Ft.-Lbs.

349. OVERHAUL LIFT CYLINDER ASSEMBLY. To service the lift cylinder components, refer to following appropriate paragraphs 350 to 353. Refer to Fig. 394 for exploded view of the lift cylinder assembly.

350. LIFT CYLINDER SAFETY VALVE. To protect the lift cylinder from excessive hydraulic pressure due to shock loads imposed by rear mounted implements, a safety relief valve is threaded into the lift cylinder; refer to Fig. 393. The safety valve will open when pressure within the cylinder reaches 2750 to 2850 psi.

As the opening pressure of the lift cylinder safety valve is above hydraulic system relief pressure, the valve can be bench tested only by suitable connectors to a hand pump and a high pressure hydraulic gage. If test equipment is not available and condition of the safety valve is questionable, renew the valve. Note: **Do not** attempt to disassemble or adjust the safety valve assembly.

On models 2000 and 3000, the safety valve is threaded into top face of cylinder casting and extends into a hole in the lift cover; the valve can be removed without removing lift cover assembly from tractor by removing the accessory cover (or selector valve or remote control valve if so equipped) and unscrewing the valve assembly with a deep well socket.

On model 4000, the safety valve is threaded into front end of cylinder casting and can be removed after removing lift cover as outlined in paragraph 347.

When reinstalling the lift cylinder

safety valve, place a new seal on the valve and tighten to a torque of 45-55 Ft.-Lbs. on models 2000 and 3000 and to a torque of 75-90 Ft.-Lbs. on model 4000. Excessive tightening torque may distort the valve body causing improper valve operation.

351. LIFT CYLINDER PISTON AND SEALS. To remove the lift cylinder piston on models 2000 and 3000, apply air pressure to cylinder pressure port. On model 4000 tractor, remove the lift cylinder safety valve and using

352. CONTROL VALVE AND BUSHING. To service the control valve, first remove the lift cylinder assembly as outlined in paragraph 348, then proceed as follows:

Unbolt and remove the retainer plate (46—Fig. 394) and remove the valve (47) and spring (48) from plate, then slide spring from valve. Remove the baffle plate (36) from front end of cylinder and remove the bushing (47B) by carefully pressing bushing out towards front (closed) end of cylinder. Note: Special care should be used in selecting a sleeve to push bushing from cylinder casting to be sure that the bore in cylinder is not damaged in any

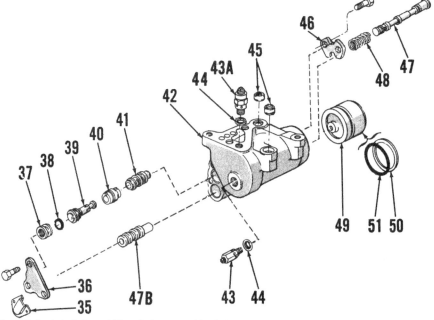

Fig. 394–Exploded view of lift cylinder assembly. Safety valve (43) for model 4000 screws into front end of lift cylinder; safety valve (43A) for models 2000 and 3000 screws into top of lift cylinder casting.

35. Baffle	40. Front bushing	44. Sealing ring
36. Front plate	41. Rear bushing	45. Hollow dowels
37. Unloading valve plug	42. Lift cylinder	46. Rear plate
38. "O" ring	43. Safety valve (4000)	47. Control valve
39. Unloading valve	43A. Safety valve (2000 & 3000)	47B. Valve bushing
48. Return spring		
49. Lift piston		
50. Backup ring		
51. "O" ring		

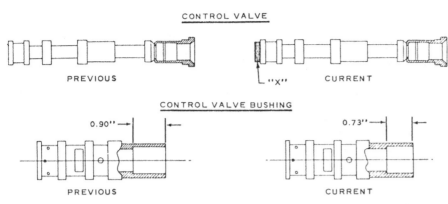

CONTROL VALVE

PREVIOUS   "X"   CURRENT

CONTROL VALVE BUSHING

0.90"   0.73"

PREVIOUS   CURRENT

Fig. 395–View showing type of control valve spool and bushing used in hydraulic lift units beginning with production date of September 16, 1969.

check valve spool fit, lubricate the valve spool and bushing and insert spool in bushing from open (rear) end of cylinder. A drag should be felt on the valve when moving it in bushing through normal range of travel. If valve moves freely through bushing, select a larger diameter valve spool; if valve sticks or binds, select a smaller diameter valve spool. The valve spool finally selected should be the largest diameter that will allow the valve to slide through the bushing from its own weight. The different valve size ranges are color coded as follows:

CONTROL VALVE SPOOLS

Size Range	Color Code
0.5917-0.5919	White
0.5919-0.5921	Blue
0.5921-0.5923	Yellow
0.5925-0.5926	Green
0.5927-0.5928	Orange

NOTE: As the color code indicates a size range only, a valve of one color code may fit correctly while other valves of same color code may fit too tight or too loose. As the color code on bushings indicates outside diameter only, do not attempt to select a control valve spool by matching spool color

way; if bore is scored during bushing removal, the cylinder must be renewed. When available, use of special bushing removal and installation tool (Nuday tool No. N-508-A) is recommended. Refer to Fig. 396.

Inspect the lands on control valve spool for erosion or scoring and renew the valve spool and bushing if either are damaged in any way, or if trouble shooting checks indicated leakage at the control valve. Neither the valve nor bushing should be renewed without renewing mating part.

NOTE: Hydraulic lift units beginning with production date of September 16, 1969, are fitted with control valves and control valve bushings of the type shown in Fig. 395. Notice that the current control valve spool is longer by the amount of the stepped extension "X" and the depth of counterbore in valve bushing has been decreased. Otherwise, dimensions of valve components are identical with earlier components and service procedures remain the same.

The control valve bushing is available in eight different outside diameter size ranges; the bushing is color coded to indicate the size range and the cylinder casting is also color coded near the bushing bore. Always renew the bushing with one of the same size range. Bushing outside diameter size ranges and color codes are as follows:

sure that bore in cylinder and bushing are clean and free of nicks or burrs. Lubricate both the control valve bushing and the bushing bore. Then, insert the bushing into the cylinder as shown in Fig. 394. Refer also to Fig. 397 for proper orientation. Press the bushing forward until the front face of bushing is flush with front machined face of lift cylinder casting.

Control valve spools are available in five different size ranges. The correct size valve can be determined by selective fit only **after** the bushing has been pressed into the cylinder casting. To

CONTROL VALVE BUSHINGS

Size Range	Color Code
1.0000-1.0002	Blue/White
1.0002-1.0004	White
1.0004-1.0006	Blue
1.0006-1.0008	Yellow
1.0008-1.0010	Green
1.0010-1.0012	Orange
1.0012-1.0014	Green/White
1.0014-1.0016	Red/White

When installing new bushing, be

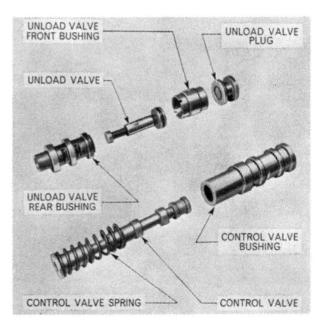

Fig. 396–Views showing use of valve bushing removal and installation tool (Nuday tool No. N-508-A). Earlier model control valve bushing is shown; however, tool can be used with all current models covered in this manual.

TOOL N508 A

Fig. 397–Exploded view of unloading valve and bushings and control valve and bushing showing relative valve and bushing location when installed in lift cylinder. Also refer to Figs. 395 and 398.

UNLOAD VALVE FRONT BUSHING

UNLOAD VALVE PLUG

UNLOAD VALVE

UNLOAD VALVE REAR BUSHING

CONTROL VALVE BUSHING

CONTROL VALVE SPRING   CONTROL VALVE

CURRENT TYPE

PREVIOUS TYPE

Fig. 398–View showing the previous (early) and current (late) type unloading valve front bushing.

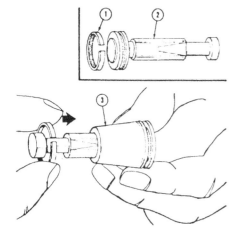

Fig. 399–View showing cast iron sealing ring (1) used on current unloading valves (2). Item (3) (Nuday tool No. SW24A) is a tapered guide used to install sealing ring on valve.

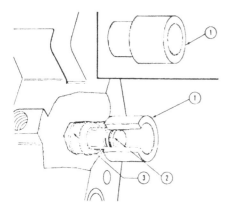

Fig. 400–View showing Nuday tool No. SW24B (1) being used to install unloading valve (2) in bushing (3).

code to bushing color code; however, it may happen that the valve spool selected and the bushing will be of the same color code.

To install control valve, slide spring over the valve and fit valve into notch in retainer plate. Lubricate the valve and spool and insert the valve. Tighten retainer plate cap screws to a torque of 25-30 Ft.-Lbs. Do not reinstall front

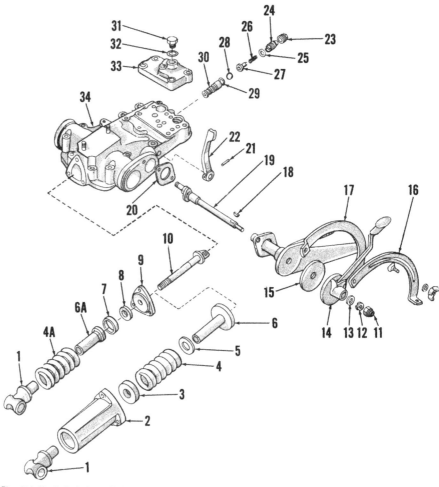

Fig. 401–Exploded view of lift cover assembly. Main draft control spring assemblies for 2000 and for models 3000 and 4000 are shown separately. Lift cover (34) may have 0.010 oversize bore for check valve bushing (30) on some tractors; refer to Fig. 403. Early type quadrant is shown.

1. Yoke	10. Plunger	18. Woodruff key	26. Spring
2. Housing	11. Self locking nut	19. Control lever shaft	27. Guide
3. Rear seat	12. Spring	20. Gasket	28. Check-valve ball
4. Main spring	13. Washer	21. Pin	29. "O" ring
5. Shims	14. Control lever	22. Selector lever	30. Check valve seat
6. Front seat	15. Friction disc	23. Plug	31. Plug
7. Rubber seal	16. Quadrant	24. Pilot	32. Sealing ring
8. Felt seal	17. Support	25. "O" ring	33. Accessory cover
9. Plate			34. Lift cover

(baffle) plate until lift cylinder has been reinstalled in cover and the control linkage is adjusted as outlined in paragraphs 345 and 346.

353. UNLOADING VALVE, BUSH-INGS AND PLUG. To remove the unloading valve, first remove the control valve as outlined in paragraph 352; it is not necessary to remove the control valve bushing if renewal is not indicated. Then, proceed as follows:

Thread a slide hammer adapter into the unloading valve plug (37—Fig. 394) and pull the plug from valve bore with slide hammer. The unloading valve (39) can then be removed by pushing the valve forward with a small screwdriver inserted through the rear bushing (41). Remove "O" ring (38) (early), or cast iron ring (late). Remove the unloading valve bushings if renewal is indicated by pressing them out towards front (closed) end of cylinder with a suitable sleeve. Note: Spe-

cial care should be taken in selecting the sleeve used to press bushings from cylinder casting to be sure not to damage bore. When available, use of special bushing removal and installation tool (Nuday tool No. N-508-A) as shown in Fig. 396 is recommended.

The unloading valve bushings (40 and 41—Fig. 394) are available in eight different outside diameter size ranges and should be renewed using parts of the same size range. It is usually not necessary to renew the unloading valve plug (37) unless damaged during removal procedure. The plug is also available in eight different size ranges and if plug, or a new plug of the same size range fits loosely in bore, a plug of the next larger size range should be installed. Unloading valve bushing and plug outside diameter size ranges and color codes for the different size ranges are the same as listed for the control valve bushing in paragraph 352.

NOTE: Changes have been made in the unloading valve and valve front bushing, however changes on the valve and front bushing occurred at different production dates.

Beginning with September 16, 1969 production the unloading valve front bushing has two notches at both ends whereas the previous front bushing had two notches in one end and one notch at the other end. The later front bushing can be installed in either direction. See Fig. 398.

Beginning with January 31, 1971 production, the unloading valve was changed to utilize a cast iron sealing ring instead of the previous "O" ring. See Fig. 399. At the same time, two special tools were made available to facilitate installation of the cast iron sealing ring onto the unloading valve (Nuday tool No. SW 24A) and the subsequent installation of unloading valve into the unloading valve bushing (Nuday tool No. SW 24B). See Figs. 399 and 400.

When installing new bushings, be sure that bore in lift cylinder is clean and free of nicks or burrs. Lubricate the bushings and bushing bore. The front (large I.D.) bushing (prior September 16, 1969 production) has one notch in one end and two notches in the other end; insert this bushing in rear (open) end of cylinder casting with end having single notch forward. Late bushings can be installed in either direction. Place the rear bushing against the rear (double notch) end of front bushing with long nose to rear. Refer to Fig. 397 for relative bushing position if necessary. Press both bushings forward into bore until rear side of land on rear bushing is flush with machined surface at rear side of lift cylinder.

Renew the unloading valve (39—Fig. 394) if scored, excessively worn or otherwise damaged. Valve is available in one size only. Lubricate the valve and insert it in bushings **without** the "O" ring (38) or cast iron ring. (Fig. 399). Valve should be a free sliding fit in bushings. Be sure that the front and rear bushing bores are concentric by checking clearance between unloading valve and inside diameter of front bushing at several different points. Remove the valve and install "O" ring or cast iron ring, lubricate the valve and ring and reinstall in bushings using Nuday tool SW 24B on those with cast iron sealing ring. The "O" ring should impart a slight drag when moving the valve back and forth in bushings. If the valve sticks or binds, or moves freely as without the "O" ring being installed, install a different "O" ring as there may be slight differences in "O" ring size. CAUTION: Do not in-

Fig. 402—Removing check valve seat with special puller (Nuday tool No. NCA-997) and adapter (NCA-997-A).

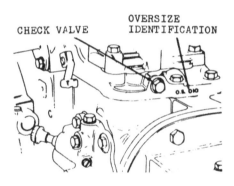

Fig. 403—Lift cover castings with 0.010 oversize bore for check valve bushing will be identified as shown.

stall an "O" ring of unknown quality or composition at this location. Some "O" ring materials may shrink or swell when subjected to hydraulic fluid and heat and therefore cause malfunction of the unloading valve.

With the unloading valve and "O" ring installed in bushings, install plug with threaded hole out and so that outer face of plug is flush with machined front surface of lift cylinder.

OVERHAUL LIFT COVER ASSEMBLY, CONTROL LINKAGE AND LIFT SHAFT

354. **HYDRAULIC SYSTEM CHECK VALVE.** The check valve for the hydraulic lift system is located in the front edge of lift cover; to remove the check valve on models 2000, 3000 and 4110 L.C.G., the lift cover must

first be removed as outlined in paragraph 347. Then, proceed as follows:

Unscrew the hex plug (23—Fig. 401 and using needle nose pliers, remove the pilot (24), spring (26), spring guide (27) and steel check valve ball. The check valve seat (30) is threaded on inside diameter. Screw the valve seat puller bolt (Nuday tool No. NCA-997 and NCA-997A adapter or equivalent) into valve seat until finger tight, then back out one turn. Refer to Fig. 402. While holding puller bolt from turning, screw puller nut down against spacer to remove the valve seat from bore in cover.

Inspect check valve seat and valve ball for worn spots, nicks, burrs or erosion and renew if any defect is noted. NOTE: On a few tractors, the check valve bore in lift cover may be 0.010 oversize requiring an oversize check valve seat; such lift covers will be identified by the code "O. S. 010" stamped in front edge of cover near the check valve bore as shown in Fig. 403. Renew the check valve spring if rusted, cracked or worn. Always renew the sealing "O" rings.

To install seat, remove spacer from valve seat puller and thread puller bolt into valve seat. Install new "O" ring on the seat, lubricate the "O" ring, seat and bore in lift cover. Then, using a steel hammer, drive valve seat into bore in lift cover until it bottoms against shoulder in bore. Insert check valve ball, spring guide and spring in bore. Install new "O" ring on check valve pilot, lubricate the "O" ring and push pilot into bore. Install and tighten retainer plug to a torque of 45-55 Ft.-Lbs.

355. **LIFT CONTROL LINKAGE AND LIFT SHAFT.** With the lift cover assembly removed as outlined in paragraph 347 and the lift cylinder assembly removed from cover as in paragraph 348, remove and disassemble lift control linkage and lift shaft as follows:

Unscrew the yoke (1—Fig. 401) from rear end of draft control plunger (10). On model 2000, remove the draft control spring (4A), then unbolt and remove the plunger locking plate (9) along with spring seat (6A), rubber seal (7) and felt (8). On models 3000 and 4000, unbolt draft control main spring housing (2) from rear end of lift cover and slide the housing, bearing (3), spring (4), shims (5) and spring seat (6) from draft control plunger.

NOTE: On models 3000 and 4000, a special washer (W—Fig. 379) is available to install between main draft control spring (4—Fig. 401) and spring seat (6) when draft control reaction from top link being

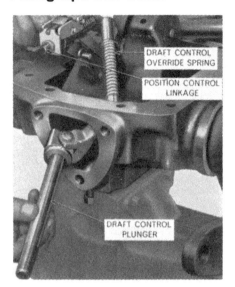

Fig. 404–Removing draft control plunger, link and override spring assembly.

under tension is not desirable. The special washer can be installed without removing any other parts.

Turn the lift shaft to move ram arm forward and remove the draft control plunger and override spring from rear opening in cover as shown in Fig. 404. Unbolt and remove the lift arms from lift shaft and disconnect piston rod from ram arm. Remove the lift shaft bushings (80—Fig. 405) and lift shaft (78) from lift cover, then remove the ram arm and thrust washer (76) located at right side of ram arm.

Remove the snap ring (69—Fig. 406) and remove the actuating lever (68) and swivel assembly as shown in Fig. 407. Unbolt and remove the quadrant (retainer) (16—Fig. 401) from support (17) (early models), then remove the control lever retaining nut (11), spring washer (12), flat washer (13), lever (14), Woodruff key (18) (early models) and the friction disc (15). Unbolt the control lever shaft support (17) from

lift cover and slide the support from shaft. Remove the cotter pin and flat washer that retain link (54—Fig. 406) to selector lever shaft (55), then remove the control lever shaft and position control linkage as shown in Fig. 409.

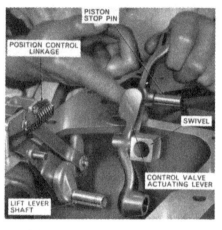

Fig. 407–Removing actuating arm assembly from lift control lever shaft.

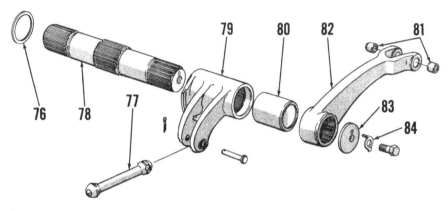

Fig. 405–Exploded view of hydraulic lift shaft assembly. Shaft (78), ram arm (79) and lift arms (82) have master splines. Spacer (76) is placed at right side (as viewed from rear) of ram arm.

76. Spacer	79. Ram arm	81. Bushings	83. Retainer
77. Piston rod	80. Bushings	82. L. H. lift arm	84. Lock
78. Lift shaft			

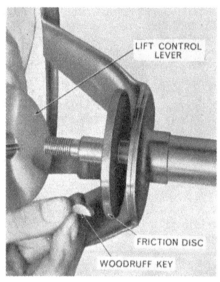

Fig. 408–Woodruff key must be removed from outer end of lift control lever shaft before support can be removed from shaft.

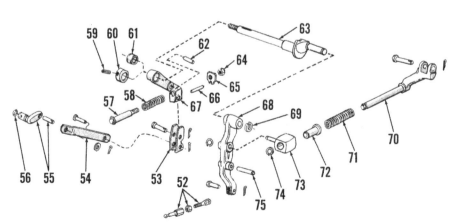

Fig. 406–Exploded view of hydraulic system control linkage. Spacer (61) is used on model 2000; cam (60) is used on models 3000 and 4000.

52. Turnbuckle	58. Spring	64. Jam nut	70. Link
53. Cam	59. Roll pin	65. Cam plate	71. Override spring
54. Link	60. Cam	66. Pin	72. Bushing
55. Selector arm	61. Spacer	67. Position control arm	73. Swivel
56. "O" ring	62. Pin	68. Actuating arm	74. Snap ring
57. Position control rod	63. Control shaft	69. Snap ring	75. Pin

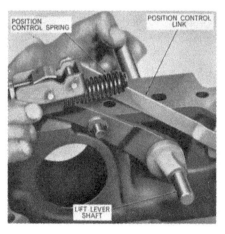

Fig. 409–Removing lift control lever shaft and linkage from lift cover.

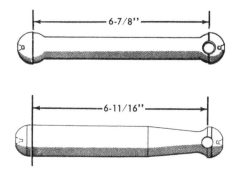

Fig. 410–Two different types of lift piston connecting rod have been used on model 4000 tractor. Correct actuating arm assembly (Fig. 411) must be used depending on type of connecting rod. Refer to text for details.

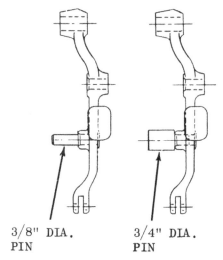

3/8" DIA. PIN   3/4" DIA. PIN

Fig. 411–Drawing of different actuating arm assemblies used on model 4000 tractor. Arm with ⅜-inch diameter pin is used with shorter (bowling pin type) connecting rod shown in Fig. 410; arm with ¾-inch diameter pin is used with longer (dog bone type) rod.

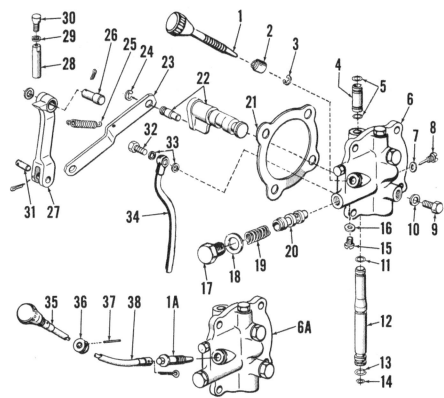

Fig. 412–Exploded view of flow control valve and linkage used on models 3000 and 4000. Items 1A, 6A and 35 through 38 are used on model 4200 only instead of items 1 and 6.

1. Spindle & knob	9. Plug	19. Plunger spring	29. Shims
1A. Spindle	10. Seal ring	20. Plunger	30. Cam follower
2. Seal	11. "O" ring	21. Gasket	31. Pin
3. Snap ring	12. Lower tube	22. Restrictor valve	32. Hollow bolt
4. Upper tube	13. Snap ring	23. Link	33. Seal washers
5. "O" rings	14. "O" ring	24. Snap ring	34. Dump tube
6. Valve plate	15. Screw plug	25. Spring	35. Knob
6A. Valve plate (4200)	16. Seal ring	26. Pivot pin	36. Collar
7. Seal	17. Plug	27. Crank lever	37. Pin
8. Retaining pin	18. Seal ring	28. Adjuster	38. Flexible shaft

On models 3000 and 4000, drive the roll pin (59—Fig. 406) from cam (60) and control lever shaft, then remove the cam and position control lever (67) from shaft. On model 2000, remove the spacer (61) and position control lever from shaft. Drive the roll pin (21—Fig. 401) from selector lever (22), remove lever from shaft (55—Fig. 406), then remove shaft and "O" ring (56) from lift cover.

NOTE: Two different lift piston connecting rods (see Fig. 410) have been used on model 4000 systems. The longer (dogbone) type connecting rod shown at top is no longer available for service and must be renewed using the shorter (bowling pin) type rod shown at bottom. For proper hydraulic system operation, the proper actuating lever must be used with each type connecting rod.

With the longer (dogbone) type connecting rod, an actuating lever with a ¾-inch diameter dowel pin must be used. When system is equipped with the newer, short (bowling pin) type rod, an actuating lever with a ⅜-inch dowel pin must be used. Refer to Fig. 411. Dowel pins are available separately from actuating lever.

Need and procedure for further disassembly will be evident on inspection of removed parts and with reference shown in exploded view in Fig. 406. To reassemble and reinstall, reverse the removal and disassembly procedure. Tighten control lever shaft support retaining cap screws to a torque of 35-47 Ft.-Lbs. Be sure that control lever is a free sliding fit over shaft and Woodruff key on early models, then install flat washer, spring washer and retaining nut; on model 2000, tighten retaining nut so that a force of 4 to 5 pounds is required to move control lever and on models 3000 and 4000, tighten nut so that 8 to 10 pounds force is required. Tighten lift arm to lift shaft retaining cap screws equally and so that all end play is removed, but

without causing excessive binding; lift arms should fall of their own weight. Bend tangs of retainers (84—Fig. 405) against cap screws when lift shaft end play is properly adjusted.

FLOW CONTROL VALVE AND OPERATING LINKAGE
Models 3000 and 4000

356. **OPERATING PRINCIPLES.** Turning the flow control knob (1 or 35 —Fig. 412) adjusts the position of the restrictor (22) in the pressure line from hydraulic pump. The flow control valve spool (20) position is then controlled by full pump pressure against inner end of spool and by reduced pressure via drilled ports from passage at upper side of restrictor aided by pressure from spring (19). Thus, according to position of restrictor, the control valve spool moves in bore to dump some oil back into the sump via the tube (34). When the 3-point hitch control lever is moved to top of quadrant (raising position), a cam on control lever shaft forces the

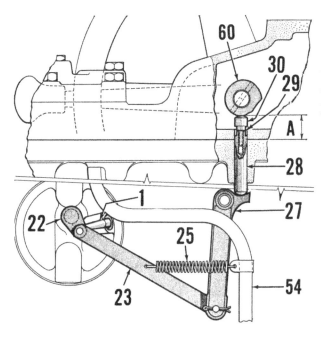

Fig. 413–Assembly drawing of flow control valve override linkage and actuating cam. Dimension "A" is from machined surface of lift cover to cam (60) on lift control lever shaft.

1. Spindle
22. Restrictor valve
23. Link
25. Spring
27. Crank lever
28. Adjuster
29. Shims
30. Cam follower
54. Return line
60. Cam

Fig. 414–View showing flow control override adjuster and cam follower being inserted in bore in edge of rear axle center housing.

plunger (28, 29 & 30) downward which moves the restrictor to full flow position through connecting linkage (23 and 27).

357. **ADJUST FLOW CONTROL VALVE LINKAGE.** The flow control restrictor is returned to full flow position by moving the lift control lever to top of quadrant (raising position); to properly adjust the linkage between cam on control lever shaft and restrictor, proceed as follows:

First, remove the hydraulic lift cover and cylinder assembly as outlined in paragraph 347. Move lift control lever to bottom of quadrant and, using a depth gage, measure distance between lower machined face of lift cover and nearest point of cam on lift control shaft (dimension "A", Fig. 413).

Refer to Fig. 414 and with flow control restrictor knob screwed fully out and flow control override adjuster inserted in bore of rear axle center housing, measure protrusion of adjuster above machined surface of rear axle center housing. The adjuster should protrude 0.010 more than measurement from face of lift cover to cam on control lever shaft (dimension "A" plus 0.010).

If necessary, unscrew the cam follower (30—Fig. 412) from top of adjuster (28) and add or remove shims (29) as required to obtain correct protrusion of adjuster. Shims are available in thickness of 0.010 only.

358. **R&R FLOW CONTROL VALVE PLATE.** To remove the flow control plate, first drain the hydraulic system (rear axle center housing) and remove the hydraulic lift cover and cylinder assembly as outlined in paragraph 347. Then, proceed as follows:

Remove snap ring (24—Fig. 412) from restrictor (22) pin and pull link (23) from pin. Remove the drilled cap screw (32), sealing washers (33) and by-pass tube (34) from inner side of valve plate. Loosen the valve plate retaining cap screws and pull upper pressure pipe (4) from bore in rear axle center housing and valve plate. On models 3000 and 4110 L.C.G., refer to Fig. 374 and remove the pressure relief valve assembly from bottom of rear axle center housing. On model 4000 with gear type hydraulic pump, pry snap ring from groove in upper end of lower pressure pipe and push or drive the pipe upward into valve plate until clear of hydraulic pump flange; pipe is located on outside of rear axle center housing. On other models (with piston type hydraulic pump), pry snap ring from groove in lower pressure pipe (12 —Fig. 412) and push or drive the pipe downward into rear axle center housing or adapter flange until pipe is clear of flow control valve plate. Then, on all models, unbolt and remove plate from side of rear axle center housing.

To reinstall the flow control valve plate, reverse the removal procedure using a new gasket (21) and all new sealing rings. Tighten the valve plate retaining cap screws to a torque of 35-37 Ft.-Lbs. and tighten the pressure relief valve to a torque of 45-55 Ft.-Lbs. Reinstall hydraulic lift cover as outlined in paragraph 347, refer to paragraph 339 or 340 for proper hydraulic fluid.

359. **OVERHAUL FLOW CONTROL VALVE AND PLATE ASSEMBLY.** Refer to exploded view of the assembly in Fig. 412 and proceed as follows:

Unscrew the plug (17) and remove the flow control valve spring (19) and valve spool (20). Inspect spring for wear, cracks or distortion, and check spring free length against new spring; renew spring if damaged in any way or if free length is less than that of new spring.

The flow control valve spool (20) should be a free sliding fit in bore of valve plate; renew spool if worn or scored. Usually, a new valve spool with the same color code of that being renewed should be selected. However, if the next size larger valve is a free sliding fit, the larger size spool should be installed. The valve should not bind or drag. Valves are color coded to indicate outside diameter as follows:

Valve Outside Dia.	Color Code
0.6670-0.6672	Red
0.6672-0.6674	Yellow
0.6674-0.6676	Blue
0.6676-0.6678	Green
0.6678-0.6680	White

If the flow control valve bore in plate is scored or excessively worn, the complete flow control valve assembly must be renewed.

Remove restrictor retaining screw (8) and pull restrictor (22) from bore. Restrictor should be renewed if scored or worn; pin is renewable separately. Renew flow control valve assembly if restrictor bore is excessively worn or scored. Inspect tip of retainer screw (8) and renew screw if tip is damaged or worn. Remove snap ring (3) from flow control knob (1) spindle and screw the knob and spindle out of plate. Renew the seal (2), lubricate threads on spindle and screw the knob back into plate and install retaining ring.

For disassembly of flow control valve restrictor operating linkage, refer to

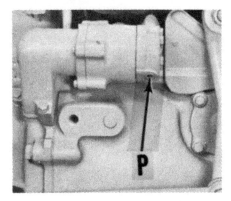

Fig. 415–Remove plug (P) to bleed piston type hydraulic pump. Refer to text for bleeding procedure.

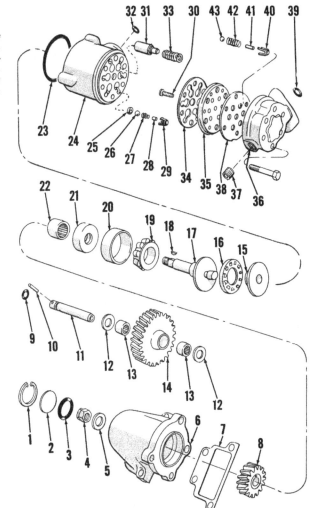

Fig. 416–Exploded view of piston type hydraulic pump assembly. Pistons (31) for model 2000 are 9/16-inch diameter; models 3000 and 4000 pump assemblies have ⅝-inch diameter pistons.

1. Snap ring
2. Plug
3. "O" ring
4. Self-locking nut
5. Washer
6. Drive housing
7. Gasket
8. Driven gear
9. "O" ring
10. Pin
11. Idler shaft
12. Washers
13. Needle bearings
14. Idler gear
15. Wobble plate
16. Thrust bearing
17. Pump shaft
18. Woodruff key
19. Bearing cone & roller
20. Bearing cup
21. Oil seal
22. Needle bearing
23. "O" ring
24. Pump body
25. Intake valve seats
26. Intake valve balls
27. Springs
28. Pins
29. Guides
30. Valve plate screw
31. Pistons
32. "O" rings
33. Return springs
34. Gasket
35. Exhaust valve plate
36. Pump cover
37. Plug
38. Gasket
39. "O" ring
40. Guides
41. Pins
42. Springs
43. Exhaust valve balls

exploded view in Fig. 412. Renew any bent, worn or otherwise damaged components. Remove oil passage sealing plugs (9) from flow control valve plate if necessary to clean the passages or to renew the plug seals (10).

HYDRAULIC PUMP
4000 Without Independent PTO, All 2000 and 3000

360. **REMOVE AND REINSTALL.** On 4000 without independent PTO and all 2000 and 3000, the piston type hydraulic pump is mounted on left rear side of engine and is driven from gear attached to rear end of engine camshaft.

To remove pump, first remove (on 2000 and 3000 only) the front steering gear cover from under left side of fuel tank. Then, on all models, thoroughly clean pump and lines and disconnect hydraulic line manifold from pump. The pump can then be unbolted and removed from engine. Place a clean rag over the open hydraulic lines to prevent entry of dirt or other foreign material.

To reinstall pump, place new sealing "O" rings in the grooves of hydraulic line manifold and install pump to engine using new gasket. Tighten the pump to engine cap screws securely, then install and tighten the bolts retaining hydraulic line manifold to pump.

361. To prime pump, proceed as follows: On diesel models, pull out the engine stop control; on gasoline models, remove distributor rotor or disconnect coil wire. Then, remove socket head plug (P—Fig. 415) from pump cover and turn engine with starter until oil flows from plug opening. Continue to turn engine with the starter until no air bubbles appear in the oil flowing from opening, then reinstall socket head plug while oil is still flowing and tighten plug securely.

362. **OVERHAUL PISTON TYPE PUMP.** With pump removed from tractor, refer to exploded view in Fig. 416 and cross-sectional view in Fig. 417 then proceed as follows:

363. DRIVE HOUSING, SHAFTS, BEARINGS AND GEARS. Clamp the pump body in a vise with drive housing (front) end up. Alternately and evenly loosen the four cap screws retaining drive housing assembly to pump body; piston return springs will force housing upward as screws are loosened. With cap screws removed, lift the drive housing from pump body taking care not to drop the wobble plate (15—Fig. 416) and thrust bearing (16).

NOTE: At this time, the pump pistons and piston return springs may be removed and inspected, and the general condition of the pump valves checked. Lift wobble plate and thrust bearing from body if not removed with drive housing and extract the pistons and return springs. Inspect pistons, springs and piston bores in pump body; renew any broken springs and check pump valves as follows: Insert return springs in bores, dip

pistons in oil and insert in bores with small diameter of pistons towards the springs. Cover intake port in pump body and actuate pistons; if vacuum is not created at intake port, outlet valves are leaking; refer to paragraph 364. Cover exhaust port and again actuate pistons; if pressure is not created, intake valves are leaking; refer to paragraph 364.

Drive pin (10) from housing, insert a ¼-inch x 20 screw in end of idler shaft and pull shaft from housing. Remove and discard "O" ring (9). Withdraw the idler gear (14) and thrust washers (12) from housing. If idler shaft shows signs of wear, renew the shaft. Inspect the needle bearings (13) in idler gear; if needle rollers are loose or have flat spots, they can be renewed if idler gear is otherwise serviceable. Press old bearing cages from gear and install new bearings. Note: Press on lettered end of bearing cages only; bearing cage should be flush or just below flush with hub of gear.

Remove the snap ring (1), plug (2) and "O" ring (3) at front end of housing and discard "O" ring. Using small punch, bend tab of washer (5) away

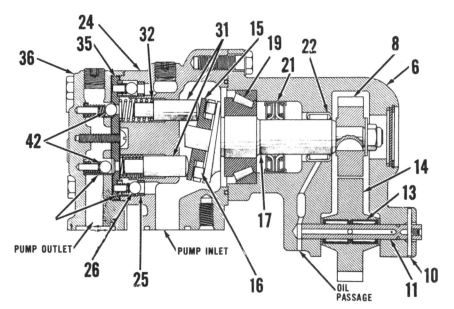

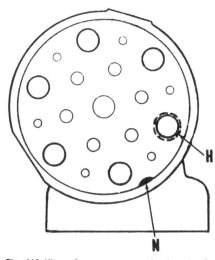

Fig. 417–Cross-sectional view of piston type hydraulic pump assembly. Note oil passage through drive housing (6); needle bearings (13 and 22) are lubricated through passage by engine oil pressure. Refer to Fig. 416 for exploded view of pump and parts identification.

Fig. 418–View of pump cover and exhaust valve plate showing proper position of alignment notches (N) for assembly. One hole (H) in cover is oversize to facilitate removal of plate from cover with punch.

from nut (4) and remove nut and washer. With suitable soft drift, tap shaft (17) out rear (open) end of housing and remove gear (8). Using OTC puller No. 943-S or equivalent tool, remove bearing cup (20). Remove and discard the shaft seal (21). With OTC step plate No. 630-3 and suitable drift, drive needle bearing (22) from housing. If bearing cone and roller assembly is worn or damaged and pump shaft is otherwise serviceable, remove bearing cone with OTC bearing puller attachment No. 951, or equivalent.

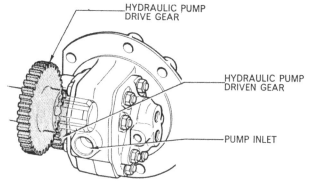

Fig. 419–Gear type hydraulic pump for model 4000 is driven from gear on transmission PTO shaft and is mounted in right side of rear axle center housing.

Carefully clean and inspect all parts and renew any parts not serviceable. Reassemble using new seal and "O" rings as follows: Using OTC step plate No. 630-3, press needle bearing (22) into housing taking particular care that the oil hole in bearing cage is aligned with the oil passage in housing. Press new seal into housing with wide lip towards needle bearing; refer to cross-sectional view for correct depth of seal in housing. Press bearing cup (20) into housing with OTC step plate No. 630-11; be sure cup is firmly seated. Press bearing cone and roller assembly onto shaft until firmly seated. Lubricate seals and bearing, place pump shaft gear in housing and Woodruff drive key in shaft; then insert shaft into housing through seals, bearing and gear. Install new washer (5) and nut (4), tighten nut to a torque of 10-15 Ft.-Lbs. and bend tab of washer against flat on nut. Install new "O" ring (3), plug (2) and snap ring (1). Note: If renewing snap ring, check suffix of casting number on housing; if suffix is "A", install a 0.060/0.064 thick snap ring, or if suffix is "B", install a 0.048/0.052 thick snap ring.

Place idler gear, with needle bearings installed, in housing. Insert thrust washer at each side of gear; use new washers if ones removed show signs of wear. Install a new "O" ring (9) in groove on idler shaft, lubricate shaft bore and insert shaft into housing through gear and thrust washers so that pin hole in shaft is aligned with pin hole in housing. Install idler shaft retaining pin.

Be sure that pistons are installed in body with rounded ends outward; push each piston in against the return spring to be sure they are free and operating properly. Place thrust bearing and wobble plate on pump shaft, insert a new "O" ring (23) in groove of pump body and, while holding thrust bearing and wobble plate, place the housing assembly down on pump body and cover. Push housing down against piston return spring pressure and insert the four housing retaining cap screws. Tighten the cap screws evenly and alternately until housing is pulled against pump body, then tighten the cap screws to a torque of 38-42 Ft.-Lbs. Turn idler gear to be sure pump oper-

ates without binding; a popping sound should be heard as the pistons actuate the inlet and outlet valves. Reinstall and prime pump as outlined in paragraphs 360 and 361.

364. PUMP BODY AND COVER ASSEMBLY. Remove the drive housing, wobble plate, thrust bearing, pistons and return spring as outlined in paragraph 363, then proceed as follows.

Clamp pump body in vise with cover end up. Remove the six cap screws retaining cover to body and pry cover assembly loose. Lift off the cover assembly, remove the six ball guides and six steel ball inlet valves from pump body and remove the six inlet valve springs from pins in valve plate attached to inside of cover. Remove the screw from center of valve plate, place cover on bench with valve plate down and tap the valve plate loose with punch inserted through the large bolt hole (See Fig. 418) in cover. Separate valve plate and cover taking care not to lose any of the six steel ball outlet valves. Remove the outlet valve springs and ball guides from the cover. Scrape all traces of gasket from both

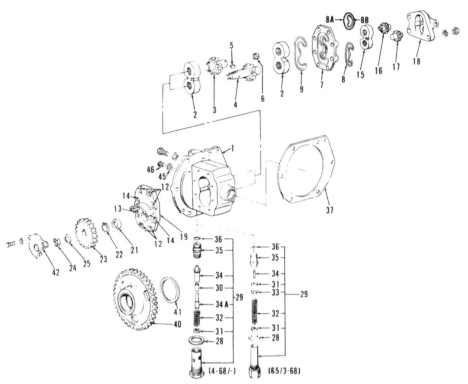

Fig. 420–Exploded view of hydraulic pump used on 4000 models with independent PTO. Note early and late type relief valves. Both relief valves can be disassembled by unscrewing valve seat (35).

1. Pump body	9. Seal ring	23. Drive gear	34A. Guide pin
2. Bearing assembly	14. Dowel bolts	24. Nut	35. Valve seat
3. Driven gear	15. Bearing assembly	25. Tab washer	36. "O" ring
4. Drive gear	16. Drive gear	28. Seal	37. Gasket
5. Woodruff key	17. Driven gear	29. Relief valve assy.	40. Idler gear
6. PTO pump coupling	18. Rear pump body	30. Seal	41. Lock ring
7. Rear cover	19. Front cover	31. Shims	42. Gear shroud
8. Seal ring (early)	21. Seal	32. Spring	45. Washer
8A&B. Seal rings (late)	22. Snap ring	33. Spring seat	46. Plug
		34. Relief valve	

sides of valve plate and the body and cover. Wash all parts in solvent and carefully inspect each part for excessive wear or damage. Renew the valve inlet and outlet springs if bent, broken or damaged in any way. Renew the valve plate if tapered seats for steel outlet balls are damaged or worn. Inspect valve seats in pump body; seats can be renewed if worn or damaged and body is otherwise serviceable. Remove old seats with special puller (Nuday tool No. NCA-600-G or equivalent and install new seats with driver (Nuday tool No. NCA-600-EA or equivalent). Be sure that seats are installed with chamfered ball seating surface out. Pins (28—Fig. 416) in valve plate or pins (41) in cover may be renewed if damaged and plate or cover is otherwise serviceable.

To reassemble, proceed as follows: Stick the gasket (38) to cover with thin film of grease; be sure all holes are aligned and that location notch is placed as shown in Fig. 418. Insert the valve cages (outlet ball guides) and outlet valve springs in the cover and place the steel outlet balls on top of the

springs. Carefully place the valve plate on top of the balls so that each ball enters a tapered seat and location notch in plate is placed as shown in Fig. 418. Thread the valve plate retaining screw into cover, insert two cover retaining cap screws through bolt holes in plate, gasket and cover; then, tighten the valve plate retaining screw. Using the light film of grease, stick gasket (34—Fig. 416) to pump body so that holes in gasket are aligned with holes in body. Install inlet valve spring on the pins in valve plate. Insert the valve cages (ball guides) and steel ball inlet valves in the valve cavities of pump body. Carefully lower the pump cover assembly onto body so that the springs enter the valve cages, then install the cover retaining cap screws and tighten the cap screws to a torque of 38-42 Ft.-Lbs. Reinstall drive housing assembly as outlined in paragraph 363 then, reinstall and bleed pump as outlined in paragraphs 360 and 361.

365. PUMP SUCTION AND PRESSURE LINES. On models with piston type pump prior to production date 10/69, formed steel tubes with an

integral manifold at each end are used to connect pump to lower right side of hydraulic sump (rear axle center housing).

Whenever the manifolds are disconnected from pump or hydraulic sump, new sealing "O" rings should be installed. Tighten the manifold to pump and manifold to rear axle center housing cap screws to a torque of 40-45 Ft.-Lbs. On model 4000, tighten the manifold to adapter cap screws to torque of 40-45 Ft.-Lbs.

On models beginning with production date of 10/69, elbows and gland nuts are used at rear axle center housing end of suction and pressure lines.

Model 4000 With Independent PTO

366. R&R HYDRAULIC PUMP. To remove the gear type hydraulic pump from side of rear axle center housing on model 4000 with independent type PTO, proceed as follows:

Drain the hydraulic system (rear axle center housing) and remove the hydraulic lift cover as outlined in paragraph 347. Disconnect brake pedal return springs and remove the right step plate (foot rest). Loosen the pump retaining cap screws, pry snap ring from groove in pressure pipe between pump and flow control valve and drive the pipe upward into flow control valve plate until clear of pump flange. Remove the pump retaining cap screws and remove pump from suction tube, PTO clutch pressure tube and from side of rear axle center housing.

While the pump is removed, the pump suction tube should be removed from center housing for cleaning. Refer to paragraph 341.

To reinstall pump, proceed as follows: Position the pump suction tube and screen in center housing, but do not bolt to pinion bearing cage at this time. Install new "O" rings on pump suction tube and PTO clutch pressure tube, lubricate the "O" rings and install new gasket on pump flange. Place the pump in opening in side of center housing, inserting the PTO clutch pressure tube as pump is moved into place. Install the pump retaining cap screws and insert suction tube into pump inlet. Install the suction tube bracket retaining cap screw and lock washer and tighten cap screw to a torque of 80-100 Ft.-Lbs. Tighten the pump retaining cap screws to a torque of 35-47 Ft.-Lbs. Reinstall lift cover assembly as outlined in paragraphs 347 and refill rear axle center housing with proper lubricant; refer to paragraph 340.

Fig. 421–Removing pump gears and bearings as an assembly.

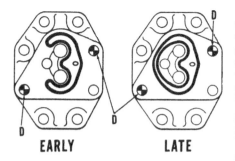

EARLY   **LATE**

Fig. 423–Early production pump rear cover had single sealing ring groove. Late production cover has grooves for two sealing rings for improved sealing to rear pump body. "D" indicates dowel bolt locations.

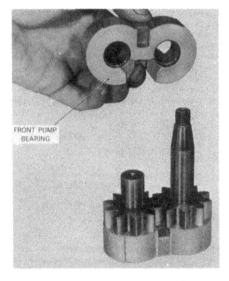

FRONT PUMP BEARING

Fig. 422–Assembling pump gears and bearings.

edge (measured at 90 degrees to gear center line) should not exceed 0.001 and the bearing journal on either side of each gear should be within 0.001 of the other side. If necessary to renew a pump gear, the mating gear must also be renewed if the face width of the new gear is more than 0.002 wider than the mating gear. Light score marks on gear bearing faces can be removed with "O" grade emery paper on a flat (lapping) plate. Remove light score marks on gear journals with "O" grade emery paper. The gear bearings (2) must always be renewed as a pair.

Thoroughly clean all parts, air dry and then lubricate with hydraulic fluid prior to reassembly. Reassemble by reversing disassembly procedure using all new seals. Fig. 422 can be used as an assembly guide; the large notch in gear bearings must be toward the intake side of pump as shown in Fig. 420 and 421. Refer to Fig. 423 for early and late seal installation on rear cover for the independent PTO pump gear bearing. Alternately and evenly tighten the nuts on pump through bolts to a torque of 40-45 Ft.-Lbs. NOTE: Be sure the two dowel bolts are reinstalled in correct position as shown in Fig. 420 and Fig. 423. Install new washer (25) and bend tang of washer against gear retaining nut (24) after tightening the nut.

(small) pump body (18) from rear cover plate (7) and remove gear bearing (15), pump drive gear (16) and driven gear (17). Remove the rear pump drive connector (6) and seal ring(s) (8 or 8A and 8B) from rear cover and separate rear cover and front cover (19) from pump body (1). Carefully remove the drive gear (4), driven gear (3) and bearings (2) from pump body as an assembly as shown in Fig. 421. Remove the seal rings (9—Fig. 420) from pump covers. Separate the gear bearings and pump gears.

Carefully clean and inspect all parts. Renew the pump body if scored or if gear track is worn more than 0.0025 deep at inlet side of body. The maximum runout across gear face to tooth

REMOTE CONTROL VALVES

367. OVERHAUL PUMP.
With hydraulic pump removed as outlined in paragraph 366, proceed as follows:

NOTE: Beginning with production date 4/68 a different relief valve has been installed as shown in Fig. 420. Shims (31) for early type relief valves were available in thicknesses of 0.010, 0.015 and 0.025; whereas the shims for the late type are available in thicknesses of 0.010 and 0.015. While valves differ somewhat, service procedures remain basically the same.

Unscrew the pressure relief valve assembly (29—Fig. 420) from pump body flange. Disassemble the valve by unscrewing valve seat (35) from valve body and removing the valve parts. Be careful not to lose any of the shims (31).

Remove the gear shroud (42), then straighten locking tab on washer (25) and remove nut (24), washer, driving gear (23) and Woodruff key (5).

Remove the nuts and lockwashers from the pump through bolts and remove the bolts. Separate the rear

SELECTOR VALVE
Models 2000, 3000 and 4000 So Equipped

368. A selector valve is available for model 2000, 3000 and 4000 tractors to provide for operation of single acting remote hydraulic cylinders, either in conjunction with or independently of the 3-point lift cylinder, from the tractor hydraulic system control valve. Refer to Fig. 424 for exploded view of the valve.

When the selector valve spool (3) is pushed fully in, the tractor hydraulic control valve will direct hydraulic pressure to the 3-point lift cylinder only. When valve is pulled fully out, hydraulic pressure is directed to the selector valve port (10) only. With valve in middle detent position, pressure is directed to both the 3-point lift cylinder and the remote cylinder port in the selector valve when control valve is in raising position. Note: The lift arms of the 3-point hitch must be slightly lowered before pulling the

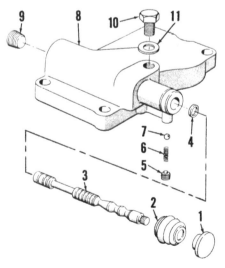

Fig. 424–Exploded view of selector valve assembly available for models 2000, 3000 and 4000. Clamp for boot (2) is not shown.

1. Knob	
2. Boot	
3. Valve spool	7. Detent ball
4. "O" ring	8. Valve body
5. Plug	9. Plug
6. Spring	10. Port sealing plug
	11. Seal ring

selector valve spool out to operate remote cylinder independently of the 3-point hitch; otherwise, the tractor hydraulic control valve spool cannot be moved to raised position to supply pressure to the remote cylinder port.

369. **R&R AND OVERHAUL.** To remove the selector valve, first place selector lever on lift cover in draft control position, move the control lever to bottom of quadrant and stand on lift arms to force all oil from 3-point hitch lift cylinder. Then, unbolt and remove the selector valve assembly from top of lift cover. Remove the sealing "O" rings from counterbores in bottom of valve body.

Remove the valve plug (5—Fig. 424), detent spring (6) and detent ball (7). Loosen clamp, pull the dust cover (2) from boss on valve body and withdraw the valve spool from bore. Remove "O" ring (4) from groove in valve spool, unscrew the knob (1) and remove dust cover. Remove plug (9) if necessary for cleaning purposes.

Carefully inspect valve spool and bore in valve body. If either part is scored or otherwise damaged beyond further service, it will be necessary to renew the complete valve assembly as neither part is available separately.

To reassemble valve, install new "O" ring in groove on valve spool, place dust cover and clamp on outer end of spool and screw valve knob tightly onto spool. Lubricate valve bore in body, valve spool and "O" ring and insert the valve into bore. Install detent ball, detent spring and retaining set screw, then push the dust cover over boss on valve body and secure clamp.

Be sure mounting surface is clean, install all new "O" rings in counterbores on bottom side of valve body and reinstall valve on top of lift cover. Tighten the retaining cap screws to a torque of 40-45 Ft.-Lbs. Reinstall plug (9) and if removed, reinstall cylinder port plug and seal.

SINGLE AND DOUBLE SPOOL REMOTE CONTROL VALVES

Single spool or double spool remote control valves, with or without detents, have been used on the model 2000, 3000 and 4000 tractors. Refer to Fig. 425 for an exploded view of the single spool, no detent valve, and to Fig. 426 for double spool, no detent valve, used prior to production date 5-70 (8-70 double spool).

For exploded views of valves used from production date 5-70 (8-70 double spool) to production date 12-70, refer to Figs. 428 and 429.

For exploded views of valves used from production date 12-70 and after, refer to Figs. 434 and 435.

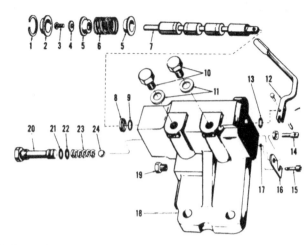

1. Snap ring
2. Plug
3. Screw
4. Washer
5. Spring cups
6. Centering spring
7. Valve spool
8. Seal retainer
9. "O" ring
10. Port plugs
11. Seal rings
12. Control lever
13. "O" ring
14. Pivot screw
15. Switch valve
16. Retainer plate
17. "O" ring
18. Valve body
19. Plug
20. Check valve plug
21. "O" ring
22. "O" ring
23. Spring
24. Check valve ball

Fig. 425–Exploded view of single spool remote control valve without spool detent mechanism.

1. Snap rings
2. Plugs
3. Screws
4. Washers
5. Spring cups
6. Centering springs
7. Valve spools
8. Seal retainers
9. "O" rings
10. Port plugs
11. Seal rings
12. Long control lever
12A. Short control lever
14. Pivot screw
15A. By-pass valve
16A. Plug
17. "O" ring
18. Valve body
19. Plug
20. Check valve plug
21. "O" ring
22. "O" ring
23. Spring
24A. Check valve
25. Plug

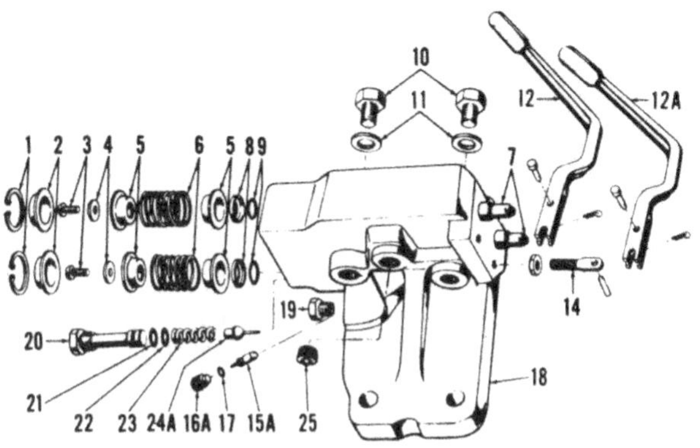

Fig. 426–Exploded view of two spool remote control valve without spool detent mechanism. Valve spools (7) are not shown removed; refer to (7–Fig. 425).

In all cases, valve body and valve spool are a select fit and are not available separately. When disassembling double spool valves, keep spools identified with their correct bores.

Valves can be identified by their exterior configuration.

Remote Control Valves Without Spool Detent (Prior to 12-70)

370. Exploded views of the single spool and double spool remote control valves without spool detent are shown in Figs. 425 and 426. Either valve can be used with single or double acting remote cylinders. On single spool valve, open (turn out) the switch valve (15—Fig. 425) to operate a single acting remote cylinder from "LIFT" port of valve. On the double spool valve, back the Allen head screw (16A —Fig. 426) out two or three turns to operate a single acting cylinder from the "No. 1 LIFT" port. The valve levers must be held in raising or lowering position as the valve spools do not have detents. If valve is held open after cylinder reaches end of stroke, the tractor hydraulic system relief valve will open at approximately 2500 pounds pressure.

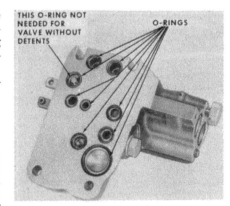

Fig. 427–View showing bottom of remote control valve assembly. On remote control valves without spool detent, "O" ring is not needed at indicated position.

371. **REMOVE AND REINSTALL.** To remove the remote control valve, first disconnect any remote hoses, place the lift system selector lever in draft control position and move 3-point hitch control lever to bottom of quadrant. Stand on lift arms to force all oil from 3-point hitch lift cylinder. Then, unbolt and remove the remote control valve assembly from lift cover.

When reinstalling the control valve, always renew the sealing "O" rings between valve body and lift cover or manifold. Refer to Fig. 427. Tighten the retaining cap screws to a torque of 40-45 Ft.-Lbs.

372. **OVERHAUL VALVE ASSEMBLY.** With the remote control valve assembly removed from the tractor as outlined in paragraph 371, refer to the following paragraphs for overhaul procedure:

373. CONTROL VALVE SPOOLS, CENTERING SPRINGS AND SEALS. Remove the control levers by removing the clevis pins retaining lever to valve spools, then pull levers from pivot pins in the adjusting screws (14—Fig. 425 or 426). Remove the snap rings (1) that retain centering spring caps (2), then push or tap on rear (lever) end of valve spools to remove the valves (7), centering spring (6) assemblies and caps from front end of valve body. NOTE: On double spool control valves, be sure to tag the valve spools as they are removed as each valve is a select fit and not interchangeable. Remove screws (3) from ends of valve spools to remove centering springs (6), spring cups (5) and washers (4). Using a slide hammer with small internal pulling attachment (Nuday tool No. 7600-E or equivalent), remove the seal retainers (8) from valve bores. Remove the "O" rings (9 and 13) from valve spool bores.

Thoroughly clean and inspect all parts. If the control valve spools or bores are seriously damaged or scored, the complete valve assembly must be renewed as spools and body are not available separately. Renew the centering springs, retainers or washers if cracked or distorted. Centering spring free length (new) is 1.103; renew spring if free length varies materially from this dimension.

To reassemble, proceed as follows: Lubricate the control valve spools and insert them into front (centering spring) end of valve body. Push the spools rearward so that rear ends are even with the "O" ring groove in valve body. Lubricate the "O" rings (13—Fig. 426) and using rear ends of valve spools as "back stops," insert the "O" rings in grooves in valve bores. Push the spools on through the "O" rings so that they extend about ¾-inch from rear of valve body. Lubricate the front "O" rings and push them into bore around the front ends of valves, then install the retainers (8) using a special driver (Nuday tool No. N-651 or equivalent). Reinstall the centering springs and cups, flat washers, retaining screws, caps, snap rings and control valve levers.

374. SINGLE SPOOL SWITCH VALVE. Loosen the locknut on lever

adjusting screw (14—Fig. 425) and swing the retainer (16) out of way. Then, using an Allen wrench, unscrew the switch valve (15) from valve body. Remove and discard the "O" ring (17). Renew the switch valve if excessively pitted or distorted. Install and lubricate new "O" ring, then screw valve into valve body. Move the retainer into position and tighten adjusting screw lock nut while holding adjusting screw square with control lever.

375. DOUBLE SPOOL SWITCH VALVE. Using an Allen wrench, remove the set screw (16A—Fig. 426). Then, using needle nose pliers, pull the valve (15A) from valve body; remove

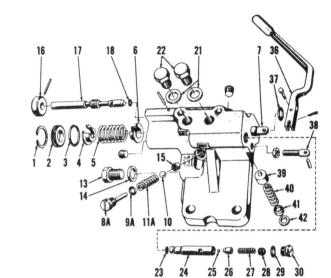

1. Snap ring
2. Plug
3. "O" ring
4. Spring cup
5. Centering spring
6. Spring cup
7. Valve spool
8A. Plug
9A. "O" ring
10. Check valve ball
11A. Spring
13. Plug
14. Seal ring
15. Check valve seat
16. Knob
17. Float valve
18. "O" ring
21. Seal rings
22. Port plugs
23. "O" ring
25. Valve body
25. Valve ball
26. Ball seat
27. Spring
28. Adjusting plug
29. "O" ring
30. Plug
39. Spool detent
40. Spring
41. Retainer
42. Retaining ring

Fig. 428–Exploded view of single spool remote control valve with spool detent (39) mechanism. Valve spool (7) is not shown removed from valve body; refer to Fig. 429. Refer to Fig. 430 for late production detent regulating valve assembly.

1. Snap rings
2. Plugs
3. "O" rings
4. Spring cups
5. Centering springs
6. Spring cups
7. Valve spools
8. Plug
9. "O" ring
10. Springs
11. Check valve balls
12. Check valve seat
13. Plug
14. Seal ring
15. Check valve seat
16. Knob
17. Float valve
18. "O" rings
19. Bypass valve
20. Plug
21. Seal rings
22. Port plugs
23. "O" ring
24. Valve body
25. Valve ball
26. Ball seat
27. Spring
28. Adjusting plug
29. "O" ring
30. Plug
31. Plug
32. "O" ring
33. Spring
34. Detent piston
35. Long valve lever
36. Short valve lever
37. "O" rings
38. Pivot screws
39. Detent piston
40. Spring
41. Retaining plug
42. Retaining ring

Fig. 429–Exploded view of two spool remote control valve assembly with spool detent (34 and 39) mechanism. Refer to Fig. 430 for view of late production detent regulating valve assembly.

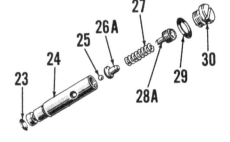

Fig. 430–Exploded view of late production detent regulating valve assembly for valves shown in Figs. 428 and 429.

23. "O" ring	27. Spring
24. Valve body	28A. Adjusting plug
25. Valve ball	29. "O" ring
26A. Spring retainer	30. Plug

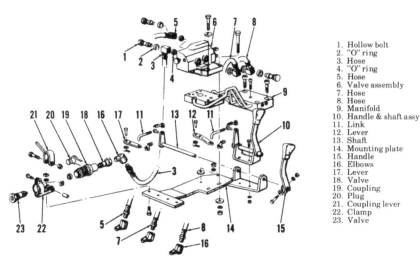

Fig. 431–Exploded view of model 4200 (rowcrop) remote control valve manifold and lines for two spool remote control valve assembly prior to production date 8/70.

1. Hollow bolt
2. "O" ring
3. Hose
4. "O" ring
5. Hose
6. Valve assembly
7. Hose
8. Hose
9. Manifold
10. Handle & shaft assy.
11. Link
12. Lever
13. Shaft
14. Mounting plate
15. Handle
16. Elbows
17. Lever
18. Valve
19. Coupling
20. Plug
21. Coupling lever
22. Clamp
23. Valve

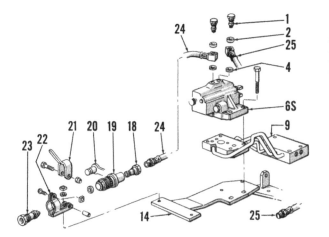

Fig. 432–Exploded view of model 4200 (rowcrop) remote control valve manifold and lines for single spool remote control valve prior to production date 5/70.

1. Hollow bolts
2. "O" rings
4. "O" rings
6S. Valve assembly
9. Manifold
14. Mounting plate
18. Valve
19. Coupling
20. Plug
21. Coupling lever
22. Clamp
23. Valve
24. Hose
25. Hose

and discard the "O" ring (17). Renew the valve if pitted or distorted. Install new "O" ring in groove on valve, lubricate the valve and "O" ring and push into place in valve body. Reinstall the Allen head plug.

376. CHECK VALVE. To remove the check valve ball (24—Fig. 425) on single spool valve or the check valve poppet (24A—Fig. 426) on double spool valve, unscrew the plug (20) and remove spring (23) and ball or poppet. Remove the sealing "O" rings (21 and 22) from plug. Renew the ball or poppet valve if damaged in any way and renew the spring if free length is not equal to that of new spring.

Install new "O" rings on the check valve plug. Drop the ball or poppet valve into bore, insert spring, then lubricate the "O" rings and screw the plug into valve body. Tighten plug securely.

Remote Control Valves With Spool Detents (Prior to 12-70)

377. Exploded views of single and double spool remote control valves with spool detents are shown in Figs. 428 and 429. Either valve may be used to operate single or double acting remote cylinders. To operate single acting remote cylinder from single spool valve or from top spool of double spool valve, turn the knob (16) counter-clockwise until switch valve (17) is screwed out against stop pin. To operate single acting remote cylinder from lower spool of double spool valve, screw the Allen head screw (20) counter-clockwise until against stop pin which will allow switch valve (19) to lift from seat. Unless float action is desired, the switch valve (17 or 19) must be closed to operate a double acting remote cylinder. Detent pistons hold the control valve spools in raising or lowering position until remote cylinder reaches end of stroke; then, increase in hydraulic pressure will open the detent regulating valve and pressurized hydraulic fluid will lift the detent pistons away from spools allowing the spool centering spring to return the spools to neutral position. The control levers can be returned to neutral position man-

ually before remote cylinder reaches end of stroke, or can be held in raising or lowering position after detent release pressure is reached. The remote control cylinder and valve are protected by the 2500 psi tractor hydraulic system relief valve when the control valve levers are manually held in raising or lowering position or if the detent regulating valve is adjusted to pressure exceeding that of the relief valve.

378. **ADJUST DETENT REGULATING VALVE.** The detent regulating valve should be adjusted to release the spool detent pistons at a pressure slightly higher than that of normal cylinder operating pressure. When so adjusted, the control valve will be returned to neutral position without an excessive pressure build up when remote cylinder reaches end of stroke.

To adjust the detent regulating valve, refer to Fig. 429 for double spool valve or to Fig. 428 for single spool valve and proceed as follows: Remove the regulating valve plug (30), then using an Allen wrench, turn the socket head adjusting screw (28) into valve body (24) to increase the detent release pressure or out to decrease release pressure. Reinstall plug with new sealing "O" ring (29) after completing the required adjustment. Late style detent regulating valve assembly is shown in Fig. 430.

379. **REMOVE AND REINSTALL.** To remove and reinstall the remote control valve with spool detents, refer to paragraph 371 which outlines the procedure for removing and reinstalling the similar remote control valve without spool detents.

380. **OVERHAUL.** First, remove the remote control valve from tractor as outlined in paragraph 371, then proceed as outlined in following paragraphs 381 through 384.

381. CONTROL VALVE SPOOLS, SPOOL CENTERING SPRINGS, SPOOL DETENT PISTONS AND SEALS. Refer to the exploded view of the double spool control valve in Fig. 429 and proceed as follows:

Remove the clevis pins retaining control levers (35 and/or 36) to valve spools (7) and remove the levers from pivot pins in adjusting screws (38). Remove the detent plug (31) and "O" ring (32) and/or the retaining ring (42) and retainer (41), then withdraw the detent spring (33 and/or 40) and piston (34 and/or 39). Remove the snap rings (1) from front end of valve bores and tap or push the valve spools forward out of valve body; the bore plugs (2)

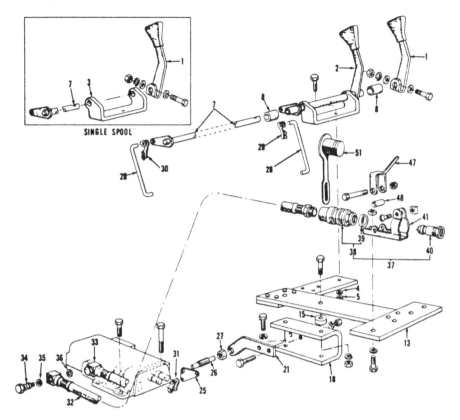

will be pushed out by the valve spools. Remove the "O" rings (3 and 37) from bore plugs and from rear of valve bores in body. Compress the centering springs (5) and remove the outer spring seats (4), then release spring pressure and remove the springs and inner spring seats (6) from the valve spools.

Thoroughly clean and inspect all parts. If the control valve spools or bores in valve body are seriously damaged, the complete valve assembly must be renewed as valve spools and body are not available separately. Renew the centering springs, retainers (spring seats) or bore caps if cracked or distorted. Centering spring free length (new) is 1.103; renew spring if free length varies materially from this dimension.

To reassemble, proceed as follows: Place rear spring seat (6) on valve spools, slide centering spring over end of spool and compress the spring far enough to install the front spring seat (4); then, release the spring. Lubricate the valve spools and insert them into valve bores from front end of valve body. Hold the spools so that rear (lever) ends are even with the "O" ring grooves in valve bores, then lubricate and install the "O" rings. Push the valves on through the "O" rings until spring seats contact shoulders in valve bores. Install new "O" rings on the bore plugs, then install the plugs and retaining snap rings. Lubricate and install the detent pistons, springs and spring retaining plug with new "O" ring or the retainer and retaining ring. Reinstall the control levers on pivot pins and connect to valve spools with clevis pins and cotter pins.

Fig. 433–Exploded view of remote control valve linkage typical of that used on model 4200 tractors with production date of 5/70 and later for single spool control valves, or production date of 8/70 and later for double spool control valves. Links (28) for model 4200 tractors are 1.83 inches long.

1. Control lever	21. Bracket	31. Pivot link	38. Coupling housing
2. Control lever	25. Pivot	32. Hose	39. Housing seal
3. Shaft support	26. Pivot shaft	33. Hose	40. Coupling half
7. Lever shaft	27. Jam nut	34. Hollow bolt	41. Coupling clamp
8. Bushing	28. Link	35. "O" ring	47. Coupling lever
13. Mounting plate	29. Clip	36. "O" ring	48. Spacer
18. Support	30. Clip	37. Coupling assy.	51. Dust plug

Fig. 434–Exploded view of single spool remote control valve after production date of 12-70. Valve shown has detent assembly (items 24 through 28). Non-detent valves, except for detent assembly, remain the same. Detent valve can be made to function as a non-detent valve by removing only detent valve spring (26).

1. Plug	11. Washer	20. Valve spool	29. Detent regulating valve
2. Check valve plug	12. Spring retainer	21. "O" ring	30. Seal
3. Seal	13. Centering spring	22. Handle pivot	31. Valve body
4. Spring	14. Spring retainer	23. Seal	32. Steel ball
5. Check valve ball	15. "O" ring	24. Detent ball	33. Spring retainer
6. Seal	16. Float valve stem	25. Poppet	34. Spring retainer
7. Plug	17. Seal	26. Spring	35. Adjusting plug
8. Retaining ring	18. Spring pin	27. Spring retainer	36. "O" ring
9. End cap	19. Valve body	28. Snap ring	37. Plug
10. Screw			

Fig. 435–Exploded view of double spool remote control valve after production date of 12-70. Except for detent assembly (items 24 through 28), non-detent valves remain the same. The inboard spool, without detents, is always double acting and cannot be converted to single acting operation. Refer to Fig. 434 for legend.

382. DETENT REGULATING VALVE. Refer to Fig. 428 for single spool valve and to Fig. 429 for double spool control valve. Then, proceed as follows:

Unscrew the regulating valve plug (30) and remove the plug and "O" ring (29). Insert an Allen wrench in the socket head plug (28) and withdraw the valve assembly from bore in valve body by prying and pulling on the wrench. If valve body (24) is stuck tightly, unscrew the plug, extract the spring (27) and pull body (24) from bore with hooked wire. Be careful not to lose the valve ball retainer (26) or valve ball (25). Note: Early production regulating valve is shown in Figs. 428 and 429, refer to Fig. 430 for exploded view of later production regulating valve. The regulating valve spring fits inside the early valve ball retainer and outside of the later valve ball retainer and spring guide (26A). The later adjusting plug

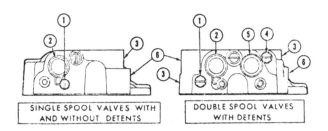

Fig. 436–View showing front side of single and double spool remote control valves.

1. By-pass valve
2. Outboard spool
3. Drop port
4. Float valve
5. Inboard spool
6. Lift port

(28A) also has a spring guide and is used as service replacement for the earlier plug (28—Fig. 428 and 429).

Inspect all parts and renew any that are worn or damaged. If necessary to renew valve body (24), the complete regulating valve assembly must be renewed. Install valve with new "O" ring (23) and lubricate the "O" ring before inserting valve in bore. Refer to paragraph 378 for valve adjustment. Install regulating valve retaining plug with new "O" ring (29) when valve is correctly adjusted. Note: It will be necessary to temporarily install the plug while checking detent release pressure.

383. SWITCH (FLOAT) VALVES. Refer to Fig. 428 for single spool valve and to Fig. 429 for double spool valve.

To remove the switch valve (17) on single spool valve or for top spool for double spool valve, drive the retaining roll pin from valve body and turn the knob (16) to unscrew valve completely. Remove "O" ring (18) from valve stem.

To remove switch valve (19) for lower spool of double spool valve, drive the retaining roll pin from valve body and unscrew plug (20) with Allen wrench. Then, using needle nose pliers, pull the valve from valve body and remove the "O" ring from valve stem.

Install the new sealing "O" ring on the valve stem, lubricate the "O" ring and push or thread valve stem into valve body. Install plug to retain switch valve for lower spool in double spool valve assembly. Install the retaining roll pins in control valve body.

384. CHECK VALVES AND SEATS. On single spool valve assembly, refer to Fig. 428 and remove the spring guide plug (8A), spring (11A) and check valve ball (10). Using a wire hook or other suitable tool, extract the valve seat (15).

On double spool valve assembly, refer to Fig. 429 and remove the plug (8), spring (10), check valve ball (11) and using a wire hook or other suitable tool, remove the seat (12). Then remove the second spring and check valve ball and extract the inner seat (15) with suitable tool.

Renew the check valve balls and seats if pitted, chipped or worn. Renew any spring if cracked, distorted or if free length is less than that of new

spring. Reinstall by reversing removal procedure and using a new "O" ring (9) on retaining plug (8 or 8A).

Remote Control Valves, With Or Without Spool Detents (12-70 and After)

385. Exploded views of single and double spool remote control valves with spool detents are shown in Figs. 434 and 435. Except for the detent assemblies (items 24 through 28), remote control valves are the same. Either control valve can be used to operate single or double acting cylinders. Refer to Fig. 436 and Table A.

Disassembly of detent and non-detent valves will be the same except the double spool valves have twice as many parts and the non-detent valves do not include detent assemblies.

386. ADJUST DETENT REGULATING VALVE. The detent regulating valve (29—Fig. 434 and 435) should be adjusted to release the spool detent poppets at a pressure slightly higher than that of normal cylinder operating pressure. When so adjusted, the control valve spool will be returned to neutral position without an excessive pressure build up when remote cylinder reaches end of stroke.

To adjust the detent regulating valve, refer to Fig. 434 or 435 and pro-

ceed as follows: Remove valve plug (37), then using an Allen wrench, turn the socket head adjusting plug (35) into valve body (31) to increase the detent release pressure, or out to decrease release pressure. Reinstall plug with new "O" ring (36) after completing the adjustment.

387. REMOVE AND REINSTALL. To remove and reinstall remote control valve, refer to paragraph 371 which outlines the basic procedure for all remote control valves.

388. OVERHAUL. Remove remote control valve as outlined in paragraph 371, then proceed as follows: Remove pin (bolt) securing handle to valve spool and disengage handle from spool and handle pivot (22—Fig. 434 or 435). Remove snap ring (28), spring retainer (27), detent spring (26), poppet (25) and detent ball (24).

NOTE: On double spool detent valves, keep poppets identified with their bores as poppets differ slightly in that one poppet has a small bleed hole.

Remove retaining ring (8) and end cap (9), then push on handle end of spool and remove spool and centering spring assembly from front of valve body (19). Remove screw (10), washer (11), spring retainer (12), centering spring (13) and spring retainer (14) from spool. Remove "O" rings (15 and 21). Remove plug (37) and "O" ring (36), pull regulating valve (29) from body, then remove adjusting plug (35), regulating valve spring (34), spring retainer (33) and ball (32) from regulating valve body (31). Remove seal (30) from valve body.

Remove socket head plug (1), then remove check valve plug (2) with seal (3), spring (4) and check valve ball (5)

Control Valve Type	Cylinder Application(s)	Float Valve	Bypass Valve	Single-Spool Valves or Outboard Spool of Double-Spool Valves		Inboard Spool of Double-Spool Valves	
				Lift Port	Drop Port	Lift Port	Drop Port
Single-Spool Without Detents	One Double-Acting	N/A	Closed	Lift Hose	Drop Hose	N/A	N/A
	One Single-Acting	N/A	Open	Lift Hose	Plug	N/A	N/A
Single-Spool With Detents	One Double-Acting	N/A	Closed	Lift Hose	Drop Hose	N/A	N/A
	One Single-Acting	N/A	Open	Lift Hose	Plug	N/A	N/A
Double-Spool Without Detents	Two Double-Acting	N/A	Closed	Lift Hose	Drop Hose	Lift Hose	Drop Hose
	One Double-Acting One Single-Acting	N/A	Open	Lift Hose	Plug	Lift Hose	Drop Hose
Double-Spool With Detents	Two Double-Acting	Closed	Closed	Lift Hose	Drop Hose	Lift Hose	Drop Hose
	One Double-Acting One Single-Acting	Open	Closed	Lift Hose	Plug	Lift Hose	Drop Hose
	Two Single-Acting	Open	Open	Lift Hose	Plug	Lift Hose	Plug

Table A–Refer to Fig. 436 for float valve and by-pass valve locations and to above table for cylinder application, valve settings and hose installation. The inboard spool of the double spool control valve without detents is always double acting and cannot be converted to single acting operation. N/A–Does Not Apply.

from valve body.

Drive out roll pin (18), then unscrew float valve (16) and remove seal (17).

Plug (7), seal (6), handle pivot (22) and seal (23) can be removed, if necessary.

Thoroughly clean and inspect all parts. If control valve spools or spool bores in valve body are excessively scored or worn, the complete valve assembly must be renewed as valve spools and body are not available separately. Centering spring free length is 2.022 for detent type control valves, or 2.157 for non-detent type control valves. Renew the centering springs if they differ substantially from given dimensions, or if they show signs of fractures or distortion. Also inspect check valve spring (4) and regulator spring (34) for signs of fracture or distortion and renew as necessary.

To reassemble, use all new "O" rings and seals, lubricate parts and reassemble by reversing the disassembly procedure.

NOTES

NOTES

NOTES

NOTES